Concrete Fracture Models and Applications

Shailendra Kumar · Sudhirkumar V. Barai

Concrete Fracture Models and Applications

Springer

Asst. Professor Shailendra Kumar
National Institute of Technology
Civil Engineering Department
Jamshedpur
India
shailendrakmr@yahoo.co.in

Prof. Sudhirkumar V. Barai
Indian Institute of Technology
Civil Engineering Department
Kharagpur
India
skbarai@civil.iitkgp.ernet.in

ISBN 978-3-642-16763-8 e-ISBN 978-3-642-16764-5
DOI 10.1007/978-3-642-16764-5
Springer Heidelberg Dordrecht London New York

Cover design: eStudio Calamar S.L, Figueres/Berlin

Printed on acid-free paper

Springer is part of Springer Science+Business Media (www.springer.com)

Dedicated to All Those who Relentlessly Endeavour to Unite a Fractured World

Foreword

Fracture mechanics as a discipline of mechanics goes back to the early years of the 20th century. It started with the description and explanation of the cracking behavior and failure of glass which could not be explained by means of the strength of materials approach. The material to which the new theory was applied had to be elastic and brittle. After glass, the failure of brittle types of metals was investigated. Later, linear elastic fracture mechanics was extended to elastic–plastic material behavior with well-established theories.

Although concrete exhibited brittleness in conventional force-controlled tensile tests, it was only in the early 1960s that fracture mechanics principles penetrated slowly into the field of concrete. First attempts were made to apply linear elastic fracture mechanics to concrete, but there was no great success. However, the idea to apply fracture mechanics to concrete and concrete structures was very important. Many researchers started to think of concrete and fracture mechanics. At the same time, the testing facilities developed enormously. It became possible to perform displacement-controlled tensile tests on concrete. One realized that concrete is not perfectly brittle but strain softening, i.e., failure in tension, occurred only after a considerable nonelastic displacement in the post-peak region. The idea of cohesive stresses in a concrete crack emerged.

After recognizing the real behavior of concrete, nonlinear fracture mechanics theories were developed. The present book presents these theories in detail. The book is a comprehensive treatment of the state of knowledge and adds some new findings to the field. The authors succeeded to write the book in a way that sometimes complicated theories can be followed with ease. I congratulate the authors to this achievement and wish that not only many teachers will use the book in their classes but also code makers will use it as a compendium of the principles of fracture mechanics of concrete in order to introduce these principles finally into the design standards.

Stuttgart, Germany
July 10, 2010

Hans-Wolf Reinhardt

Preface

Concept of linear elastic fracture mechanics was first applied to pre-cracked concrete elements in the early 1960s. Thereafter, extensive experimental and numerical research investigations proved that the classical form of linear elastic fracture mechanics cannot be applied to normal size concrete members. From the past research and studies it also became clear that the modified form of linear elastic fracture mechanics or nonlinear fracture mechanics can be useful and powerful tools for the analysis of the growth of distributed cracking and its localization in concrete if the softening behavior of the material is taken into account. The nonlinear fracture models coupled and introduced the tension-softening constitutive law in the fracture mechanics concepts to study the crack initiation and its propagation in concrete and concrete structures.

In due course of time, a number of nonlinear fracture models have been proposed and used to predict the nonlinear fracture behavior of cementitious materials. These are fictitious crack model or cohesive crack model, crack band model, two-parameter fracture model, size-effect model, effective crack model, K_R-curve method based on cohesive force distribution in the fracture zone, double-K fracture model, and double-G fracture model. Fracture mechanics concept introduced energy approach for crack development and its growth which can avoid the unobjectivity in the results, predict the post-peak response with a less complexity and exhibit size-effect behavior. The brittleness of the material can quantitatively be defined and more uniform safety of factors can be achieved in the structural design with the help of fracture mechanics concept.

It is a well-known phenomenon that the fracture parameters of concrete depend on the softening function of concrete, concrete strength, specimen size, specimen geometry, geometrical factors like relative size of notch length and the loading condition. The literature reports extensive numerical and experimental investigation on nonlinear fracture behavior of concrete. All the important nonlinear fracture models are widely applied to characterize the related fracture parameters and these studies are available in scattered literature. This book attempts to present the theoretical development and applications of various nonlinear concrete fracture models in a unified manner using different fracture parameters. In this regard, the authors investigated the behavior of fracture parameters of concrete at different phases of crack propagation phenomena of concrete.

There are six major chapters in the textbook which are mainly based on the recent research and studies carried out by the authors in the recent years. The detailed introduction of the book is mentioned in the opening chapter. In the subsequent chapters, cohesive crack model for three-point bending test, four-point bending test, and compact tension specimens using important softening functions of concrete are developed. The numerical results are compared with the experimental results available in the literature. Further, a systematic study on the different cohesive crack fracture parameters is carried out. Introduction of weight function method is explained to determine the double-K fracture parameters and the K_R-curve method based on cohesive stress distribution. Furthermore, attempts are made to put forward some new developments regarding behavior of different fracture parameters using the cohesive crack model as the reference. A comprehensive comparison between the double-K and double-G fracture criteria is presented. Emphasis on the effect of various parameters including specimen geometry, size effect, and loading condition on the double-K and the K_R-curve method is also focused. Finally, a comparative study among different fracture parameters obtained from important nonlinear models is presented. Hence, the textbook presents results of a comprehensive study on the crack initiation and its growth in concrete-like materials using various fracture models. At last, the flowcharts of various fracture models are presented in Appendix.

In this book, the authors have taken a small step to present a basic introduction on the various nonlinear concrete fracture models considering the respective fracture parameters. It can be helpful to undergraduate and postgraduate students who are studying this subject. An immense help to the beginners and researchers in the area of fracture mechanics of concrete is expected from this book which will provide a sound basis on the relevant subject to carry out further innovative research work in the future. Appendix can be of much use to the readers for computing different fracture parameters using computer programs.

At last, the authors would be thankful to the readers and their invaluable suggestions or comments for the further improvement of the book.

Jamshedpur, India Shailendra Kumar
Kharagpur, India Sudhirkumar V. Barai

Acknowledgement

Experimental and numerical investigations showed that the classical form of linear elastic fracture mechanics (LEFM) cannot be applied to normal size concrete members because of the presence of large and variable size of fracture process zone. However, when the softening behavior of material was taken into account, nonlinear fracture mechanics models emerged as a powerful tool for analyzing growth of distributed cracking and its localization in concrete. There is a scattered literature exploring nonlinear fracture behavior of concrete. It consists of numerical and experimental applications of all important nonlinear fracture models to characterize related fracture parameters of concrete. This inspired us to attempt to present a unified view of the theoretical development and applications of these models on concrete.

In the journey of writing this book we came across many legendary researchers who have actively contributed directly or indirectly toward the content of this book.

First of all we would like to thank Prof. Dr.-Ing. Hans-Wolf Reinhardt, University of Stuttgart, Germany, who has been constant source of inspiration and support to our book and kindly agreed to write *Foreword* for this book.

We wish to extend our sincere thanks to Prof. Per-Erik Petersson (Sweden), Prof. Zdenek P. Bažant (USA), Prof. B.L. Karihaloo (U.K.), Prof. Jaime Planas (Spain), Prof. Alberto Carpinteri (Italy), Prof. S P Shah (USA), Prof. Shilang Xu (China), Prof. Grzegorz Glinka (Canada), Prof. John P. Dempsey (USA), Prof. Gustavo V. Guinea (Spain), Prof. Goangseup Zi (South Korea), Prof. B.K. Raghuprasad (India), Prof. P.C. Pandey (India), Prof. T. Kant (India), Prof. B. Pradhan (India), Prof. S.K. Bhattacharyya (India), Prof. L.S. Ramachandra (India), Prof. D. Deb (India), Prof. N Dhang (India), Prof. Yanhua Zhao (China), Prof. Jeff Roesler (USA), Prof. Ravindra Gettu (India), and Dr. Xiufang Zhang (China) for enriching inputs and feedbacks.

We are indeed grateful to Dr. Thomas Lehnert, senior editor engineering Springer-Verlag GmbH, Berlin, Germany, who helped us in publishing this book with Springer-Verlag.

Shailendra is thankful to National Institute of Technology, Jamshedpur, India, for providing a conducive environment for working on this book. Further, he would like to express a word of appreciation to his lovely wife, Neelam Prabha, for her unprejudiced encouragement and his sons Somil and Shankhin for bearing with all difficulties during this exciting journey of writing the book.

Sudhir is grateful to Indian Institute of Technology, Kharagpur, India, for providing stimulating environment for working on this book. Behind every accomplishment, the family plays an important role in sacrificing the precious time of togetherness. He would like to thank his wife – Parama – and lovely daughters – Sristi and Shailey – for all their warmth and support and helping him to sail through during difficult times.

Above all, we would like to express profound gratitude and appreciation to Almighty.

Jamshedpur, India
Kharagpur, India
10 July 2010

Shailendra Kumar
Sudhirkumar V. Barai

Contents

List of Symbols and Abbreviations

List of Symbols

a	crack length
a_o	initial crack length
a_c	effective crack length at peak (critical) load
a_e	effective crack length at peak (critical) load from ECM
B	width of the test specimen
b_i	boundary force distribution defined on A_T
$[C]$	a vector of crack opening displacement at node i for unit external load
c_1, c_2	material constants for nonlinear softening function
c_f	effective extension of crack length for infinitely large specimen
$\{C_g\}$	a vector of crack opening displacement at node i due to self-weight of specimen
CTOD_{cs}	critical value of crack-tip opening displacement from TPFM
c_p	length of fully developed fracture process zone
d_a	maximum size of aggregates
D	depth of the test specimen (characteristic dimension)
D_g	load point deflection due to self-weight of the specimen
D_L	load point deflection when external load is unity
$\{D_p\}$	a vector of load point deflection when $\{p\} = \{1\}$
E	modulus of elasticity of concrete
f	a function
F_1, F_2	geometric factors for different loading cases
$F(x/a, a/D)$	the standard Tada Green's function for edge cracks subjected to pair of forces normal to the crack face
f_c	cylinder compressive strength
f_{ck}	characteristics strength of concrete
f_{cm}	mean compressive strength
f_{cu}	cube compressive strength of concrete
f_{tm}	mean tensile strength

F_i	the body force
f_t	uniaxial tensile strength of concrete
G_f	initial fracture energy of concrete, defined as the area under curve of the horizontal intercept of the initial tangent of the softening curve
G_F, G_{FC}	fracture energy of concrete, defined as the area under curve of the post-peak stress and crack opening displacement relation
$g_f(x)$	Hillerborg's local cohesive breaking energy at any location x along the FPZ
G_{FB}	fracture energy of concrete for infinitely large specimen
G_{I-coh}	the average value of cohesive breaking energy per unit length
${G_F}^{ini}$	crack initiation fracture energy obtained using LEFM equation from known value of ${K_{IC}}^{ini}$
${G_F}^{un}$	unstable fracture energy obtained using LEFM equation from known value of ${K_{IC}}^{un}$
${G_{IC}}^{ini}$	the crack initiation fracture energy release
${G_{IC}}^{un}$	the unstable fracture energy release
${G_{IC}}^{C}$	the critical value of the cohesive breaking energy
H_o	thickness of the clip gauge holder
$[K]$	a symmetric matrix and the value of K_{ij} is the COD at node i by a unit opening nodal force applied at node j
K_I	stress intensity factor in mode I fracture
${K_I}^{COH}$	the cohesive toughness during crack extension
${K_C}^{t}$	total crack extension resistance
${K_{IC}}^{ini}$	the crack initiation toughness
${K_{IC}}^{un}$	the unstable fracture toughness
${K_{IC}}^{C}$	the cohesive toughness at critical condition
${K_{IC}}^{e}$	critical value of SIF from ECM
${K_{IC}}^{s}$	critical value of SIF from TPFM
${K_{IC}}^{b}$	critical value of SIF from SEM
$\overline{K_{IC}^{ini}}$	the effective crack initiation toughness
$\overline{K_{IC}^{un}}$	the effective unstable fracture toughness
K_{INu}	stress intensity factor for P_u and a_o
k_P	stress intensity factor at the process zone tip due to unit applied load
K_σ	the stress intensity factor at the process zone tip due to cohesive stress
K_{IC}, K_C	the critical stress intensity factor
$[{K_{IC}}^{C}]_{NI}$	${K_{IC}}^{C}$ determined using numerical integration method
$[{K_{IC}}^{ini}]_{NI}$	${K_{IC}}^{ini}$ determined using numerical integration method
$[{K_{IC}}^{C}]_{WF4}$	${K_{IC}}^{C}$ determined using weight function method with four terms

$[K_{IC}^{ini}]_{WF4}$	K_{IC}^{ini} determined using weight function method with four terms
$[K_{IC}^{C}]_{WF5}$	K_{IC}^{C} determined using weight function method with five terms
$[K_{IC}^{ini}]_{WF5}$	K_{IC}^{ini} obtained using weight function method with five terms
K_{IC}^{un*}	K_{IC}^{un} evaluated by Xu and Reinhardt (1999b, c) in Chap. 4
K_{IC}^{ini*}	the K_{IC}^{ini} evaluated by Xu and Reinhardt (1999b, c) in Chap. 4
$k(\alpha)$	a geometric factor
l_{ch}	characteristic length of material
$m(x,a)$	weight function
M_1, M_2, M_3, M_4	parameters of weight function
$\{p\}$	a vector of nodal forces
P_{ini}	initial cracking load
P_u	maximum applied load
$P_{u,exp}$	maximum applied load obtained from experiment
$P_{u,CCM}$	maximum applied load obtained using cohesive crack model
$P_{u,LEFM}$	maximum load obtained using LEFM concept
r_u	undamaged length of the ligament during crack propagation
S	span of the test beam
u_i	an admissible displacement field for the system
W	the strain energy density function defined in the body
w	crack opening displacement ahead of the crack tip
w_1	the horizontal intercept of the initial tangent of softening
w_t	crack opening displacement at the initial notch tip
w_c	maximum COD when the cohesive stress becomes zero
w_g	the self-weight per unit length of the beam
w_s	COD at slope change in the bilinear softening
α	*a/D* ratio
β_B	the brittleness number
δ	load point deflection
Δa_c	effective crack extension at critical load
ε_{ij}	strain field the surface potential defined in the part of boundary A_P
υ	Poisson's ratio of concrete
$\sigma_s(CTOD_c)$	cohesive stress at $CTOD_c$
$\sigma(w_t)$	cohesive stress at the initial notch tip
σ	cohesive stress ahead of the crack tip
σ_{Nu}	peak nominal stress
σ_s	cohesive stress at slope change in the bilinear softening
σ_{tip}	stress value generated at the crack tip during loading
Π	the total potential

List of Abbreviations

CBM	crack band model
CCM	cohesive crack model
CMOD	crack mouth opening displacement
$\mathrm{CMOD_c}$	critical value of crack mouth opening displacement
COD	crack opening displacement
$\mathrm{COD_c}$	critical value of crack opening displacement
CT	compact tension
CTOD	crack-tip opening displacement
$\mathrm{CTOD_c}$	critical value of crack-tip opening displacement
DGFM	double-*G* fracture model
DKFM	double-*K* fracture model
ECM	effective crack model
FCM	fictitious crack model
FEM	finite element method
FPBT	four-point bending test
FPZ	fracture process zone
LEFM	linear elastic fracture mechanics
LHS	left-hand side
RHS	right-hand side
SEM	size-effect model
SIF	stress intensity factor
TPBT	three-point bending test
TPFM	two-parameter fracture model
WST	wedge splitting test

Chapter 1
Introduction to Fracture Mechanics of Concrete

1.1 General

Fracture mechanics is a branch of solid mechanics which deals with the behavior of the material and conditions in the vicinity of a crack and at the crack tip. While the concept of linear elastic fracture mechanics has been well developed for more than past 40 years and successfully applied to metallic structures, several civil engineering materials such as cementitious materials, rocks, and fiber-reinforced composites commonly known as *quasibrittle* need a different fracture mechanics approach to model the fracture process. Cementitious materials can be modeled at various scales like the nano-, micro-, meso-, and macrolevels. At mesolevel, they can be considered as a two-phase particulate composite, i.e., the matrix and the reinforcement. In the cement pastes, mortar, and concrete, the matrices can be considered as the hydrated cement gels, cement paste, and mortar, respectively, whereas the reinforcements in the corresponding materials can be taken as unhydrated cement particles, fine aggregates, and coarse aggregates.

Concrete is made up of many ingredients such as cement, fine aggregates, coarse aggregates, water, and admixtures in complex arrangements. Apart from the two-phase particulate composite, multitudes of internal voids ranging up to several millimeters are present in the hardened concrete (Shah et al. 1995). These voids including pores in cement, cracks at matrix–aggregate interface, and shrinkage cracking have significant influence on the mechanical behavior of concrete. The presence of defects plays an important role and gives rise to a microcracked zone ahead of the tip of a macro-crack, resulting in a progressive material failure as the crack grows. Preceding to unstable or critical load, a sub-critical crack grows when a pre-cracked concrete specimen is loaded. The pre-critical crack growth is often termed as fracture process zone or damage process zone or slow crack growth.

During 1960–1970s, several experimental and numerical investigations proved that the classical form of linear elastic fracture mechanics cannot be applied to normal size concrete members. The inapplicability of linear elastic fracture mechanics was discovered and reasoned as the presence of large and variable size of fracture process zone. From the past research and studies it became clear that the fracture mechanics can be a useful and powerful tool for the analysis of the growth

S. Kumar, S.V. Barai, *Concrete Fracture Models and Applications*,
DOI 10.1007/978-3-642-16764-5_1, © Springer-Verlag Berlin Heidelberg 2011

of distributed cracking and its localization in concrete if the softening behavior of the material is taken into account. The actual application of tension-softening constitutive law was unknown until about late 1970s. Then using nonlinear fracture mechanics, Hillerborg and co-workers (1976) put forward a pioneer work in which the development of fictitious crack model based on cohesive crack model (Dugdale 1960, Barenblatt 1962) for the crack propagation study of unreinforced concrete beam was introduced. Thereafter, a number of fracture models have been proposed and used to predict the nonlinear fracture behavior of cementitious materials. The other nonlinear models are crack band model (Bažant and Oh 1983), two-parameter fracture model (Jenq and Shah 1985), size-effect model (Bažant 1984, Bažant et al. 1986), effective crack model (Nallathambi and Karihaloo 1986), K_R-curve method based on cohesive force (Xu and Reinhardt 1998, 1999a), double-K fracture model (Xu and Reinhardt 1999a), and double-G fracture model (Xu and Zhang 2008). More appropriately, the cohesive crack and crack band models are attributed to numerical approach such as finite element or boundary element technique, whereas the others models employ the concept of linear elastic fracture mechanics in its modified form.

Design of concrete structures has already passed through two important phases in its historical evolution. The first phase of development took place until 1930s with the elastic no-tension analysis and in its second phase during 1940–1970, the plastic limit theory was introduced. Since concrete structures have been designed and successfully built according to national codes of practice without using the concept of fracture mechanics, it might seem unnecessary to change the current design practice (ACI Committee-446 1992, Bažant and Planas 1998). However, a reasonable consensus has appeared from the researchers and professionals around the world to explore the possibility of introducing the fracture mechanics into the design practices. There are good reasons to believe that the third phase of evolution in the design of concrete structures may probably come by introduction of fracture mechanics parameters. The most important reasons (ACI Committee-446 1992, Karihaloo 1995, Bažant and Planas 1998) for the need of evoking fracture mechanics are briefly mentioned as below:

- Lack of energy criterion for the crack development and its growth in the existing national design codes.
- Application of fracture mechanics will avoid the unobjectivity in the results.
- Due to lack of yield plateau and presence of material softening damage or cracking behavior, the plastic hinges do not form at isolated locations. Instead, the failure is associated with the propagation of the failure zone throughout the structure.
- The existing failure theory cannot give information on the post-peak response in order to obtain the energy absorption under the complete load–deflection curve. This information can be conveniently determined using fracture mechanics theory.
- The concrete structures exhibit size effect due to various reasons including boundary layer or wall effect, diffusion phenomena, hydration heat, statistical

size effect, fracture mechanics size effect, and fractal nature of crack surfaces (Bažant and Planas 1998). Out of these, the most important size effect is due to the release of stored energy of the bulk of the structure into the fracture surface energy (fracture mechanics size effect). This particular behavior can be easily modeled using the fracture mechanics principle.

- Besides above, the brittleness of the material can be obtained quantitatively with the help of fracture mechanics concept. This parameter may be useful for measuring the ductility property of concrete, especially for high-strength concrete, which shows relatively more brittle failure. Moreover, more uniform safety of factors can be achieved in the structural design.

The preceding investigations show that the nonlinear fracture behavior is significantly influenced by softening function of concrete. Many shapes of softening curve have been proposed by different groups of researchers. In real sense, the actual shape of the softening can be determined from a direct tension test on the concrete specimen. However, indirect methods using notched specimens have been also proposed in the past to avoid the intricacy involved in the true (direct) tensile tests. In the indirect method, the load–displacement curves obtained from the experiment and theoretical prediction are compared until they are up to a limited error tolerance. Trial-and-error or optimization method is adopted to gain an accurate softening curve until the theoretically predicted and experimentally observed load–displacement curves are in a good agreement within the acceptable error range.

For practical engineering applications of the fracture mechanics-based concept to the design of concrete structures, some kind of simplicity in the structural analysis is always welcomed. The nonlinear fracture models based on the numerical approach are relatively more involved in the computations. For this reason, probably, the fracture models based on the modified linear elastic fracture mechanics may bridge the gap between the computational efficiency and the model predictive capability of results, because they are relatively more computationally efficient but have limited capacity to predict the fracture parameters. Nevertheless, recently proposed double-K fracture model, K_R-curve method based on cohesive stress distribution in fracture process zone and the double-G fracture model belonging to the modified linear elastic fracture mechanics concept can predict more number of fracture parameters without needing closed-loop testing system in the experiments. Precisely, they can predict the important stages of fracture processes like crack initiation, stable crack propagation, and unstable fracture. Moreover, with the use of K_R-curve method, the complete fracture process can be analyzed. However, calculation method of the fracture models, particularly double-K fracture model and K_R-curve method, needs specialized numerical technique because of singularity problem at the integral boundary. To avoid this, a simplified method (Xu and Reinhardt 2000) has been proposed in the past which uses two empirical relationships for determining the double-K fracture parameters.

One of the major reasons which possibly inhibit the wide application of the fracture parameters to a common practice is that the concrete fracture parameters are influenced by many factors including softening function of concrete, concrete strength, specimen size, specimen geometry, geometrical factor like relative size of notch length and the loading condition.

The literature reports extensive numerical and experimental investigation on nonlinear fracture behavior of concrete. All the important nonlinear fracture models are widely applied to characterize the related fracture parameters. However, a systematic study using cohesive crack model with important softening functions of concrete may be a useful supplement. Further, closed-form equation solutions for determining the double-K fracture parameters and the K_R curve associated with cohesive stress in the fictitious fracture zone will lead in computational efficiency and avoid the need of the specialized numerical technique. From the observations based on the experimental studies (Xu and Reinhardt 1999a, b), it was reported that the double-K fracture parameters are almost independent of the specimen size. Though the size-effect study of double-K fracture parameters based on experimental results is available in the literature, no such numerical study has been attempted in the past. The relationship between double-K and double-G fracture parameters using experimental tests has been demonstrated, the same using numerically obtained data has not been presented in the literature. The various studies including influence of specimen geometry and the type of loading condition on the double-K fracture parameters and the K_R curves are not available in the literature. It is interesting to carry out those studies. The load–displacement curves are the preliminary requirements for the above studies. For the investigation of the influence of specimen geometry, the results of load–displacement curves tested under the similar conditions for different specimen geometries are needed. Similarly, for the effect of loading condition, the load–displacement plots tested under the similar conditions except for different loading conditions are required. Most of the experimentally measured load–displacement curves available in the literature are not favorable to investigate the influence of specimen geometry and loading condition on the above fracture parameters. In other way, these investigations should be carried out with the load–displacement curves generated using nonlinear fracture models like the cohesive crack model or the crack band model. The advantage of gaining such load–displacement curves is that they may be free from unseen errors due to experimental measurements. The above lacunas were considered in the doctoral research program of Kumar (2010), who presented some of the behavior of fracture parameters based on numerical studies. In this text book, most of the contents are taken from the works of Kumar (2010) and Kumar and Barai (2008a–c, 2009a–f, 2010a, b) but more in an elaborative manner. In the subsequent chapters, cohesive crack model for three-point bending test and compact tension specimens using important softening functions of concrete are developed. The numerical results are compared with the experimental results available in the literature. Further, a systematic study on the different fracture parameters is carried out. Weight function method is introduced to determine the double-K and the K_R-curve method based on cohesive stress distribution. Application of universal form of weight function enables one to obtain closed-form equation for the cohesive toughness of the material during crack extension. The validity of the formulations is established by comparing results obtained from existing analytical or simplified methods using both the experimental and numerical input data. Furthermore, attempts are made to put forward a possible new equation of brittleness of concrete with the help of double-K fracture parameters and formulize the size-effect predictions from the double-K fracture using the cohesive crack model as the reference. A comprehensive comparison between the double-K and double-G fracture criteria is presented. Emphasis on the effect of various parameters including specimen geometry, size effect, and loading condition on the double-K and the K_R-curve method is also focused.

1.2 Organization of the Book

The textbook presents results of a comprehensive study on the crack initiation and its growth in concrete-like materials using various fracture models. A brief outline of the book consisting of six chapters as shown in Fig. 1.1 is given below:

- Chapter 1 presents the general introduction of the book briefly.
- Chapter 2 gives a state-of-the-art review on various aspects of the material behavior and development of different concrete fracture models. The critical observations on the available literature and the scope of the present book are also highlighted.
- Chapter 3 presents the development of cohesive crack or fictitious crack model for two standard tests: (i) three-point bending test and (ii) compact test specimens using commonly used softening functions of concrete. The results are analyzed and compared with the experimental results available in literature. The size-effect study between the size-effect model and the cohesive crack model is also carried out.
- Chapter 4 contains the extensive study on the double-K and double-G fracture parameters. Initially, the weight function approach is introduced for determining the double-K fracture parameters. This approach is compared with the existing *analytical method* and *simplified approach*. Prediction of size effect from the double-K fracture parameters is formulized. Study of the influence of many factors including specimen geometry, loading condition, size effect, and softening function on the double-K fracture parameters is presented. At last, a comparative study between double-K and double-G fracture parameters is carried out.
- Chapter 5 presents the application of weight function approach for determining the K_{R} curve associated with cohesive stress distribution in the fracture zone. This method is validated with the existing *analytical method* using test data available in the literature. Further, effect of specimen geometry, loading condition, size effect, and softening function on the K_{R} curve is investigated.
- Chapter 6 presents a comparative study between various fracture parameters obtained from cohesive crack model, two-parameter fracture model, size-effect model, effective crack model, double-K fracture model, and double-G fracture model. Finally this chapter ends with final closing remarks.

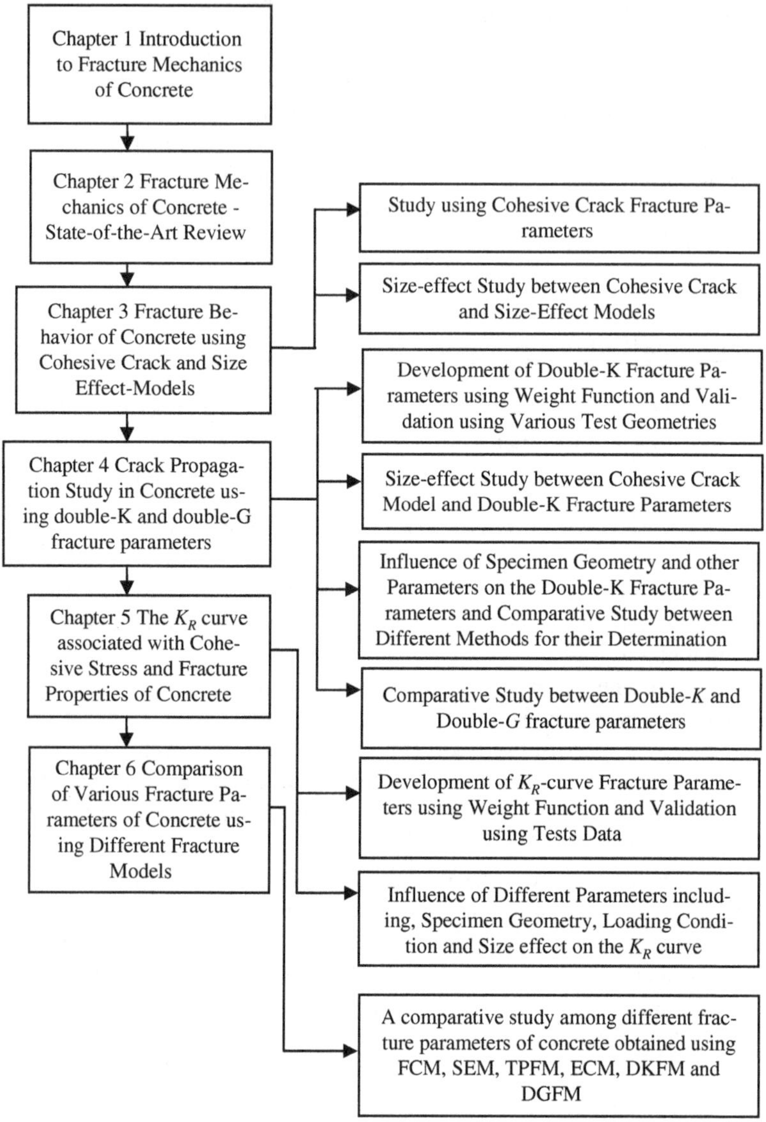

Fig. 1.1 Outline of the book

1.3 Closing Remarks

A brief application of the fracture mechanics concepts to concrete was addressed in this chapter. The important reasons why fracture mechanics to concrete can be applied were focused. At the end, organization of the book was addressed.

References

ACI Committee 446, Fracture Mechanics (1992) Fracture mechanics of concrete structures. Part I, State-of-Art Report (Edited by Z.P. Bažant), Elsevier Applied Science, London and New York: 1–140.

Barenblatt GI (1962) The mathematical theory of equilibrium cracks in brittle fracture. Adv Appl Mech 7 (1): 55–129.

Bažant ZP (1984) Size effect in blunt fracture: Concrete, rock, metal. J Eng Mech ASCE 110 (4): 518–535.

Bažant ZP, Oh BH (1983) Crack band theory for fracture of concrete. Mater Struct 16 (93): 155–177.

Bažant ZP, Kim J-K, Pfeiffer PA (1986) Determination of fracture properties from size effect tests. J Struct Eng ASCE 112 (2): 289–307.

Jenq YS, Shah SP (1985) Two parameter fracture model for concrete. J Eng Mech ASCE 111 (10): 1227–1241.

Bažant ZP, Planas J (1998) Fracture and size effect in concrete and other quasibrittle materials. Florida, CRC Press.

Dugdale DS (1960) Yielding of steel sheets containing slits. J Mech Phys Solid 8 (2): 100–104.

Hillerborg A, Modeer M, Petersson PE (1976) Analysis of crack formation and crack growth in concrete by means of fracture mechanics and finite elements. Cem Concr Res 6: 773–782.

Karihaloo BL (1995) Fracture mechanics and structural concrete, concrete design and construction series. Longman Scientific & Technical, Harlow, Essex.

Kumar S (2010) Behaviour of fracture parameters for crack propagation in concrete. Ph.D. Thesis submitted to Department of Civil Engineering, Indian Institute of Kharagpur, India.

Kumar S, Barai SV (2008a) Influence of specimen geometry and size-effect on the K_R-curve based on the cohesive stress in concrete. Int J Fract 152: 127–148.

Kumar S, Barai SV (2008b) Influence of specimen geometry on determination of double-K fracture parameters of concrete: A comparative study. Int J Fract 149: 47–66.

Kumar S, Barai SV (2008c) Cohesive crack model for the study of nonlinear fracture behaviour of concrete. J Inst Eng (India), CV 89 (Nov.): 7–15.

Kumar S, Barai SV (2009a) Determining double-K fracture parameters of concrete for compact tension and wedge splitting tests using weight function. Eng Fract Mech 76: 935–948.

Kumar S, Barai SV (2009b) Effect of softening function on the cohesive crack fracture parameters of concrete CT specimen. Sadhana-Acad Proc Eng Sci 36 (6): 987–1015.

Kumar S, Barai SV (2009c) Equivalence between stress intensity factor and energy approach based fracture parameters of concrete. Eng Fract Mech 76: 1357–1372.

Kumar S, Barai SV (2009d) Size-effect of fracture parameters in concrete: A comparative study. Comput Concr An Int J (under review).

Kumar S, Barai SV (2009e) Weight function approach for determining crack extension resistance based on the cohesive stress distribution in concrete. Eng Fract Mech 76: 1131–1148.

Kumar S, Barai SV (2009f) Influence of loading condition and size-effect on the K_R-curve based on the cohesive stress in concrete. Int J Fract 156:103–110.

Kumar S, Barai SV (2010a) Determining the double-K fracture parameters for three-point bending notched concrete beams using weight function. Fatigue Fract Eng Mater Struct 33(10): 645–660. DOI: 10.1111/j.1460-2695.2010.01477.x.

Kumar S, Barai SV (2010b) Size-effect prediction from the double-K fracture model for notched concrete beam. Int J Damage Mech 9: 473–497.

Nallathambi P, Karihaloo BL (1986) Determination of specimen-size independent fracture toughness of plain concrete. Magn Concr Res 38 (135): 67–76.

Shah S P, Swartz SE, Ouyang C (1995) Fracture Mechanics of Concrete: Applications of Fracture Mechanics to Concrete, Rock and Other Quasi-Brittle Materials. Wiley, New York, NY.

Xu S, Reinhardt HW (1998) Crack extension resistance and fracture properties of quasi-brittle materials like concrete based on the complete process of fracture. Int J Fract 92: 71–99.

Xu S, Reinhardt HW (1999a) Determination of double-*K* criterion for crack propagation in quasi-brittle materials, part I: Experimental investigation of crack propagation. Int J Fract 98: 111–149.

Xu S, Reinhardt HW (1999b) Determination of double-K criterion for crack propagation in quasi-brittle materials, part II: Analytical evaluating and practical measuring methods for three-point bending notched beams. Int J Fract 98: 151–177.

Xu S, Reinhardt HW (2000) A simplified method for determining double-K fracture meter parameters for three-point bending tests. Int J Fract 104: 181–209.

Xu S, Zhang X (2008). Determination of fracture parameters for crack propagation in concrete using an energy approach. Eng Fract Mech 75: 4292–4308.

Chapter 2
Fracture Mechanics of Concrete – State-of-the-Art Review

2.1 Introduction

The genesis of the development of fracture mechanics goes back to the beginning of 20th century when Inglis (1913) published a pioneer work on stress analysis for an elliptical hole in an infinite linear elastic plate loaded at its outer boundaries, in which a crack-like discontinuity was modeled and stress singularity was observed at the crack tip by making the minor axis very much less than the major axis. The actual development in this field could not occur until a new approach was postulated by Griffith (1921). Since then, it took around another four decades when for the first time the concept of fracture mechanics was applied to cementitious materials. Thereafter, the study of crack propagation in cement-based materials and structures attracted interest of a large number of researchers around the world until today. In this chapter, a state-of-the-art review on various aspects of fracture process of concrete-like materials now-a-days called as *quasibrittle* materials is presented.

2.2 Linear Elastic Fracture Mechanics

From analysis of a sharp crack in a sheet of brittle material (glass) subjected to a constant remotely applied stress, Griffith (1921) presented the first explanation of the mechanism of brittle fracture using a new energy-based failure criterion. According to this criterion, a certain amount of the accumulated potential energy in the system must decrease to overcome the surface energy of the material in order for the crack to propagate. It was shown that the stresses near the crack tip tend to approach infinity.

The Griffith's theory for ideally brittle materials was extended to account for the limited plasticity near the crack tip in majority of the engineering materials (Irwin 1955, Orowan 1955). It was postulated that the resistance to crack extension is taken as sum of the elastic surface energy and the plastic work. For ductile materials, the plastic dissipation energy is much greater than the elastic energy; therefore, resistance to crack growth is mainly governed due to plastic work.

S. Kumar, S.V. Barai, *Concrete Fracture Models and Applications*,
DOI 10.1007/978-3-642-16764-5_2,

Later, based on the existing mathematical procedures (Westergaard 1939), a series of linear elastic crack stress fields was developed (Irwin 1957) and it was shown that the stress field near a sharp crack tip shows a fundamental singular variation, i.e., the stress near the crack tip decreases in proportion to the square root of the distance to the crack tip r ($r << a$, and a is the crack length). In a cracked body, the singularity is independent of the boundary conditions, geometry, loading, and the type of cracking mode for which three fundamentally different types of failure modes such as mode I, mode II, and mode III as shown in Fig. 2.1 exist. For mode I cracking, the stresses and displacements near the crack tip as shown in Fig. 2.2 and expressed by Eqs. (2.1) and (2.2) are given in polar coordinates with origin at the crack tip, at a distance r and angle θ:

$$\begin{aligned}\sigma_{xx} &= \frac{K_{\mathrm{I}}}{\sqrt{2\pi r}}\cos\frac{\theta}{2}\left(1-\sin\frac{\theta}{2}\sin\frac{3\theta}{2}\right)\\ \sigma_{yy} &= \frac{K_{\mathrm{I}}}{\sqrt{2\pi r}}\cos\frac{\theta}{2}\left(1+\sin\frac{\theta}{2}\sin\frac{3\theta}{2}\right)\\ \tau_{xy} &= \frac{K_{\mathrm{I}}}{\sqrt{2\pi r}}\cos\frac{\theta}{2}\sin\frac{\theta}{2}\cos\frac{3\theta}{2}\end{aligned} \tag{2.1}$$

$$\begin{aligned}u &= \frac{K_{\mathrm{I}}(1+\upsilon)}{E}\sqrt{\frac{2r}{\pi}}\cos\frac{\theta}{2}\left(\frac{k-1}{2}+\sin^2\frac{\theta}{2}\right)\\ v &= \frac{K_{\mathrm{I}}(1+\upsilon)}{E}\sqrt{\frac{2r}{\pi}}\sin\frac{\theta}{2}\left(\frac{k-1}{2}-\cos^2\frac{\theta}{2}\right)\end{aligned} \tag{2.2}$$

The parameter k is $(3-\upsilon)/(1+\upsilon)$ and $(3–4\upsilon)$ in plane stress and plane strain conditions, respectively. The value K_{I} is called as the stress intensity factor (SIF) for mode I situation. Similar expressions for the stresses and displacements in the mode II and mode III failure conditions can be written in which the stress intensity factors are denoted by K_{II} and K_{III}, respectively.

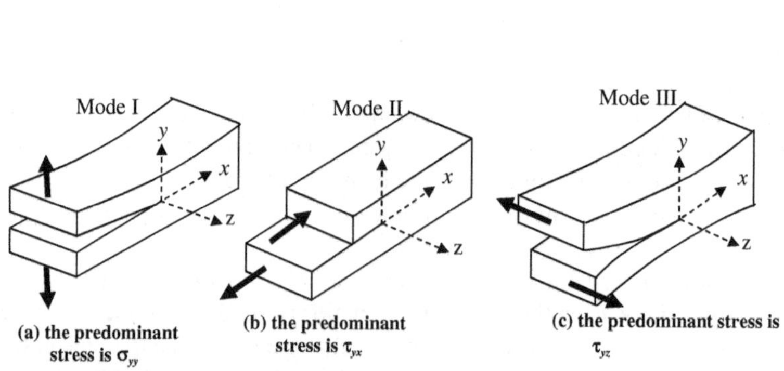

Fig. 2.1 Cracking modes: (**a**) mode I (or opening mode), (**b**) mode II (or sliding or in-plane shear mode), (**c**) mode III (or the tearing or anti-plane shear mode)

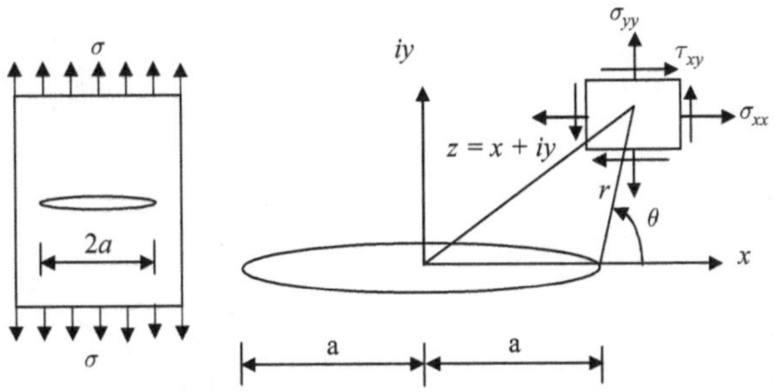

Fig. 2.2 Stress field in the vicinity of crack tip for mode I fracture using Westergaard complex coordinates

2.2.1 Significance of Stress Intensity Factor

The term stress intensity factor is different from the term stress concentration factor, which is used to characterize the ratio between actual and average or nominal stress at a geometric discontinuity. On the other hand, the stress intensity factor defines the amplitude of the crack-tip singularity, that is, the stress field in the vicinity of the crack tip increases proportionally to stress intensity factor. When a structural component is stressed in tension or bending, the developed stress field in the vicinity of a crack tip under elastic conditions shows a singularity following an inverse square root relationship with distance from the crack tip. In other words, stress intensity factor describes the strength of this singularity. Since the stress and displacement field in the vicinity of the crack tip is controlled by the stress intensity factor, it may be further assumed that critical stress or displacement condition at the crack tip can be explained using a critical value of stress intensity factor for any modes of failure. Hence the concepts of linear elastic fracture mechanics may be reasonably characterized using a single parameter, that is, stress intensity factor. Furthermore, local yielding occurs in engineering materials that relieves the singularity and hence the size of the plastic zone can be directly related to the stress intensity factor.

2.2.2 Concept of R Curve

In early 1960s, the *R*-curve approach (Irwin 1960, Krafft et al. 1961) based on energy balance was proposed. In the *R*-curve concept, an energy release rate *G* as a measure of the energy available for an increment of crack extension was postulated. According to definition, crack extension occurs when strain energy release rate *G* is equal to *R*, where *R* is called the material resistance to crack extension. The *R* curve is represented by a plot between the crack extension resistance expressed in terms of either strain energy release rate *G* or stress intensity factor *K* and the corresponding crack extension Δa as shown in Fig. 2.3. It is generally called as G_R curve and K_R curve depending upon the unit of the crack resistance parameters strain energy

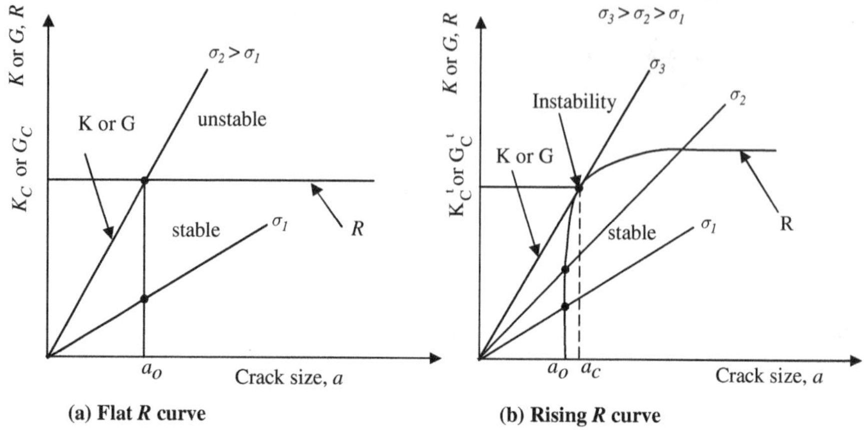

Fig. 2.3 Representation of *R* curve

release rate and stress intensity factor, respectively. The shape of the *R* curve determines the onset of crack instability. For most of the engineering materials except truly brittle material, a rising *R* curve is observed due to slow stable crack extension. As one crack extends, a rising *R* curve produces several small cracks because the required stress intensity to propagate the crack will increase to a point where it is favorable to propagate a different crack. On the other hand, a flat *R* curve for ideal brittle materials will give rise to a single and possibly catastrophic crack. Depending on how *G* and *R* vary with the crack size, the crack growth may be stable or unstable.

In Fig. 2.3a for a flat *R* curve, the crack is stable between the stress levels σ_1 and σ_2 and beyond the stress σ_2, the crack propagates unstably. In this case the total value of critical energy release is G_{C}. In Fig. 2.3b for a rising *R* curve, the crack extension is stable between the stress levels σ_2 and σ_3 and beyond the stress σ_3, the crack propagates unstably. At the onset of unstable crack growth, the driving force is the tangent to the *R* curve and corresponding value of the total crack extension resistance at the point of tangency between *G* and *R* curves is $G_{\mathrm{C}}{}^{\mathrm{t}}$ at the critical effective crack extension a_{c}.

The fracture criterion and instability of a propagating crack in cracked solid structures represented using the K_{R} curve can be expressed mathematically using Eq. (2.3):

$$\frac{\partial K}{\partial a} < \frac{\partial K_R}{\partial a} \quad \text{The crack extension takes place in stable manner}$$

$$\frac{\partial K}{\partial a} = \frac{\partial K_R}{\partial a} \quad \text{Critical condition is reached at onset of unstable crack extension}$$

$$\frac{\partial K}{\partial a} \geq \frac{\partial K_R}{\partial a} \quad \text{The crack propagates uncontrollably} \tag{2.3}$$

2.3 Elastic–Plastic Fracture Mechanics

Linear elastic fracture mechanics (LEFM) is applicable to the materials in which the nonlinear behavior is confined to a small region near the crack tip. There are many materials, however, for which the applicability of LEFM is impossible or at least suspicious. Moreover, the *R*-curve concept describes the fracture behavior of a material of limited ductility in the vicinity of a crack tip for plane stress situation. Therefore, an alternative elastic–plastic fracture mechanics is applied for too ductile material behavior which exhibits time-independent, nonlinear behavior (plastic deformation). In the nonlinear behavior at the crack tip, the fracture criterion is characterized by two parameters:

1. crack-tip opening displacement (CTOD) criterion (Wells 1962) and
2. the *J*- integral approach (Rice 1968a, b).

Critical values of CTOD and *J*-integral give nearly size-independent measures of fracture toughness, even for relatively large amount of crack-tip plasticity. However, there are still limits to the applicability of *J*-integral and CTOD criteria, but these limits are much less restrictive than the validity requirements of LEFM (Anderson 2005).

2.3.1 The CTOD Criterion

The stability analysis for crack growth in the CTOD criterion is based on the characteristic critical value of crack-tip opening displacement CTODc. According to this approach the crack is stable as long as CTOD $\leq$ CTOD$_c$ (Wells 1962).

2.3.2 The J-Integral Approach

A path-independent contour integral called *J*-integral was introduced (Rice 1968a, b) for the analysis of crack extension. The value of *J*-integral is equal to the energy release rate in a nonlinear elastic cracked body and can be viewed as both an energy parameter and a stress intensity parameter. Considering an arbitrary counter-clockwise path Γ as shown in Fig. 2.4 around the crack tip, the path-independent *J*-integral is given by

$$J = \int_{\Gamma} \left(W\,\mathrm{d}y - T_i \frac{\partial u_i}{\partial x} \mathrm{d}s \right) \tag{2.4}$$

where W is the strain energy density $= \int_0^{\varepsilon_{ij}} \sigma_{ij}\,\mathrm{d}\varepsilon_{ij}$; T_i is the component of the traction vector $= \sigma_{ij} n_j$, being the components of the unit vector normal to Γ;

Fig. 2.4 The definition of the J-integral

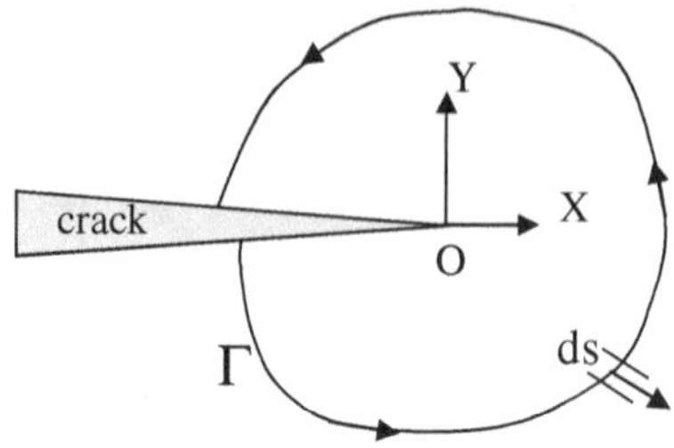

u_i is the displacement vector components; ds is the length increment along the contour Γ. The characteristic critical value of the J-integral J_c is considered to define the following form of the crack initiation criterion:

$$J = J_c \tag{2.5}$$

where J_c is considered as a material fracture parameter.

2.4 Early Research Using LEFM to Concrete

The application of LEFM concept to concrete was first attempted in early 1970s by Kaplan (1961). Notched concrete beams tested in three-point bending and four-point bending configurations were used to measure the critical strain energy release rate of concrete. The results indicate that the critical strain energy release rate of concrete depends on the concrete mix proportion, type of loading condition, relative size of initial notch length, and specimen dimension. It was pointed out that the proper account of slow crack growth prior to fast fracture could be introduced in the process of fracture analysis. However, the Griffith concept of a critical energy rate being a condition for rapid crack propagation may be applicable to concrete after suitable modification in analytical and experimental procedures. Thereafter, many experimental and numerical investigations have been performed to predict fracture behavior of concrete. Researchers in the 1960s used LEFM to cementitious materials similar to the experimental determination of critical stress intensity factor K_{IC} in metals.

Glucklich (1963) examined the fracture of concrete using fracture mechanics approach and revealed that the strain energy is converted mainly to surface energy but the surface involved is much larger in area than the surface of the effective crack. The rate of energy absorption at the crack tip in concrete is mainly due to entire highly stressed zone in the form of microcracking and not to plastic flow. The increase of the microcracked zone and the heterogeneity of the composite materials contribute to the relatively high value of the strain energy release rate. The driving force (strain energy release rate) increases with the crack length in tension, whereas it is a constant value in compression fracture.

Naus and Lott (1969) conducted experimental investigation to determine the effects of several concrete parameters: water–cement ratio, air content, sand–cement ratio, curing age, and size and type of coarse aggregate on the effective fracture toughness of concrete using three-point bending test. A regular dependency of the effective fracture toughness was observed with the variation of concrete parameters.

Shah and McGarry (1971) concluded that hardened Portland cement paste is a notch-sensitive material, whereas mortar and concrete with the normally used amounts and volume of stone aggregates are notch-insensitive materials with notch length lower than at least a few centimeters. The critical notch length for mortar and concrete depends on the volume, type, and size of aggregate particles.

Walsh (1971) presented test results of fracture toughness of geometrically similar notched concrete beams tested in three-point bending geometry. The test results of nominal strength and their sizes were plotted on double logarithmic scale. From the graphs, it was found that the results did not follow the straight line of slope −1/2, which inferred that the LEFM was not applicable to concrete. He suggested that the crack propagation load obtained using LEFM depends upon the size of the specimen. Further, in order to apply the LEFM to concrete, a minimum value of ligament length 150 mm for a beam depth 225 mm should be used for fracture testing.

Brown (1972) used two methods: a notched-beam technique and a double-cantilever beam to measure the effective fracture toughness of cement pastes and mortars. Tests of both pastes and mortars showed that the fracture toughness of cement is independent of crack growth but that the toughness of mortar increases as the crack propagates.

Kesler et al. (1972) performed the experimental investigation on a large number of cracked cement paste, mortar, and concrete specimens with an objective to determine the applicability of the LEFM to these materials. Based on the analysis of the test results, the authors concluded that the concepts of LEFM cannot be directly applicable to the cementitious materials having sharp cracks.

Brown and Pomeroy (1973) determined the effective fracture toughness on cement paste and mortar using both a notched-beam and a double-cantilever beam method. The influence of size and quality of aggregates on the effective fracture toughness was reported in the experimental research. It was found that the addition of aggregate not only increases the toughness but also results in a progressive increase in toughness with crack growth. The higher the proportion of aggregate, the larger the increase in toughness: fine aggregate is possibly more effective than coarse in this respect.

Fracture mechanics studies on plain and polymer-impregnated mortars (Evans et al. 1976) showed that macro-crack propagation resistance of these materials is not significantly influenced by water/cement ratio and curing time but is greatly enhanced by polymer impregnation. The fracture parameters are independent of the crack length for cracks larger than ~2 cm. Acoustic emission measurements showed that the susceptibility to microcracking is substantially restarted by polymer impregnation.

In a subsequent paper, Walsh (1976) showed the influence of specimen size on the fracture strength parameter. The investigation indicates that specimen size can have considerable effect upon the validity of extrapolations to real structures from the performance of laboratory of laboratory-scale tests. The validity of LEFM depends upon the extent of microcracking, slow crack growth, or other inelastic behavior near the crack tip. If specimen is large enough, the zone of stress disturbance can be considered to be surrounded by an area in which the stresses are substantially in accordance with the ideal stress distribution and the LEFM can be applied to obtain the cracking load. On the other hand, if the specimen is small in relation to the zone of microcracking, the LEFM cannot be applied as failure criterion.

Mindess and Nadeau (1976) performed experiments on mortar and concrete notched specimens using three-point bending test to show whether the critical stress intensity factor depends upon the length of the crack front (specimen width). The test results indicated that within the size range studied, there is no dependence of fracture toughness upon the length of the crack front and therefore, it appears that there is no significant plastic zone for mortar and concrete.

Gjϕrv et al. (1977) investigated experimentally the notch sensitivity and fracture toughness of concrete with different types, volumes, sizes, and strengths of aggregate using small three-point bending test. Results showed that both mortar and concrete are notch sensitive materials although not as sensitive as cement paste. If the fracture of cement paste is governed by LEFM, it appears that this theory is inapplicable to small-size specimens of mortar and concrete. Lightweight concrete showed the same fracture properties as the cement paste.

Hillemeier and Hilsdorf (1977) conducted experimental and analytical investigations to determine fracture mechanics characteristics of hardened cement paste, aggregates, and aggregate–cement paste interfaces using a suitable testing method of wedge-loaded compact tension specimens. The fracture toughness for cement paste and interfaces were found to be decreasing as the initial crack length increased and for further increase in the value of crack length, the fracture toughness attained constant values. Finally, it was found that hardened cement paste, aggregates, and interfaces exhibit unique values of fracture toughness which are independent of initial crack length. In additional test series, the ductility of different model concretes was studied.

Cook and Crookham (1978) examined the behavior of premix and impregnated polymer concrete. The notched-beam technique was used to determine fracture parameters, together with compliance measurements to obtain the slow crack growth. Experimental results indicated that the thermal treatment to polymerize the monomer in the premix concrete has a marked influence upon fracture toughness. On the other hand, irradiation has a little influence. The plain and the premix polymer concretes are partially notch sensitive, whereas the impregnated polymer concrete is notch sensitive.

Strange and Bryant (1979) carried out fracture tests on concrete, mortar, and cement paste using bending and tension specimens. They showed that concrete is notch sensitive and fracture toughness is dependent on specimen size. The slow crack growth was also evident in the tests. These phenomena indicated that concrete cannot be regarded as an ideal elastic homogeneous material: a region of nonelastic behavior must exist in the crack-tip region.

In the past, many authors (Radjy and Hansen 1973, Naus et al. 1974, Carpinteri 1982) reviewed the preceding works carried out on the fracture testing and analysis on the cementitious materials. Until the mid of 1970s, many research attempts were made to apply the linear elastic fracture mechanics and elastic–plastic fracture mechanics to the cementitious materials for crack propagation study. And gradually, it was accepted that no single fracture mechanics parameter could uniquely quantify the resistance to crack propagation. For instance, different researchers found widely varying values of the critical stress intensity factors of cementitious materials mainly depending upon the specimen geometry, specimen size and dimension, and type of measurement technique. However, it became clear that a large portion ahead of initial crack tip is surrounded by microcracking zone and other inelastic phenomena which cause slow crack growth behavior before reaching to the instability condition. The source of showing such behavior is recognized to be the softening damage of the material in a large fracture process zone (FPZ). In addition, more than one parameter is required to quantify the fracture characteristics and crack propagation study of cementitious or *quasibrittle* materials.

2.5 Tensile Behavior of Concrete

2.5.1 Strain Localization Effect

Understanding of complete behavior of concrete specimen subjected to tensile load is inevitable in the fracture analysis because the material in the vicinity of crack tip experiences tensile stress. In 1960s, many researchers (Hughes and Chapman 1966, Evans and Marathe 1968) obtained a stable and complete stress–strain diagram of concrete in direct tension using a very stiff tensile testing machine. A typical stress–displacement curve of concrete under direct tensile test is shown in Fig. 2.5. Test

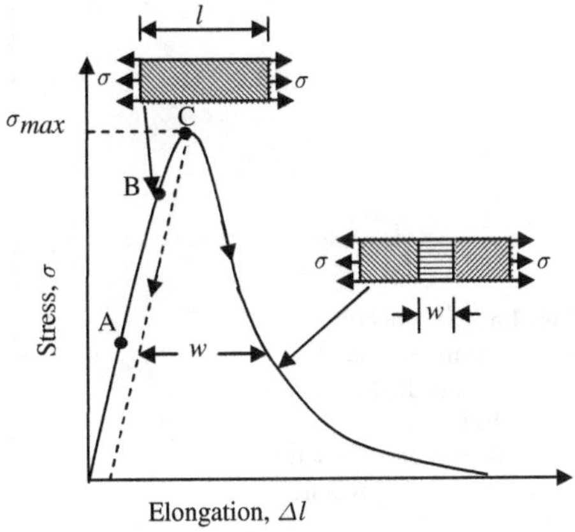

Fig. 2.5 A typical stress–displacement curve of concrete

results (Heilmann et al. 1969) show interesting results of local deformations at direct tensile tests. Furthermore, it has been confirmed from experimental techniques such as acoustic emission and electron speckle pattern interferometry that concrete shows a substantial nonlinearity before the peak load is attained (Shah and Ouyang 1994). Thus, Fig. 2.5 can be explained with three salient points: A, B, and C. Point A corresponds to about 30% of the peak load up to which propagation of microcracks or internal voids is negligible. Point B corresponds to about 80% of the peak load. The microcracks propagate between the points A and B and are isolated and randomly distributed over the specimen volume. In this region, the tensile stress is uniformly distributed in the direction of loading over the specimen length. Between the load points B and C, the microcracks start to localize into a macro-crack and the distribution of tensile strain in the loading direction is no longer uniform over the specimen. Up to the peak load (point C), a stable crack growth takes place, i.e., the crack propagates only when the load increases. At the peak load, the localized damage begins to form a band called fracture zone. Beyond the peak load, the tensile strain within the fracture zone or localized damage band continuously increases, whereas the material outside the fracture zone starts unloading. Therefore, stress–displacement relation for the fracture zone can be obtained by subtracting the elastic displacement from the total displacement as shown in Figs. 2.5 and 2.6.

Therefore, the behavior of concrete in tension may be described by two characteristic curves: the stress–strain curve up to the peak load and the stress–displacement curve after the peak load as shown in Fig. 2.6. A simple model of concrete tension specimen with initial length l is shown in Fig. 2.7.

After the maximum load, the elongation of the specimen Δl is written as

$$\Delta l = \varepsilon_0 l + w \tag{2.6}$$

If ε_0 is the uniform strain in the material outside the fracture zone and w is the width of the fracture zone, then the average strain ε_m is expressed as

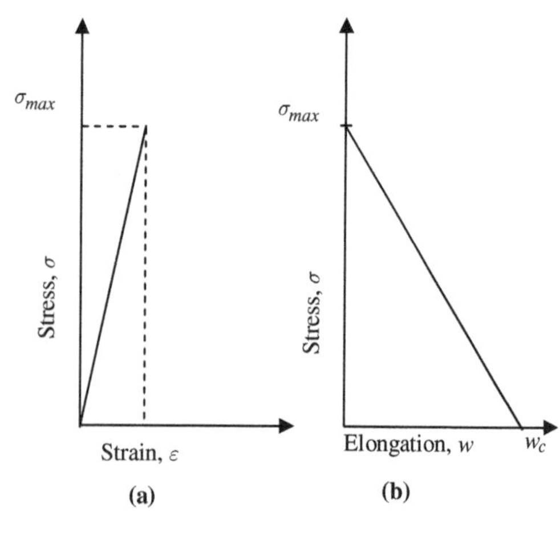

Fig. 2.6 (**a**) Linear stress–strain relation of concrete outside the fracture zone; (**b**) linear stress–displacement relation of concrete in the fracture zone

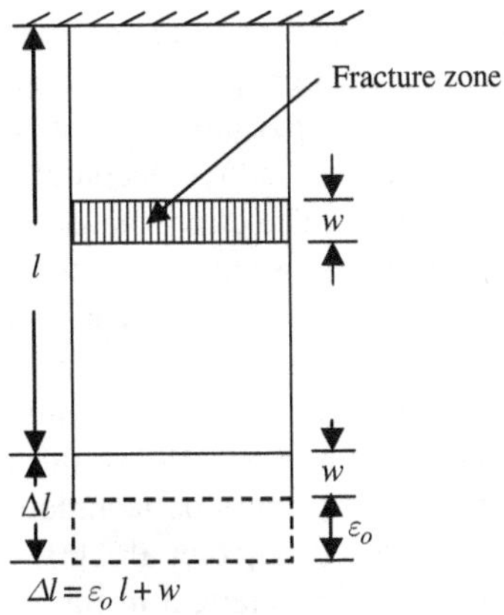

Fig. 2.7 Localization of strain in a concrete specimen under direct tension test

$$\varepsilon_{\mathrm{m}} = \frac{\Delta l}{l} = \varepsilon_0 + \frac{w}{l} \tag{2.7}$$

It is clear from Eq. (2.7) that after the maximum stress, the average strain depends on the specimen length; therefore, the stress–strain curve is not a characteristic of the material. This phenomenon is termed as the *strain localization effect* (Gdoutos 2005).

2.5.2 Fracture Process Zone

A large and variable size of damage zone exists ahead of the macro-crack when the specimen is loaded. This damage zone is known as fracture process zone (FPZ). The FPZ has a capability to still transfer the closing stress across the crack faces which decreases at increasing deformation as shown in Fig. 2.8. Many micro-failure

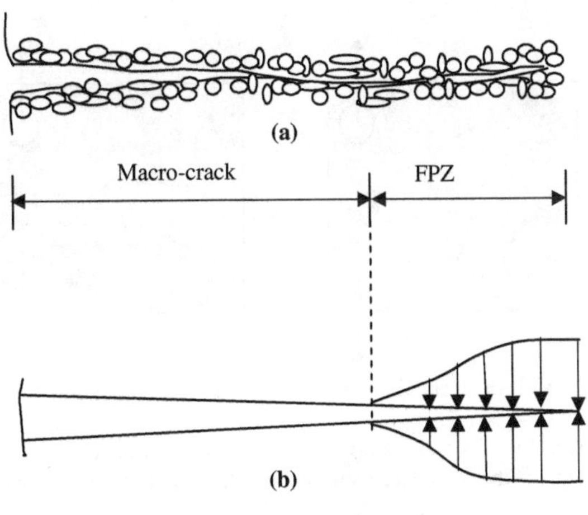

Fig. 2.8 (**a**) Fracture process zone in concrete; (**b**) distribution of cohesive stress in the FPZ

mechanisms such as matrix microcracking, debonding of cement–matrix interface, crack deflection, grain bridging, and crack branching which consume energy during the crack propagation are responsible for the stress transfer.

Direct and indirect techniques have been applied to measure the shape and size of the FPZ. The direct methods are optical microscopy (Derucher 1978), scanning electron microscopy (Mindess and Diamond 1980, 1982), and high-speed photography (Bhargava and Rehnström 1975), whereas the indirect methods are laser speckle interferometry (Ansari 1989), compliance and multicutting techniques (Hu and Wittmann 1990), penetrating dyes (Lee et al. 1981), ultrasonic measurement (Sakata and Ohtsu, 1995), infrared vibro-thermography (Dhir and Sangha 1974), and acoustic emission technique (Maji and Shah 1988, Maji et al. 1990, Ouyang et al. 1991, Hadjab et al. 2007). The typical size of FPZ is in the order of 50 cm for normal concrete, 3 m for concrete dam with extra large aggregate, 10 m for a grouted soil mass, and 50 m for a mountain with jointed rock. On the other extreme, its length is in the order of 1 cm for drilling into an intact granite block between two adjacent joints, 0.1 mm for a fine-grained silicon oxide ceramic, and 10–100 nm in a silicone wafer (Bažant 2002).

2.5.3 Nonlinear Behavior of Concrete

The basic difference between fracture principles applied to different kinds of material is shown in Fig. 2.9 (Karihaloo 1995). The nonlinear zone ahead of the crack tip is composed of fracture process zone and hardening plasticity. LEFM is applied to the material in which the zone is very small (Fig. 2.9a). Figure 2.9b shows the second case of nonlinear behavior for ductile material in which the zone with nonlinear material behavior or hardening plasticity is large and the fracture zone is small. The third case (Fig. 2.9c) is applicable to nonlinear behavior of cementitious materials in which the fracture zone is large and hardening plasticity is small.

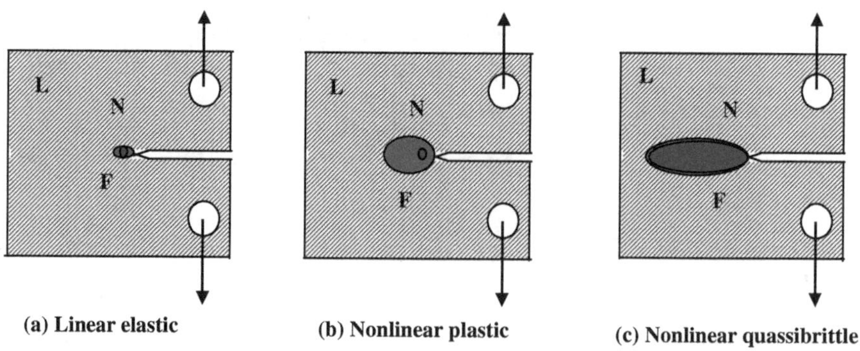

Fig. 2.9 Types of nonlinear zones in different types of materials, L denotes linear elastic, N denotes the nonlinear material behavior due to plasticity, and F denotes the fracture process zone

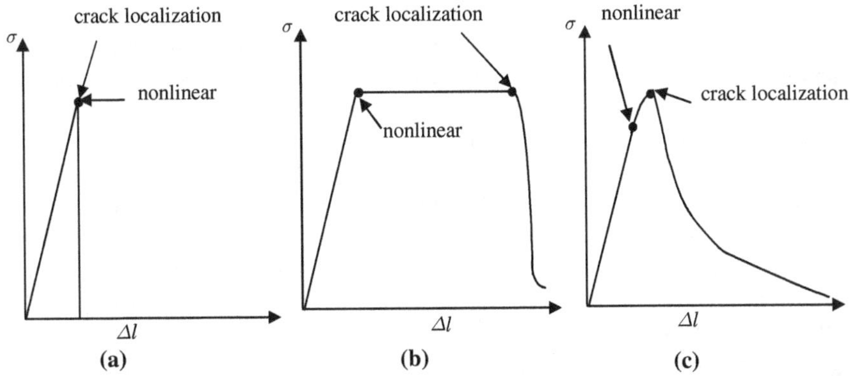

Fig. 2.10 Stress–displacement behavior under uniaxial tension: (**a**) brittle, (**b**) ductile, and (**c**) quasibrittle materials

In the other way, the nonlinear behavior of three different types of materials subjected to uniaxial tension can be shown using stress deformation relations (Fig. 2.10).

The brittle material shows a linear elastic behavior almost up to the peak load, and a catastrophic crack will propagate through the specimen just after the peak (Fig. 2.10a). The ductile material behavior shown in Fig. 2.10b is characterized by a pronounced yield plateau and the nonlinear behavior starts much before the onset of crack localization. In the *quasibrittle* material (Fig. 2.10c), the nonlinear behavior starts before the peak load and then upon arrival at the peak load, crack localization will occur. Consequently, the stress transfer capability of the material starts to reduce.

2.6 Specimen Geometry for Fracture Test of Concrete

Different specimen geometries have been used in the past to perform stable fracture testing of *quasibrittle* material like concrete. Uniaxial tensile test with displacement controlled as direct method can be used to characterize the fracture behavior of concrete. However, this test is difficult to carry out. Therefore, indirect method with various geometrical shapes is generally used to determine the fracture parameters. The commonly used shapes of test geometry for conducting indirect tests are three-point bending test (TPBT), compact tension (CT) test, and wedge splitting test (WST). The three-point bending geometry is a common specimen used to determine the fracture parameters applying the different fracture models. Fracture testing on TPBT specimen has an advantage that the testing can be performed with the standard testing machines and it is easier to perform the stable bending test on precracked beams. RILEM Technical Committee 50-FMC (1985) has recommended the guidelines for determination of fracture energy of cementitious materials conducting TPBT on notched beam. However, for large-size structures, the self-weight of the beam possibly negates the use of the specimen for fracture testing due to not only handling problems of testing specimens but also requiring special care in fracture analysis. In addition, it is not practically possible to use a drilled specimen from the construction sites or existing structures. Alternatively, CT test and WST have been used by many researchers in the past to determine the fracture parameters of concrete. The standard CT configuration is a single-edge notched plate loaded in tension. The ASTM Standard E-399 (2006) has specified the general proportions and standard configuration of a CT specimen. Many researchers (Wittmann et al. 1988, Brühwiler and Wittmann 1990) have used CT test to determine the specific fracture energy G_F of concrete. The CT specimen also suffers from some drawbacks: (i) there are no flexibilities in specimen shapes, so drilling of specimen from existing structures is difficult, (ii) there may be inconvenience with testing arrangement in directly applying the load on the specimen, and (iii) the testing can be carried out only in crack opening displacement (COD)-controlled test setup.

The wedge splitting test is a special form of the so-called compact tension test, in which a small specimen with a groove and a notch is split into two halves while monitoring the load and crack mouth opening displacement (CMOD). Linsbauer and Tschegg (1986) initially introduced the use of WST geometry for performing stable fracture tests on *quasibrittle* material and subsequently Brühwiler and Wittmann (1990) modified the testing arrangements. The use of WST specimen for stable fracture testing of concrete is now gaining more and more practical importance because of several advantages:

- The specimen is small and compact requiring less material.
- It has relatively larger ligament area-to-concrete volume ratio.
- Test results are least affected by self-weight of the specimen.
- It provides better stability during fracture test.
- Similar to TPBT geometry, the experiment on WST specimen is either controlled by constant rate of displacement of the wedges or in closed-loop COD control.
- Geometry of the specimen may be of cubical or cylindrical shapes. Hence, it is suitable for fracture testing from existing structures using drilled concrete cores.

2.6.1 Dimensions of Test Specimens

2.6.1.1 Three-Point Bending Test (TPBT)

The standard dimensions of the TPBT are shown in Fig. 2.11a in which the symbols *B*, *D*, and *S* are the width, the depth, and the span, respectively, for TPBT geometry with $S/D = 4$.

2.6.1.2 Compact Tension (CT) Test

The dimensions and configuration of standard CT specimen according to the ASTM Standard E-399 (2006) are shown in Fig. 2.11b. The dimensioning should comply $D_1 = 1.25D$, $H = 0.6D$, $H_1 = 0.275D$, and the specimen thickness $B = 0.5D$.

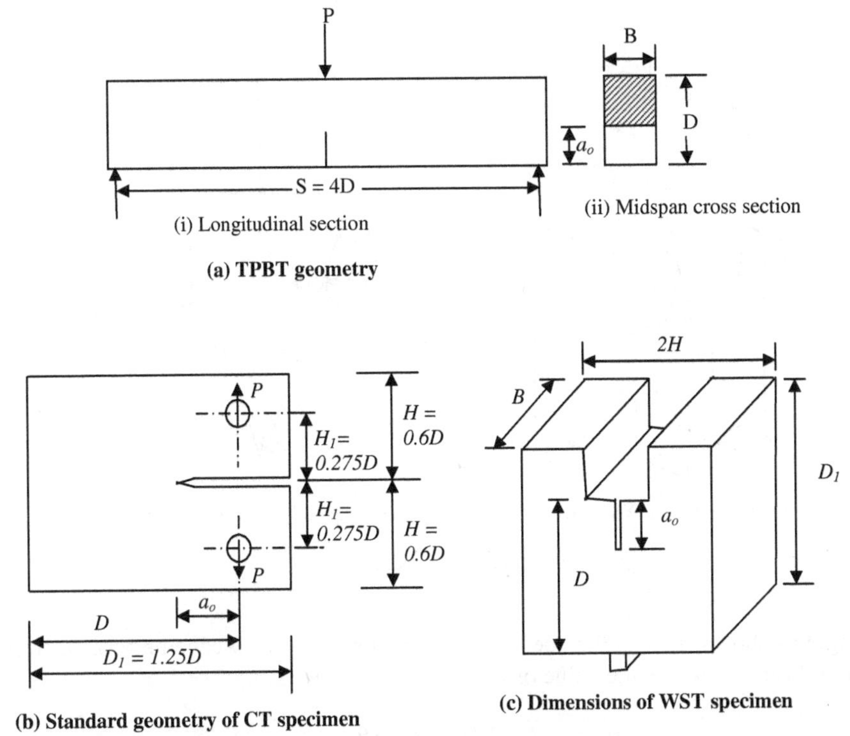

Fig. 2.11 (**a**) Dimensions of three-point bending test (TPBT), (**b**) compact tension (CT), and (**c**) wedge splitting test (WST) specimens

2.6.1.3 Wedge Splitting Test (WST)

The only special requirement for WST geometry is to use additional two massive steel-loading devices equipped with the roller bearings on each side of the specimen. The shape and dimensions of commonly used WST specimen are shown in Fig. 2.11c, whereas the testing arrangement and forces acting on the specimen are shown in Fig. 2.12.

The dimensions of D_1, $2H$, and B of WST specimens made of concrete may be slightly different from the standard CT specimen due to practical requirements and convenience. Nevertheless, because of the similarity in specimen geometry and loading condition between wedge splitting and compact tension tests, the LEFM formula to calculate the fracture parameters of wedge splitting specimens is the same as applied to the CT specimens. In wedge splitting test, the vertical load P_V and the crack opening displacement (COD) are recorded and the horizontal force P_H is computed using Eq. (2.8), whereas the horizontal force is directly recorded in CT specimen.

In Fig. 2.12b, P_V is the applied vertical load, N is the normal reaction, F_f is the friction force, P_H is the horizontal force acting on the specimen, and θ is the wedge

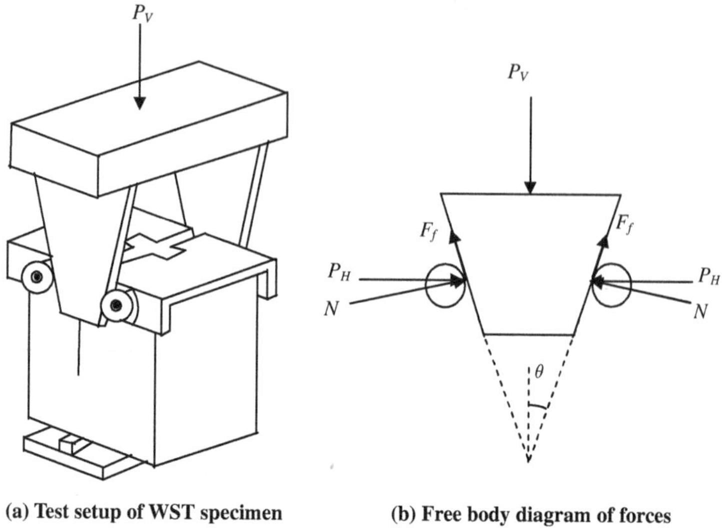

Fig. 2.12 Details of wedge splitting test (WST) specimen

angle as shown. If μ is the coefficient of friction between wedge and roller, from equilibrium of forces the value of P_{H} can be expressed as

$$P_{\mathrm{H}} = \frac{1 - \mu \tan \theta}{2(\mu + \tan \theta)} P_{\mathrm{V}} \tag{2.8}$$

Neglecting the friction force being very small and considering small wedge angle, viz. $\theta = 15°$, Eq. (2.8) reduces to

$$P_{\mathrm{H}} \approx \frac{P_{\mathrm{V}}}{2 \tan \theta} = 1.866 P_{\mathrm{V}} \tag{2.9}$$

Equation (2.9) shows that if even a smaller vertical force is subjected on the specimen, it is able to produce magnified horizontal force which reduces the need of higher capacity testing machine. For convenience, the horizontal load acting on WST specimen will be denoted as *P* from now onward.

2.7 Nonlinear Fracture Mechanics for Concrete

The nonlinear fracture models are based on two basic approaches: first using finite element method or boundary element method and second using modified LEFM concept. The cohesive crack model (CCM) or fictitious crack model (FCM) and the crack band model (CBM) fall under the former category, while two-parameter fracture model (TPFM), size-effect model (SEM), effective crack model (ECM), K_{R} curve based on cohesive force, double-*K* fracture model (DKFM), and double-*G* fracture model (DGFM) belong to the latter group. The fracture models adapting to LEFM do not directly use the initial crack length a_o. However, a critical effective crack length a_c is calculated based on experiments assuming that when peak load is attained, the far field may be described by an equivalent elastic crack, $a_c = a_o + \Delta a_c$, where Δa_c is the equivalent extension in stress-free crack length at peak load, and the corresponding stress intensity factor $K_I = K_{IC}$ is determined using LEFM principles. A brief review of each fracture model is given in the subsequent sections.

2.7.1 Cohesive Crack Model (CCM) or Fictitious Crack Model (FCM)

The cohesive crack method was first proposed by Barenblatt (1959, 1962) and Dugdale (1960). While Barenblatt applied cohesive crack to analyze the brittle fracture behavior, Dugdale introduced it to model ductile fracture behavior of a material. Under discrete cracking approach (Ngo and Scordelis 1967), Hillerborg et al. (1976) initially applied cohesive crack method (or fictitious crack model) to simulate the softening damage of concrete structures. In the pioneer work they showed that analysis of crack formation and propagation as well as failure analysis can be done with cohesive crack model even if coarse finite element is used. This method thereby eliminates the mesh sensitivity. Subsequently, the cohesive crack method has been modified and used by many researchers over the period of time.

Petersson (1981) employed a boundary integral method in which the kernel was discretized a priori using finite element formulation. Wecharatana and Shah (1983) presented a theoretical model similar to cohesive crack model (Barenblatt 1962, Dugdale 1960, Hillerborg et al. 1976) to predict the extent of nonlinear zone around the crack tip which depends on the size of the aggregate and the geometry of the specimen. The model is based on simple and approximate extensions of the concepts of LEFM. The model was successfully used to analyze the results of the experiments on double-cantilever, double–torsion, and the notched-beam specimens.

Ingraffea et al. (1984) predicted numerically the bond-slip behavior of a reinforcing bar embedded in concrete. The radial secondary cracking was modeled using a nonlinear discrete crack, interface, and finite element approach having a tension-softening element. As an example, a reinforced concrete tension diaphragm was analyzed.

In a finite element code, the fictitious crack model was extended by Ingraffea and Gerstle (1984) to simulate mixed-mode crack propagation analysis, employing six-node interface elements. However, their crack propagation algorithm based on stress intensity factors and their nonlinear algorithm were inefficient.

Barker et al. (1985) reported a series of tests on concrete crack-line wedge-loaded double-cantilever beams using a replica technique that provides an easy and accurate method for determining the total extent of cracking in concrete specimens. The process zone length ahead of a macro-crack tip is generated by the formation and coalescence of numerous microcracks. It was shown that the process zone, characterized by the two critical crack-tip opening displacements for determining the necessary condition for crack growth, continues to increase in length with increasing load. From the analysis of the results, it was concluded that LEFM techniques are not suited to predicting unstable cracking in concrete members of the proportions normally used in buildings.

Li and Liang (1986) suggested that process zone length in concrete and fiber-reinforced concrete is not to be a material property but depends on the specimen and the loading configuration. The overall mechanical behavior of these materials could be strongly influenced by the stress–separation constitutive relation.

Carpinteri (1989a–c, 1990) and Carpinteri and Colombo (1989) modified the Petersson's method of cohesive crack method to study the size effect of concrete fracture on the global structural behavior which can range from ductile to brittle when strain softening and strain localization are taken into account. The size-scale transition from ductile to brittle behavior is governed by the nondimensional brittleness number. They showed that for extremely brittle cases (e.g., initially uncracked specimens, large and/or slender structures, low fracture toughness, and high tensile strength), a snapback in the equilibrium path occurs and the load–deflection softening branch assumes a positive slope. Both load and deflection must decrease to obtain slow and controlled crack propagation. If the loading process is deflection controlled, the load-carrying capacity yields a discontinuity with a negative jump. It was proved that such a catastrophic event tends to reproduce the classical LEFM instability for small fracture toughness, high tensile strengths, and/or for large structure sizes.

Bocca et al. (1991) extended the cohesive crack model to mixed-mode crack propagation problem along with an experimental confirmation by testing four-point shear specimens of concrete. A constant crack mouth sliding displacement rate was imposed so that it was possible to control and detect the snapback load vs. deflection branches. The behavior of the larger specimens was found to be more brittle. On the other hand, the brittleness in the numerical simulations was controlled through the crack length, which was certainly a monotonic increasing function during the irreversible fracture process. The experimental load vs. deflection diagrams as well as the experimental fracture trajectories were captured satisfactorily by the numerical model. The mixed-mode fracture energy results tend to be of the same order of magnitude as the mode I fracture energy.

Planas and Elices (1991) introduced an enhanced algorithm for the solution of nonlinear simultaneous equations in the cohesive crack model. In the enhanced algorithm, the system of equations can be partitioned and a partial solution of the system is carried out during the crack propagation study. For very large specimen sizes, an asymptotic analysis was developed by the authors that allowed an accurate treatment of the cohesive zones and provided a powerful framework for theoretical development.

Gopalaratnam and Ye (1991) developed a fictitious crack model based on noniterative numerical scheme to study the development and growth of fracture process zone of different specimen sizes and two specimen geometries such as three-point bending test and compact tension test using linear and nonlinear softening laws. It was observed that the process zone reaches a steady-state length, which is dependent on specimen size and geometry. The numerical results reproduce many of the other experimentally observed characteristics in the fracture of plain concrete.

Employing a mixed-mode extension of the Hillerborg fictitious crack model, a practical interface finite element approach was used by Gerstle and Xie (1992) to model discrete crack propagation. The concept of fracture process zone is broadened to interface process zone, which includes the nonlinear behavior of the fracture process zone as well as the crack face interference behind the fracture process zone. The fictitious crack is represented by interface elements with linearly varying displacements; however, the traction distribution is arbitrary. In finite element code, an explicit incremental iterative procedure using dynamic relaxation was adopted which made the software simple. Examples demonstrate that the assumed assumptions allow the fictitious crack to be modeled using few degrees of freedom without compromising accuracy.

Liang and Li (1991a, b) developed a boundary element numerical approach to simulate crack propagation study using fictitious crack model. The model can determine the peak load solution in a nondimensional form without knowing the material properties in advance. The numerical parametric study examined the effects of specimen size, loading configuration, and the initial notch size on the nonlinear characteristics of concrete. It was also shown that the normalized energy release rate (*R* curve) is dependent on specimen size, initial notch length, and the loading configuration of mode I concrete fracture. Li and Liang (1992, 1993) proposed a new mathematical formulation of a boundary eigenvalue problem for the fictitious crack model using linear softening law. The theoretical foundation of the model was based on the second variation of the potential energy principle to displacement to describe the equilibrium and stability behavior of the cohesive crack model. Further, a new numerical algorithm for the same problem with linear softening law was developed in the framework of the finite element method (Li and Liang 1994). Li and Bažant (1994) reduced the cohesive (or fictitious) crack model into the eigenvalue problem to calculate the maximum load of geometrically similar structures of different sizes without calculating the entire load–deflection curve. They incorporated the nonlinear power function for the softening law.

Xie and Gerstle (1995) presented an energy-based approach for the FEM modeling of mixed-mode cohesive crack propagation. A two-dimensional automatic mixed-mode discrete crack propagation modeling program was developed for both nonlinear and linear elastic crack propagation problems. The numerical efficiency and convergence behavior of the approach was demonstrated through two examples: a three-point bending problem and a single-edge notched shear beam.

Saleh and Aliabad (1995) extended the dual boundary element method for modeling the fictitious crack model to simulate the pure mode I and mixed-mode crack propagation in concrete. The numerical simulation is controlled through the crack length as a monotonic increasing function during the fracture process. Three-point bending and four-point shear specimens were used to check the numerical results.

Mulmule and Dempsdey (1997) developed a fictitious crack model using weight function approach to study the fracture behavior of saline ice. The model was validated using the benchmark problem of concrete specimen of a center-cracked panel of finite width subjected to remote uniform tensile stress using a linear stress–displacement softening law (Li and Liang 1986).

For a *global* mixed-mode loading of concrete fracture, the authors (Gálvez et al. 1998, Cendón et al. 2000) showed that the crack grows with a predominantly *local* mode I fracture. Based on the cohesive crack concept of the local mode I approach, a simple and efficient numerical procedure for mixed-mode fracture of *quasibrittle* materials was presented by Cendón et al. (2000). This technique predicts crack trajectories as well as load–displacement responses. The numerical results agree quite well with different experimental sets of mixed-mode fracture of concrete beams.

Gálvez et al. (2002) introduced the formulation of the classical plasticity for a numerical simulation for mixed-mode fracture of *quasibrittle* materials using the cohesive crack approach. The numerical results agreed well with the experimental results of mixed-mode fracture of concrete beams.

Planas and Elices (1992) presented a method to analyze the evolution of a cohesive crack, particularly appropriate for asymptotic analysis. This method further improved the solution algorithm of the system of equations and later this method was modified by Bažant (1990), Bažant and Beissel (1994), and Bažant and Zi (2003) as smeared-tip superposition method.

The derivation of the cohesive crack problem was based on an energetic variational formulation. The energy formulation also provides the condition for the loss of stability of a structure with a propagating cohesive crack. Bažant and Li (1995a, b) obtained an eigenvalue problem for a homogeneous Fredholm integral equation, with the structure size as the eigenvalue exploiting the criterion of stability limit. The preceding eigenvalue method was further extended by Zi and Bažant (2003) to the cohesive crack model with a finite residual stress. The extended method allows an efficient calculation of the size-effect curves for the cohesive crack model with residual stress, for both positive and negative–positive specimen geometries.

A discrete computational model CoMICS (computational model for investigation of cracks in structural concrete) for study of crack formation and crack growth in plain and reinforced concrete plane stress members was presented by Prasad and Krishnamoorthy (2003) in a more generalized way. The model CoMICS required successive regeneration of the mesh in the vicinity of the crack tip in which the mesh modifications were limited to a selected patch of elements at the crack tip. With regard to plain concrete member, the global response of load deformation plots for uncurved and curved mode I cracks using CoMICS was compared with those experimental results and numerical models available in the literature. The study was focused to analyze reinforced concrete (RC) structural members with regard to the formation and growth of cracks. Nonlinear behavior due to tensile cracking of concrete and steel–concrete interface bond was considered using the bond-slip model for bimaterial contact. The use of plain concrete crack model in conjunction with bond-slip model gives good correlation with experimental data of RC structures.

Chudnovsky and collaborators (Issa et al. 1993, 2003) have characterized the fracture surfaces of concrete using simple nondestructive modified slit-island experimental technique. They showed that the main crack can deflect around the aggregates and/or penetrates through the aggregates depending upon the strength of the cement paste and the aggregate location with respect to the crack front. This phenomenon results in the energy consumption and affects the apparent toughness of the material. It was observed that the fracture surfaces of concrete appeared to be relatively rougher for the larger size aggregates than for the smaller size aggregates. Extensive experimental studies using wedge splitting concrete specimens of different sizes, made with varying sizes of aggregates and numerical studies (Issa et al. 2000a, b), revealed that the energy release rate increases with crack extension (*R*-curve behavior), specimen size, and size of aggregates. Fracture energy also increases with the specimen size and the size of aggregate particles. The crack trajectory deviates from the rectilinear path more in the specimens with larger aggregate sizes. Fracture surfaces with larger size aggregate show higher roughness than do those with smaller size aggregates. For similar concrete specimens, the crack tortuosity is greater for the larger size specimens.

The cohesive zone model is a strictly uniaxial method considering zero width of the process zone. The general application in structures in which triaxial stresses arise is questionable. This method is not suitable for large-scale analyses. Despite these limitations, the cohesive crack model is an idealized approximation of a physical localized fracture zone. This model has a simple mathematical procedure and it retains the real physical process. The basic essence of this model is the ability to describe the nonlinear material behavior in the vicinity of a crack and at the crack tip.

A brief overview of the cohesive crack model, its numerical aspects, advantages, limitations, and challenges can be found in the recent literatures (Guinea 1995, Elices and Planas 1996, Bažant 2002, Elices et al. 2002, de Borst 2003, Planas et al. 2003, Carpinteri et al. 2003, 2006).

The application of cohesive crack model in the recent time is also many. Kim et al. (2004) applied cohesive crack model on wedge splitting concrete specimen at early ages to compare the load–crack mouth opening displacement (P-CMOD) curves obtained between the finite element analysis and the experiments for the characterization of bilinear softening law. The derived fracture parameters and bilinear softening curves at early ages of concrete may be used for fracture criterion and input data in finite element analyses.

Raghu Prasad and Renuka Devi (2007) modified appropriately the fictitious crack model to make it applicable for plain concrete beams with vertical tortuous cracks. To model the roughness of fracture surface of concrete, a number of tortuous cracks were generated considering the crack deviations as random variable based on the maximum aggregate size. Plain concrete beams with such tortuous cracks were analyzed to study the effect of the tortuosity of the cracks on the various fracture parameters.

Roesler et al. (2007) developed finite element-based cohesive zone mode using bilinear softening to predict the monotonic P-CMOD plot of three-point bending test specimens having different sizes. Experimental results on the companion beams were used to compare the numerical results of the fracture behavior of concrete. The authors (Park et al. 2008) further characterized the bilinear softening law by using experimental results for predicting the fracture behavior of concrete. The location of the kink point of the softening law was proposed to be the stress at the critical crack-tip opening displacement. This location of the kink point is similar to the modified bilinear softening function proposed by Xu and Reinhardt (1999a) and it compares well with the range of kink point locations reported in the literature. The proposed bilinear softening law was further verified from the numerical simulations with the cohesive zone model.

In an experimental and numerical analyses, the authors (Zhao et al. 2008, Kwon et al. 2008) presented experimental results of a three-point bending test for a notched beam and a wedge splitting test with different size specimens for 10 different concrete mixes to investigate the effect of specimen size and geometry on the fracture energies. It was observed that the fracture energy increases with an increase in specimen size for both the specimen geometries. In order to find out the softening curve using inverse analysis that optimally fits the measured P-CMOD curve, a cohesive crack model (Gopalaratnam and Ye 1991) and an optimization technique were incorporated. The effect of size-dependent fracture energy on the softening curve was then investigated. It was found that the first steep branch of the softening curve is very similar for different size specimens, whereas the tail of the curves becomes longer with increase in the specimen size.

Elices et al. (2009) performed fracture tests on six types of simple concrete made with three types of spherical aggregates and of two kinds of matrix interface to study the influence of the matrix–aggregate interface and of the aggregate strength on the softening curve. A cohesive crack model with bilinear softening function was used to model the fracture behavior of concrete. The highest specific fracture energy values were determined with strong aggregates well bonded to the matrix because the crack path avoids the aggregates and wanders through the matrix resulting in the increased value of the fracture area. In general, the cohesive strength of all the concretes was found to be lower than that of the matrix. The critical crack opening value appeared to be almost insensitive to the type of interface and matrix of concrete in the case of strong aggregates were used, whereas this opening was larger than the corresponding matrix in the case of weak aggregates. The shape of the softening function was almost the same for the concrete and the matrix when strong aggregates were used and the same was different for the concrete when the weak aggregates were used. Therefore, weak interfaces provide larger initial slope of softening function, larger critical opening, and smaller cohesive strength than do strong interfaces.

2.7.1.1 Material Properties for CCM or FCM

Three material properties such as modulus of elasticity E, uniaxial tensile strength f_t, and specific fracture energy G_F are required to describe the cohesive crack model. Difficulty with FCM lies in being able to experimentally determine the values of G_F, f_t, and w_c for a material. The G_F is defined as the amount of energy necessary to create one unit of area of a crack. Hillerborg (1985a) and RILEM Technical Committee 50-FMC (1985) proposed a method using three-point bending beams to obtain the

Fig. 2.13 Work of fracture as an area between load–displacement (P–δ) curve

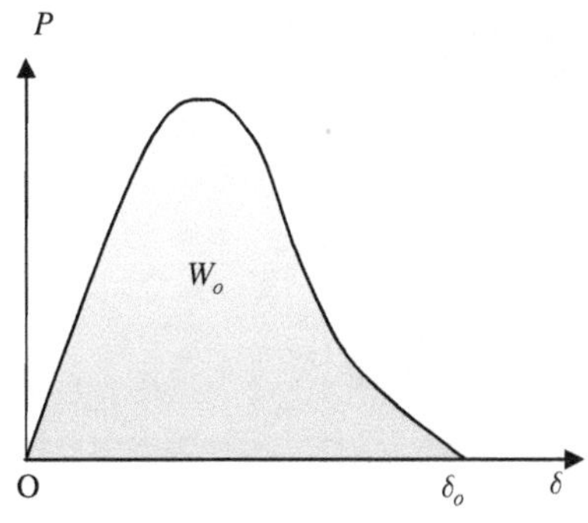

values of G_F. According to this method the area between the load–displacement ($P-\delta$) curve obtained experimentally is corrected for eventual nonlinearities at low loads, self-weight of the specimen, and other loading due to fixtures. The fracture energy is calculated from the equation

$$G_F = \frac{W_o + W_s\delta_o}{A_{lig}} \tag{2.10}$$

in which W_o is the area according to Fig. 2.13, W_s is the sum of the weight of the specimen between the supports and twice that of the part of the loading arrangement, and δ_o is the deformation at the final failure of the beam.

Hillerborg (1985b) conducted a study on the determination of the fracture energy from the tests data of about 700 mortar and concrete beams obtained from different laboratories of the world. This study showed that fracture energy is apparently dependent on the specimen size. Subsequently, Planas and co-workers (Guinea et al. 1992, Planas et al. 1992, Elices et al. 1992) carried several investigations on the measurement of the fracture energy using three-point bending tests and they showed that if different sources of energy dissipation during the test and analysis are taken into account, the value of G_F appears to be almost size independent.

The nonlinearity in the cohesive crack model is automatically introduced by using stress–crack opening displacement relations (also known as softening function) across the crack faces near the crack tip, which makes the SIF at the propagating crack tip to be zero. Hence, it is one of the fundamental material properties used as input in cohesive crack model. This constitutive law relates the cohesive stress (σ) across the crack faces and the corresponding crack opening displacement w, i.e., $\sigma = f(w)$. The terminal point of the $\sigma - w$ curve w_c is the maximum widening of the fracture zone when it is still able to transfer stress. From the definition of softening law, $f(0) = f_t$ = direct tensile strength of concrete. The G_F is graphically equal to the area under the $\sigma - w$ curve as shown in Fig. 2.14.

Hillerborg et al. (1976) initially used linear softening law in the finite element analyses considering f_t and G_F as material properties. Later Petersson (1981) proposed bilinear (two-line approximation) softening for notched concrete beam

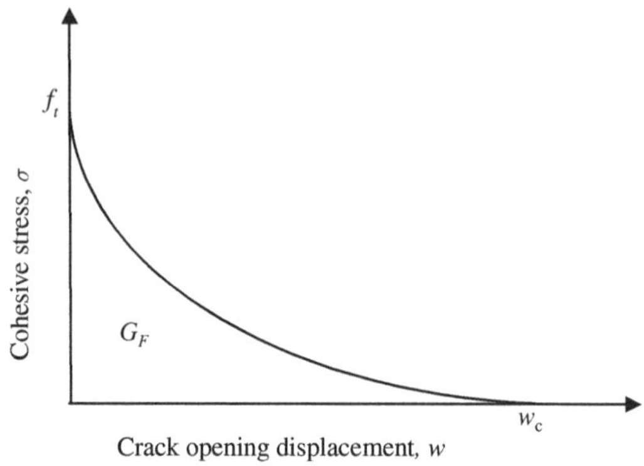

Fig. 2.14 Representation of softening function of concrete

which was in good agreement with the experimental results. Thereafter with some modifications, the bilinear softening curve and its parameters have been widely discussed and used by many researchers (Alvaredo and Torrent 1987, Wittmann et al. 1988, Hilsdorf and Brameshuber 1991, Guinea et al. 1994, Bažant 2002, Bažant and Becq-Giraudon 2002). On the basis of theoretical considerations and available experimental data, Hilsdorf and Brameshuber (1991) verified the fracture properties and stress deformation characteristics of concrete as adopted in CEB-FIP Model Code-1990 (1993). Various other fundamental softening relationships such as trilinear (Cho et al. 1984), exponential (Gopalaratnam and Shah 1985, Karihaloo 1995), nonlinear (Reinhardt et al. 1986), and quasi-exponential (Planas and Elices 1990) have also been developed. Xu (1999) proposed a formulized method to determine the parameters of bilinear softening, nonlinear softening, and exponential softening curves for different concrete grades and the maximum sizes of aggregates.

2.7.2 Crack Band Model (CBM)

Strain-softening behavior of concrete in the smeared-cracking framework was first introduced by Rashid (1968). Gradually, extensive research showed that the classical smeared-cracking concept is unobjective showing spurious mesh sensitivity, no size effect, and convergence to an incorrect failure mode with zero energy dissipation because of localizing to a zone of zero volume. Adapting to Hillerborg et al. (1976), Bažant and Oh (1983) developed the crack band model in which the fracture process zone is modeled as a system of parallel cracks that are continuously distributed (smeared) in the finite element. The material behavior is characterized by the constitutive stress–strain relationship. The width of the fracture process zone (h_c) is assumed to be constant. For normal concrete it is assumed to be three times the aggregate size. The width of the crack band is held constant in order to avoid spurious mesh sensitivity. This assures that the energy dissipation due to fracture per unit length of crack is a constant which is equal to the fracture energy of the material (G_F). In this model, the crack is modeled by changing the isotropic elastic

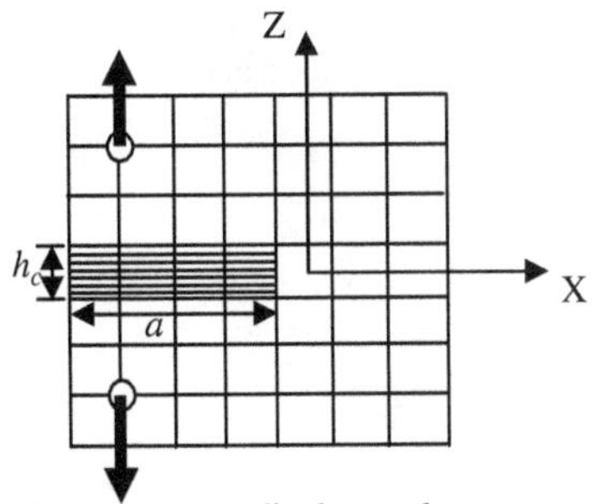

Fig. 2.15 Cartesian coordinate of crack band model

moduli matrix to an orthotropic one, thereby reducing the stiffness in the direction normal to the cracking plane. The softening behavior of concrete is modeled by superimposing the fracturing strain ε_f on the elastic strain.

A brief mathematical background (Bažant and Oh 1983) of this model is given below by considering a system of Cartesian coordinates as shown in Fig. 2.15. If concrete is idealized as homogeneous material, the triaxial stress–strain relationship can be expressed as follows:

$$\begin{Bmatrix} \varepsilon_x \\ \varepsilon_y \\ \varepsilon_z \end{Bmatrix} = \frac{1}{E}\begin{bmatrix} 1 & -\nu & -\nu \\ -\nu & 1 & -\nu \\ -\nu & -\nu & 1 \end{bmatrix}\begin{Bmatrix} \sigma_x \\ \sigma_y \\ \sigma_z \end{Bmatrix} + \begin{Bmatrix} 0 \\ 0 \\ \varepsilon_f \end{Bmatrix} \tag{2.11}$$

where σ_x, σ_y, and σ_z are the principal stresses and ε_x, ε_x, and ε_x are the principal strains; ε_f is the fracture strain by the opening of microcracks, E is the Young's elastic modulus of concrete, and is its Poisson's ratio. As the microcracks develop in the material, the effect of these opening microcracks does not cause any effect on the strains in "X"- and "Y"-directions ("Y"-axis perpendicular to the plane, see Fig. 2.15). The width of the fracture front h_c is assumed to be a material constant which can be determined from experiments and is proportional to the aggregate size for plain concrete. The fracturing strain ε_f is determined by summing all the deformation or openings of individual microcracks, $\delta_f = \sum_i \delta_f^i$ is the sum of the openings of individual microcracks intersecting axis "Z," over the width h_c. The relation of normal stress σ_z and the relative displacement δ_f across a line crack is identical to the relation of σ_z to the displacement:

$$\delta_f = \varepsilon_f h_c \tag{2.12}$$

The fracture starts when the stress at the crack tip reaches the tensile strength; at that point, ε_f is still zero. As the crack opens, δ_f starts to increase and σ_z starts to gradually decline. A simple choice for this modeling could be a linear function as shown in Fig. 2.16. Fracturing strain ε_f can be represented as a function of stress σ_z as follows:

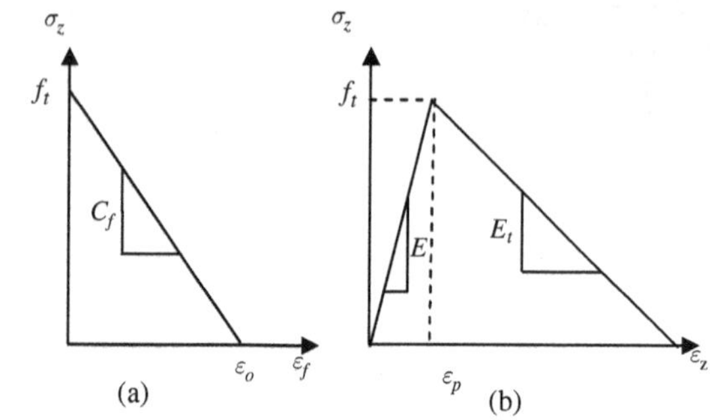

Fig. 2.16 The stress–strain relationship in crack band model

$$\varepsilon_{\mathrm{f}} = f(\sigma_{\mathrm{z}}) = \frac{1}{C_{\mathrm{f}}}(f_{\mathrm{t}} - \sigma_{\mathrm{z}}) \tag{2.13}$$

where $C_{\mathrm{t}} = f_{\mathrm{t}}/\varepsilon_{\mathrm{o}}$, ε_{o} is the strain at the end of strain softening (Fig. 2.16a) at which the microcracks coalesce into a continuous crack and σ_z vanishes. From Eqs. (2.13) and (2.11), the relation (2.11) is modified as

$$\begin{Bmatrix} \varepsilon_x \\ \varepsilon_y \\ \varepsilon_z \end{Bmatrix} = \frac{1}{E}\begin{bmatrix} 1 & -\nu & -\nu \\ -\nu & 1 & -\nu \\ -\nu & -\nu & \frac{E}{E_{\mathrm{t}}} \end{bmatrix}\begin{Bmatrix} \sigma_x \\ \sigma_y \\ \sigma_z \end{Bmatrix} + \begin{Bmatrix} 0 \\ 0 \\ \varepsilon_{\mathrm{o}} \end{Bmatrix} \tag{2.14}$$

in which the post-peak tensile stress–strain relationship is described by the tangent modulus E_{t}, which is defined as

$$\frac{1}{E_{\mathrm{t}}} = \frac{1}{E} - \frac{1}{C_{\mathrm{f}}} \leq 0 \tag{2.15}$$

The fracture energy G_{F} that is defined as the energy absorbed in creating (opening) of all cracks is given by

$$G_{\mathrm{F}} = h_{\mathrm{c}} \int_{0}^{\varepsilon_{\mathrm{o}}} \sigma_z(\varepsilon)\mathrm{d}\varepsilon \tag{2.16}$$

The integral represents the area under the stress–strain curve after the stress has reached f_{t}. If a linear relationship between stress–strain is assumed, then

$$G_{\mathrm{F}} = h_{\mathrm{c}}\frac{1}{2}f_{\mathrm{t}}\varepsilon_{\mathrm{o}} \text{ or } G_{\mathrm{F}} = \frac{f_{\mathrm{t}}^2}{2C_{\mathrm{f}}}h_{\mathrm{c}} \tag{2.17}$$

If G_{F}, f_{t}, and h_{c} are known from experiments, then the basic parameters of the stress–strain relation can be calculated as follows:

$$C_{\mathrm{f}} = \frac{f_{\mathrm{t}}^2 h_{\mathrm{c}}}{2G_{\mathrm{F}}}, \ \varepsilon_{\mathrm{o}} = \frac{f_{\mathrm{t}}}{C_{\mathrm{f}}} = \frac{2G_{\mathrm{F}}}{f_{\mathrm{t}} h_{\mathrm{c}}} \tag{2.18}$$

Alternatively, it is also possible to express the fracture energy as the total area under the stress–strain relationship. Using Eq. (2.15), it can be shown that

$$G_{\mathrm{F}} = \frac{1}{2}\left(\frac{1}{E} - \frac{1}{E_{\mathrm{t}}}\right) f_{\mathrm{t}}^2 h_{\mathrm{c}} \tag{2.19}$$

The crack band model is a smeared-cracking method based on the concept of classical continuum mechanics in which constitutive relations are the stress–strain curve with softening along with the condition of the minimum size of the localization zone. This length scale is numerical in nature which has to be accurately ascertained according to a particular problem.

The crack band model suffers from many limitations (Bažant and Lin 1989) such as (1) refinement of mesh size cannot be smaller than the cracking zone width, (2) zigzag crack band propagation needs special mathematical treatment, (3) a typical square mesh introduces a certain degree of directional bias, and (4) possible variations of the cracking width and hence fracture energy cannot be taken into account. To improve such deficiencies, a nonlocal smeared-cracking formulation has been extended in the framework of damage mechanics and metal plasticity by Bažant and co-workers to introduce the length scale in a more systematic approach. And the available literature is so vast that a complete review is not carried out because majority of studies presented in the book are beyond this scope. Further, some overview of the concrete fracture model based on distributed cracking can be found elsewhere (Bažant 1986, 2002, Planas et al. 1993a, de Borst 2003, de Borst et al. 2004).

In smeared context, the strong discontinuity approach (Simo et al. 1993, Oliver 1995, Jiraśek and Zimmermann 2001a, b) has been also applied in recent times for cohesive zone modeling. The concept of discontinuous displacement has been introduced within a finite element analysis using partition-of-unity property which can avoid the numerical difficulties associated with the smeared cohesive zone modeling. This method is known as the extended finite element method (XFEM) (Wells and Sluys 2001, Mos and Belytschko 2002, Mariani and Perego 2003, Zi and Belytschko 2003, Mergheim et al. 2005, Meschke and Dumstorff 2007). In the XFEM, unlike discrete crack models, cracks are not limited to element boundaries but can be located in the FE mesh.

2.7.3 Two-Parameter Fracture Model (TPFM)

The two-parameter fracture model was developed by Jenq and Shah (1985a, b). In this model, the actual crack is replaced by an equivalent fictitious crack. The model involves two valid fracture parameters for cementitious materials: the critical stress intensity factor $K_{\mathrm{IC}}^{\mathrm{s}}$ at the tip of the equivalent crack length at peak load and the corresponding value of the crack-tip opening displacement (CTOD) known as

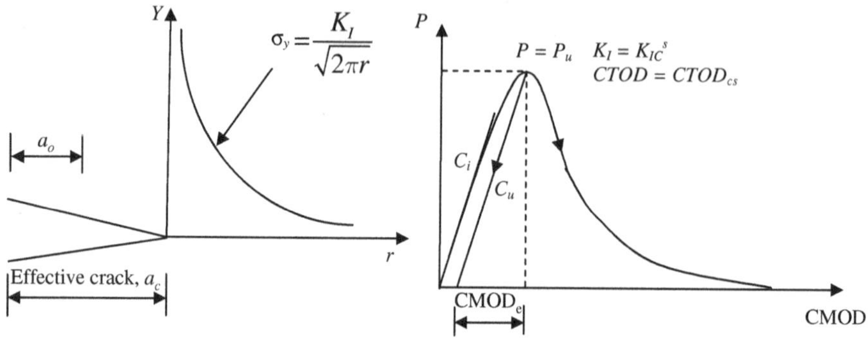

(a) (b)

Fig. 2.17 (**a**) Equivalent Griffith crack; (**b**) typical load–CMOD curve for TPFM

critical crack-tip opening displacement $CTOD_{cs}$. The loading and unloading crack mouth opening displacement compliances of standard three-point bending specimen were used to determine the value of a_c (Fig. 2.17).

A brief procedure involved in the calculation of the two parameters K_{IC}^s and $CTOD_{cs}$ for three-point bending specimens is mentioned in the RILEM formula (RILEM Technical Committee 89-FMT, 1990a). The standard LEFM formula given by Tada et al. (1985) is used to calculate the critical value of stress intensity factor. The main advantage of this method is that only a single size of three-point bending specimen is needed in the experiment.

The value of K_{IC}^s obtained from this model has been shown to be essentially not dependent on the geometries of the specimens. The results obtained by Jenq and Shah (1988a) from compact tension tests, wedge splitting cube tests, and large tapered double-cantilever beams showed that two parameters K_{IC}^s and $CTOD_{cs}$ might be considered as geometry-independent fracture parameters. However, an investigation done by Jenq and Shah (1988b) indicated a significant influence of geometry. Other researchers reported substantial size effect both on K_{IC}^s and $CTOD_{cs}$ (Planas and Elices 1989, 1990). In determination of two parameters K_{IC}^s and $CTOD_{cs}$, the inelastic part of the total CMOD part is neglected which may possibly lead to an underestimated value of a_c because the nonlinear fracture characteristic of concrete is associated with the stable crack propagation. This method needs a closed-loop testing system to determine the fracture parameters.

2.7.4 Size-Effect Model (SEM)

All important nonlinear fracture models capture adequately the structural size effect over the useful range of applicability. There are two extremes of size-effect law: (i) strength criteria and (ii) LEFM size effect. The former yields no size effect, whereas the latter shows the strongest size effect, i.e., nominal strength is inversely

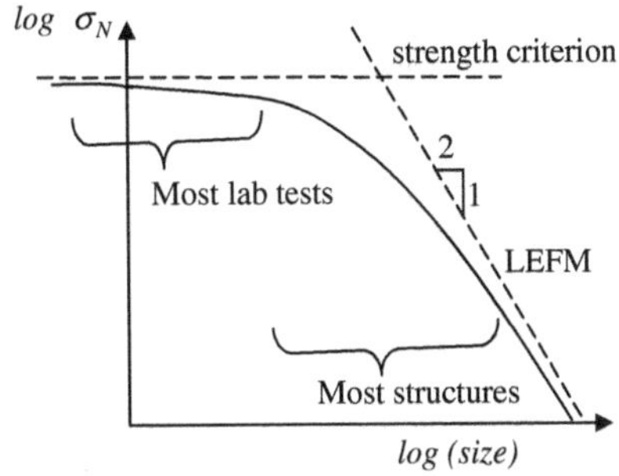

Fig. 2.18 Size effects as a plot of nominal strength vs. size on a bilogarithmic scale

proportional to the square root of the structural dimension, as shown in Fig. 2.18. A *quasibrittle* material like concrete exhibits a transitional size effect between the two extremes of size effects. Bažant and co-workers (Bažant 1984, 2002, Bažant et al. 1986, 1991, Bažant and Kazemi 1990, 1991, Gettu et al. 1990, Bažant and Jirăsek 1993) introduced size-effect model, which describes the material fracture behavior using two parameters: the fracture energy G_f (or denoted here as G_{FB}) and critical effective crack length extension c_f at peak load for infinitely large test specimen. The fracture parameters are determined from the maximum loads of geometrically similar notched specimen of different sizes. Bažant and Planas (1998) have presented a simple and elegant approach showing the transitional size effect. Assume a notched plate of width D and thickness B (Fig. 2.19) in which during the infinitesimal crack extension Δa, the approximate energy release $2\,k(a_o+c_f)B\Delta a\sigma_N^2/2E$ must be equal to energy required for crack extension, $BG_{FB}\,\Delta a$. Therefore

$$2\,k(a_o+c_f)B\Delta a\sigma_N^2/2E = BG_{FB}\,\Delta a \tag{2.20}$$

which yields

$$\sigma_N = \frac{B' f_t}{\sqrt{1+\beta}} \tag{2.21}$$

in which $\beta = (a_o/c_f) = (D/D_o)$ =constant, f_t is the tensile strength of the material, B' is the dimensionless constant, D_o is the a constant having dimension of length, c_f is a material constant in terms of elastic equivalent crack extension (for present derivation), and G_{FB} is the fracture energy, a material constant.

Equation (2.21) represents the transitional size effect for most practical size range of concrete structures and it explains the two extremes of size effects exhibited by plastic limit analysis and LEFM. The reason of size effect in concrete structures recognized is the existence of large and variable length of FPZ ahead of the crack tip. Determination of the nonlinear fracture properties using size-effect model is described in RILEM method (1990b). According to this procedure, for TPBT specimen (Fig. 2.11a), the maximum load P_u including self-weight per unit of beam w_g is expressed as

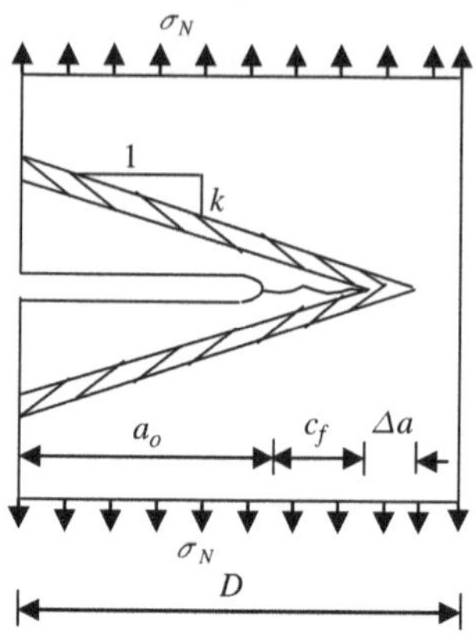

Fig. 2.19 Energy transfer during infinitesimal crack extension in a slit-like process zone

$$P_u = P_{u,test} + w_g S/2 \tag{2.22}$$

Expressing $\sigma_N = P_u/BD$, $\sigma_{Nu} = c_n P_u/BD$ and expanding Eq. (2.21)

$$\left(\frac{1}{\sigma_N}\right)^2 = \left(\frac{D}{D_o(B'f_t)^2}\right) + \left(\frac{1}{B'f_t}\right)^2 \tag{2.23}$$

$$Y_i = \left(\frac{1}{\sigma_{Ni}}\right), \quad C = \left(\frac{1}{B'f_t}\right)^2, \quad X_i = D_i \tag{2.24}$$

$$Y = AX + C \tag{2.25}$$

Plotting the values of Y and X on X–Y plot as shown in Fig. 2.20, the constants of Eq. (2.25) are determined using linear regression in which the coefficient of variation of the slope of the regression line, coefficient of variation of the intercept, and relative width of the scatter band should not exceed about 0.1, 0.2, and 0.2, respectively. Comparing Eqs. (2.23) and (2.25), the coefficients are determined as

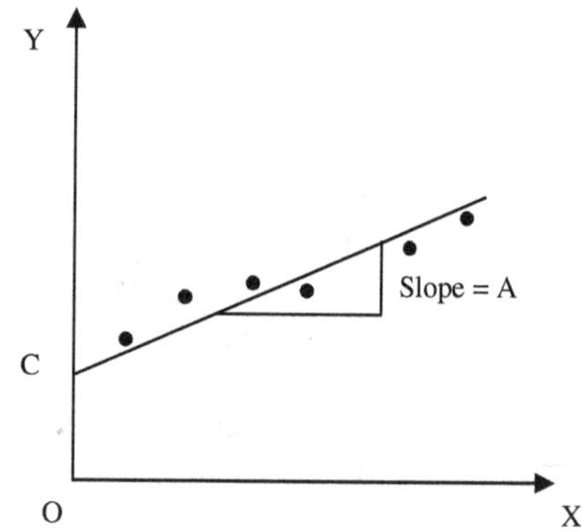

Fig. 2.20 Linear regression of test data

$$C = \left(\frac{1}{B' f_t}\right)^2, \quad A = \frac{C}{D_o} \tag{2.26}$$

Then, fracture energy, elastic equivalent crack extension, and brittleness are obtained as

$$G_{FB} = \frac{g(\alpha_0)}{AE} = \frac{[c_n k(\alpha_0)]^2}{AE} \tag{2.27}$$

$$c_f = \frac{g(\alpha_0)}{g'(\alpha_0)}\left(\frac{C}{A}\right) \tag{2.28}$$

$$\beta = \frac{D}{D_0} = \frac{g(\alpha_0)D}{g'(\alpha_0)c_f} = \frac{\overline{D}}{c_f} \tag{2.29}$$

The R curve and the crack extension during crack propagation are determined as

$$G(\alpha) = R(c) = G_{FB}\frac{g'(\alpha)}{g'(\alpha_0)}\frac{c}{c_f} \tag{2.30}$$

$$\frac{c}{c_f} = \frac{D}{D + d_0}\frac{g'(\alpha_0)g(\alpha)}{g'(\alpha)g(\alpha_0)} \tag{2.31}$$

The load–deflection curve of the structure or the specimen with a propagating crack can be determined using Castigliano's second theorem. In the SEM, the fracture energy G_{FB} by definition is independent of test specimen size, although this is true only approximately since the size-effect law is not exact. The G_{FB} is also independent of the specimen shape. This becomes clear by realizing that the fracture process zone occupies a negligibly small fraction of the specimen's volume in an infinitely large specimen. Therefore, most of the specimen is elastic, which implies that the fracture process zone at its boundary is exposed to the asymptotic near-tip elastic stress and displacement fields which are known from LEFM and are the same for any specimen geometry. Here, the fracture process zone must be in the same state regardless of the specimen shape.

The simplest method to obtain size-independent material fracture properties is perhaps provided by extrapolating to infinite size on the basis of the size-effect law. This law approximately describes the transition from the strength criterion for which there is no size effect to LEFM criterion for which the size effect is the strongest possible. An important advantage of the size-effect method is its simplicity which requires only the maximum load values of geometrically similar specimens that are sufficiently different in size which can be obtained without specialized need of closed-loop testing system. Neither the post-peak softening response nor the true crack length needs to be determined. Another advantage of the size-effect method is that it yields not only the fracture energy of the material but also the effective length of the fracture process zone, from which one can further obtain R curve and the critical effective crack-tip opening displacement.

2.7.5 Effective Crack Model (ECM)

Swartz and Go (1984) proposed that the effective crack length could be estimated at point of instability lying on the descending portion of the P-CMOD curve and at 0.95 of the maximum load (Fig. 2.21). However, for determination of effective crack length, Rafai and Swartz (1987) used maximum load calibration curves based on experimental results.

Nallathambi and Karihaloo (1986) and Karihaloo and Nallathambi (1989a, b, 1990, 1991) introduced effective crack model to evaluate Δa_c based on compliance calibration approach. The basic principle of determining the effective crack extension is to obtain the midspan deflection of the standard three-point beam test using secant compliance. According to the effective crack model, the fracture in the real structure sets in when the stress intensity factor becomes critical at crack length equal to a_e. Consider a typical load–deflection plot up to the peak load P_u and corresponding deflection is δ_u as shown in Fig. 2.22. P_i denotes an arbitrary load level in the initial (linear) portion of the load–deflection plot and the corresponding deflection is δ_i and stress intensity factor is K_I.

For an elastic notched three-point bending concrete beam containing a notch length a_o, the deflection and SIF are written as

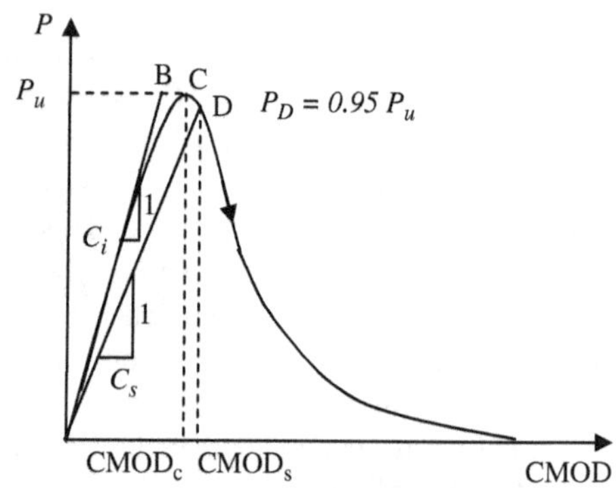

Fig. 2.21 Determination of a_c using CMOD measured at the point D ($P = 0.95\,P_u$) on P-CMOD curve by Swartz and Go (1984)

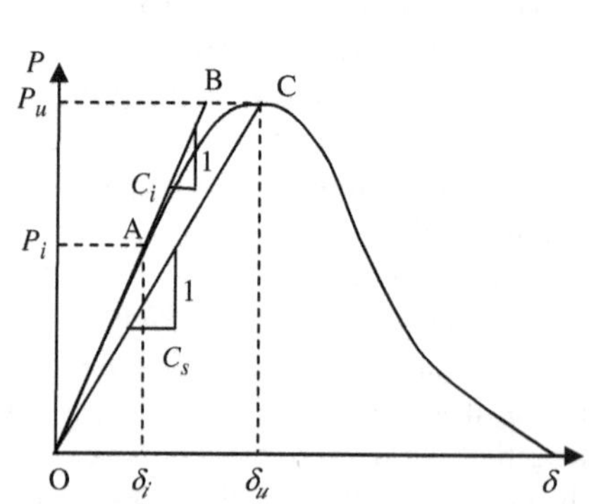

Fig. 2.22 Determination of effective crack length from P–δ curve

$$\delta_i = \frac{P_i}{4BE}\left(\frac{S}{D}\right)^3\left[1+\frac{5w_gS}{8P_i}+\left(\frac{D}{S}\right)^2\left\{2.70+1.35\frac{w_gS}{P_i}\right\}-0.84\left(\frac{D}{S}\right)^3\right]$$
$$+\frac{9P_i}{2BE}\left(1+\frac{w_gS}{2P_i}\right)\left(\frac{S}{D}\right)^2F_2(\alpha_o) \tag{2.32}$$
$$F_2(\alpha_o) = \int_0^{\alpha_o}\beta F^2(\beta)\mathrm{d}\beta$$
$$K_I = \sigma_N\sqrt{a}F(\alpha)$$

where $F(\alpha)$ is the geometric factor which depends on the *S/D* ratio of the beam. The w_g is the self-weight of the beam per unit length. Equation (2.32) is used to calculate the value of E using P_i and δ_i if it is not known from separate test. The value of a_c is computed using the following equation:

$$\delta_u = \frac{P_u}{4BE}\left(\frac{S}{D}\right)^3\left[1+\frac{5w_gS}{8P_u}+\left(\frac{D}{S}\right)^2\left\{2.70+1.35\frac{w_gS}{P_u}\right\}-0.84\left(\frac{D}{S}\right)^3\right]$$
$$+\frac{9P_u}{2BE}\left(1+\frac{w_gS}{2P_u}\right)\left(\frac{S}{D}\right)^2F_2(\alpha_c) \tag{2.33}$$
$$F_2(\alpha_c) = \int_0^{\alpha_c}\beta F^2(\beta)\mathrm{d}\beta$$

The same group of researchers also proposed an empirical equation using regression analysis to calculate the value of a_c from the load–deflection plot.

$$\frac{a_e}{D} = B_1\left(\frac{\sigma_{Nu}}{E}\right)^{B_2}\left(\frac{a_o}{D}\right)^{B_3}\left(1+\frac{d_a}{D}\right)^{B_4} \tag{2.34}$$

where d_a is the maximum size of coarse aggregate, $\sigma_n = 6M/bD^2$ is the nominal tensile stress for P_u including self-weight, and $B_1 = 0.088\pm0.004$, $B_2 = -0.208\pm 0.010$, $B_3 = 0.451\pm0.013$, and $B_4 = 1.653\pm0.109$. If the value of E is known from separate tests like compressive cylinder specimens, the same regression equation with coefficients $B_1 = 0.198\pm0.015$, $B_2 = -0.131\pm0.011$, $B_3 = 0.394\pm0.013$, and $B_4 = 0.600\pm0.092$ is used to evaluate the value of a_c. The critical value of stress intensity factor using this method is $K_{IC}{}^{e}$:

$$K_{IC}^{e} = \sigma_{Nu}\sqrt{a_e}F(\alpha_c) \tag{2.35}$$

2.7.6 Double-K Fracture Model (DKFM)

Recently, Xu and Reinhardt (1999a) presented the three stages of crack propagation in concrete: crack initiation, stable crack propagation, and unstable crack propagation based on tests of the large-size CT specimens and small-size TPBT specimens. Test results of five numbers of TPBT and four numbers of CT test specimens of different sizes were analyzed. A brief review of the test results on TPBT and CT test specimens is outlined for ready reference. In the first case, two sizes of TPBT specimens of dimensions $700 \times 150 \times 100$ mm and $515 \times 100 \times 100$ mm (span×depth×thickness) with initial notch length/depth $a_0/D = 0.4$ and span/depth ratio $S/D = 4$ made of concrete mix of 1:1.87:3.36:0.57 (cement:sand:aggregate:water by weight) with maximum aggregate size of 20 mm were tested between 28 and 32 days of maturity. The measured cube compressive strength, splitting tensile strength, and modulus of elasticity E of concrete were found to be 35.2, 2.45, and 31,000, respectively. The starting point of stable crack propagation, the critical unstable point of crack propagation, and the $\mathrm{CTOD_c}$ were measured using laser speckle interferometry experimental technique. In the second case, three sizes of CT test specimen geometry having dimensions $3000 \times 3000 \times 200$ mm, $2500 \times 2500 \times 200$ mm, and $1500 \times 1500 \times 200$ mm (length×depth×thickness) with $a_0/D = 0.4$ with the same concrete mix were tested. These specimens were tested at maturity ages between 180 and 360 days to avoid aging effect on the test results. The measured cube compressive strength, splitting tensile strength, and the value of E of concrete at age of 360 days were found to be 38.5, 2.78, and 32,400, respectively. The measurement of crack propagation was done using two approaches, viz. photoelastic coating and strain gauges.

While the fracture models, viz. TPFM, SEM, and ECM, adapting to LEFM to concrete structures can only predict the unstable fracture of concrete structures, the numerical methods using cohesive crack model and crack band model show the crack initiation, crack propagation, and failure during a fracture process until the maximum load is reached. Xu and Reinhardt (1999a) then advocated a new fracture model which can represent all the three stages of cracking phenomena in the fracture process of concrete. The mechanism of crack imitation and its growth is shown in Fig. 2.23.

The figure shows a typical P-CMOD or P–δ curve obtained from mode I fracture test of a concrete specimen. In-between the region BC, a deviation of the tangent with respect to the initial tangent (linear segment OB) is observed. Such nonlinearity is due to aggregate bridging action in the fracture process zone and it is considered that point B is the beginning of the fracture process zone and a macro-crack begins to occur in the mortar matrix. Till the point B, the solid body behaves in a linear elastic manner and the initial crack length does not start to grow which means that the crack-tip opening displacement is also zero. With this assumption, the load corresponding to point B is known as initial cracking load and the stress intensity factor is termed as initial cracking toughness $K_{\mathrm{IC}}^{\mathrm{ini}}$ of the material. Further, an unstable condition is reached at maximum load P_{u} at which both the crack length and CTOD become the critical values. The fracture criterion is named as double-K fracture

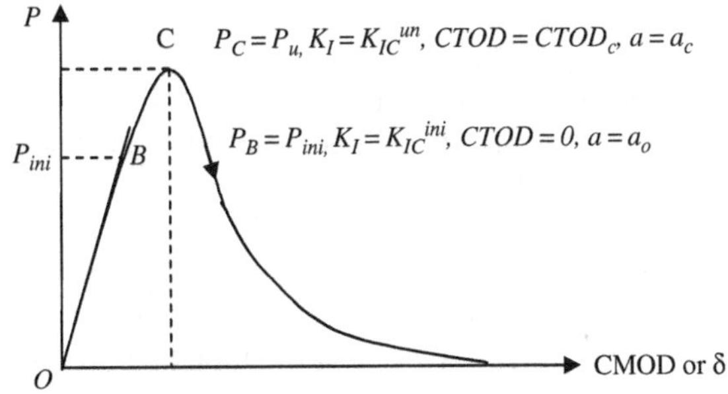

Fig. 2.23 Graphical representation of salient points on P-CMOD or *P–δ* curve

model. Hence, the double-*K* fracture criterion can predict the crack initiation, stable crack propagation, and unstable fracture process during the crack propagation in concrete structures. According to this criterion, the two size-independent parameters termed as initial cracking toughness $K_{\mathrm{IC}}^{\mathrm{ini}}$ and unstable fracture toughness $K_{\mathrm{IC}}^{\mathrm{un}}$ can predict different stages of concrete fracture process.

The first parameter $K_{\mathrm{IC}}^{\mathrm{ini}}$ is directly calculated by the initial cracking load and initial notch length using LEFM formula. The initiation toughness is defined as the inherent toughness of the materials, which holds for loading at crack initiation when material behaves elastically and microcracking is concentrated to a small scale in the absence of main crack growth. The other parameter $K_{\mathrm{IC}}^{\mathrm{un}}$ can be obtained by peak load and corresponding effective crack length using the same LEFM formula. That means the total toughness at the critical condition is known as unstable toughness $K_{\mathrm{IC}}^{\mathrm{un}}$ which is regarded as one of the material fracture parameters at the onset of the unstable crack propagation. In the tests (Xu and Reinhardt 1999a), the values of $K_{\mathrm{IC}}^{\mathrm{ini}}$ and $K_{\mathrm{IC}}^{\mathrm{un}}$ were then determined for both the specimen geometries. From the test results it was found that mean value of $K_{\mathrm{IC}}^{\mathrm{ini}} = 0.509\,\mathrm{Mpa\,m}^{1/2}$ (ranging 0.393–0.585 MPa $\mathrm{m}^{1/2}$) and 0.458 MPa $\mathrm{m}^{1/2}$ (ranging 0.455–0.459 MPa $\mathrm{m}^{1/2}$) for TPBT and CT test specimens, respectively. The mean values of $K_{\mathrm{IC}}^{\mathrm{un}}$ for these specimen geometries were found to be 2.428 MPa $\mathrm{m}^{1/2}$ (ranging 1.518–3.454 MPa $\mathrm{m}^{1/2}$) and 2.525 MPa $\mathrm{m}^{1/2}$ (ranging 2.323–2.837 MPa $\mathrm{m}^{1/2}$), respectively. It was reported from the test results that the values of $K_{\mathrm{IC}}^{\mathrm{ini}}$ and $K_{\mathrm{IC}}^{\mathrm{un}}$ are independent of the specimen sizes in the tested ranges and also that these values of fracture parameters are quite close for two specimen geometries made with concrete having almost close values of physical properties. The evaluated values of $K_{\mathrm{IC}}^{\mathrm{ini}}$ and $K_{\mathrm{IC}}^{\mathrm{un}}$ for TPBT geometry have larger dispersion because of the loading procedure in the laser speckle interferometry experimental method. Hence, it is revealed from the test results that both the parameters $K_{\mathrm{IC}}^{\mathrm{ini}}$ and $K_{\mathrm{IC}}^{\mathrm{un}}$ are dependent on material properties and not on specimen geometry and size. Xu and Reinhardt (1999b, c) developed analytical methods for three-point bending test and compact tension or wedge splitting tests to determine the values of double-*K* fracture parameters. From the available experimental results, it was also shown that double-*K* fracture parameters $K_{\mathrm{IC}}^{\mathrm{ini}}$ and $K_{\mathrm{IC}}^{\mathrm{un}}$ are not dependent on size of the specimen. Further, the evaluated values of the $\mathrm{CTOD_c}$ showed that this value appears to be size dependent (Xu and Reinhardt 1999b). The parameters K_{IC}^{ini} and K_{IC}^{un} computed from fracture tests on the small-size wedge splitting specimens show that these are independent of the relative initial notch length, slightly dependent on the size and independent of the thickness of the specimens (Xu and Reinhardt 1999c).

Both the fracture parameters K_{IC}^{ini} and K_{IC}^{un} have their own advantages. For some important structures such as high concrete dam, concrete pressure vessel, and nuclear reactor, both the parameters K_{IC}^{ini} and K_{IC}^{un} are required to be determined. The accurate prediction of crack initiation is also sometimes needed to specialized structural elements such as liquid-retaining concrete structures using the initial cracking toughness. Therefore, the K_{IC}^{ini} can be used to assess the initial cracking toughness, while the K_{IC}^{un} can be useful to describe the critical unstable fracture of cracked structures. At the same time, determination of double-*K* fracture parameters is experimentally easier because the unloading and reloading of the structure is not needed. This averts the need of closed-loop testing arrangement.

Xu and Reinhardt (2000) later proposed a simplified approach to obtain the double-*K* fracture parameters. In this approach, two empirical formulae to determine the crack mouth opening displacement and stress intensity factor due to cohesive stress in the fictitious fracture zone were proposed. It was shown that the results of double-*K* fracture for three-point bending test obtained using simplified approach are close to the corresponding results obtained using the analytical method (Xu and Reinhardt 1999b).

Zhao and Xu (2002) reported numerical experiments on three-point bending test specimens with the objective to study the influence of span/depth ratio, specimen size, and strength of concrete on the double-*K* fracture parameters. It was shown that the unstable fracture toughness is independent of span/depth ratio and dependent on the material property. The initiation fracture toughness not only varies with the strength of material but also presents size effects in terms of span/depth ratio and the depth of specimens, which require a further theoretical investigation.

Zhang et al. (2007) carried out the fracture tests on the total of 43 concrete specimens (three-point bending beam and wedge splitting specimen) with the small-size aggregates being of the maximum size of 10 mm for analyzing the double-*K* fracture parameters. The initial cracking load was determined by means of graphical method or electrical resistance strain gauge method. It was found that the ratio of initial cracking load to maximum load is 0.67–0.71 and the ratio of initial fracture toughness to unstable fracture toughness is 0.45–0.50. Further, the initial fracture toughness and the unstable fracture toughness values are approximately close to the constants when the depth of specimens is larger than 200 and 400 mm for three-point bending beams and wedge splitting specimens, respectively. The critical crack-tip opening displacement of each specimen was calculated and it was noticed that this value is different for different test geometries.

Xu and Zhu (2009) conducted the fracture tests on different sizes of TPBT specimens made of hardening cement paste and mortar of different strengths. The test results showed that fracture behavior of all cementitious composites, even hardening cement paste and mortar, was nonlinear. In addition, double-*K* fracture model was applied to calculate the fracture parameters and it was concluded that this fracture criterion can be applied to even hardening cement paste and mortar.

2.7.7 The K_R Curve Associated with Cohesive Stress Distribution in the FPZ

The basis of crack extension resistance curve (K_R curve) associated with the cohesive stress distribution in fictitious crack extension was introduced with the help of test data on CT test specimens (Xu and Reinhardt 1999a). It was shown that there existed two definite relationships: the first one between the determined values of K_{IC}^{un} and the tangency K_C^t (the value of SIF at the onset of unstable crack extension) of the K curves (SIF curve) and the second between the measured values of K_{IC}^{ini} and the onset of the stable crack propagation on the K_R curves. Hence, double-K fracture model was employed to describe the K_R curve for complete fracture process because of two distinguished characteristic parameters of concrete. The distribution of cohesive stress along the fictitious fracture zone at different stages of loading conditions is taken into account in order to evaluate the K_R curve. This approach of K_R curve differs from the principle of R curve originally proposed by Irwin (1960) and Kraft et al. (1961) in early 1960s and the conventional R-curve methods (Wecharatana and Shah 1982–1983, Bažant and Cedolin 1984, Mai 1984, Hilsdorf and Brameshuber 1984, Karihaloo 1987, Bažant et al. 1986, Ouyang et al. 1990, Bažant and Kazemi 1990–1991, Bažant and Jirăsek 1993, Elices and Planas 1993, Planas et al. 1993b).

An analytical method was proposed (Xu and Reinhardt 1998) to compute the crack growth resistance $K_R(\Delta a)$ based on the cohesive stress distribution in fictitious fracture zone during the crack propagation using P-CMOD curve obtained from test results of standard TPBT specimen available in the literature. The K_R curves for eight numbers of TPBT specimens of different sizes and geometrical parameters such as initial notch length/depth ratio a_o/D were determined and analyzed with regard to stability criterion using double-K fracture criterion. The values of K_{IC}^{ini}, K_{IC}^{un}, Δa_c, CTOD$_c$, the length of small micro-region Δa_B, and P_{ini} could be determined graphically using stability analysis. The variation of CTOD with respect to Δa obtained using analytical method also conformed to the experimental observations. Finally, it was shown that the K_R curves are independent of specimen size and a_o/D ratio.

The main requirement for determining crack extension resistance curve based on cohesive force distribution during crack propagation is to know the load–crack opening displacement (P-COD) curve a priori. Some characteristics of K_R curve were investigated numerically on standard TPBT specimen for different concrete strength and specimen sizes (Reinhardt and Xu 1999). Commercial computer program for nonlinear finite element analysis of reinforced concrete structures based on crack band model (Bažant and Oh 1983) was used to obtain P-CMOD curve for the test specimens. It was observed that K_R curve is dependent on compressive strength of concrete and has almost the same S shape. It increases with increasing crack length and increasing strength of concrete. The obtained K_R curves are almost independent of the specimen sizes. However, it was found that some difference could be noticed on the gained K_R curve by using bilinear (CEB-FIP Model Code-1990 1993) and nonlinear (Reinhardt et al. 1986) softening functions of concrete. The influence of specimen geometry on the K_R curves is not reported in the previous research work and as far as authors' knowledge is concerned, this study is also not found in the literature.

2.7.8 Double-G Fracture Model (DGFM)

Zhao and Xu (2004) discussed the local fracture energy and the average energy consumption due to the cohesive forces during crack extension. The experimental results obtained from the wedge splitting tests on specimens with different initial notch lengths are employed to investigate the property of the local and the average values of fracture energy.

Xu et al. (2006) introduced two fracture energy quantities, the stable fracture energy and unstable fracture energy, to predict energy consumption in the fracture process zone. Wedge splitting tests were carried out on concrete specimens with different ligament length to identify all three fracture energy quantities. The experimental results showed that the stable fracture energy remains almost a constant irrespective of specimen size, while the unstable fracture energy increases with increasing ligament length.

Zhao et al. (2007) investigated local fracture energy and its average value along the crack by considering the influence of specimen boundaries on the development of fracture process zone.

Zhang and Xu (2007) calculated the fracture resistance induced by aggregate bridging $G_{\text{I-bridging}}$ in the fracture process zone using both numerically simulated results and experimental results. The computed results indicate that the variation of $G_{\text{I-bridging}}$ with the development of crack firstly increases and then tends to invariable value once the initial crack-tip opening displacement reaches the characteristic crack opening displacement w_c. Also, it was found that the constant value of $G_{\text{I-bridging}}$ is comparable to the fracture energy. In this investigation, $G_{\text{I-bridging}}$ showed no size effect. These findings provide new understanding of the meaning of fracture energy.

Recently, Xu and Zhang (2008) proposed the double-G fracture criterion based on the concept of energy release rate consisting of two characteristic fracture parameters: the initiation fracture energy release $G_{\text{IC}}^{\text{ini}}$ and the unstable fracture energy release $G_{\text{IC}}^{\text{un}}$. The value of $G_{\text{IC}}^{\text{ini}}$ is defined as the Griffith fracture surface energy of concrete mix in which the matrix remains still in elastic state under the initial cracking load P_{ini} and the initial crack length a_0. Until the initial cracking load, P–δ or P–CMOD curve is linear and the crack-driving energy is less than the energy required to create a new crack surface. Once the load value P on the structure is increased beyond the value of P_{ini}, a new crack surface (macro-cracking) is formed and the cohesive stress along the new crack surface starts to act. The cohesive stress provides additional resistance to the stable crack propagation in terms of cohesive breaking energy $G_{\mathrm{I}}^{\mathrm{C}}$ until the critical condition is achieved. At the onset of unstable crack propagation, the total energy release $G_{\mathrm{IC}}^{\mathrm{un}}$ consists of initiation fracture energy release $G_{\mathrm{IC}}^{\mathrm{ini}}$ and the critical value of the cohesive breaking energy $G_{\mathrm{IC}}^{\mathrm{C}}$. Extensive test results using two specimen geometries, namely three-point bending test of size range 150–500 mm and wedge splitting test of size range 200–1000 mm, on determination of double-*G* fracture parameters and double-*K* fracture parameters were presented by Xu and Zhang (2008). Concrete mixes for different test geometries were different. From test results it was found that some scatter in the values of characteristic fracture parameters was observable even for the same specimen size of particular specimen geometry. In general, it was noticed that the obtained fracture parameters are dependent on the specimen size. Further, within certain scatter range in the test results, it was concluded that the double-*G* fracture parameters were size independent over the size range of 200 mm. The values of $G_{\mathrm{IC}}^{\mathrm{un}}$ and $G_{\mathrm{IC}}^{\mathrm{ini}}$ were found to be approximately 50 and 11 N/m, respectively, for three-point bending geometry, whereas those values for wedge splitting specimens were approximately 90 and 16 N/m, respectively. The double-*G* fracture parameters were converted to the effective initiation toughness $\overline{K_{\mathrm{IC}}^{\mathrm{ini}}}$ and effective unstable fracture toughness $\overline{K_{\mathrm{IC}}^{\mathrm{un}}}$ equivalent to the double-*K* fracture parameters using the relationship $K = \sqrt{EG}$. It was found that the values of equivalent fracture parameters in terms of stress intensity factors at the onset of crack initiation and the onset of unstable fracture using double-*G* fracture criterion and double-*K* fracture criterion are very close. The experimentally observed discrepancy in double-*G* fracture parameters due to specimen geometry and size effect paves the way for further systematic investigation in case the input data could be obtained numerically.

2.8 Comparative Study and Size-Effect Behavior

Karihaloo and Nallathambi (1989a) used the test data of three-point bending specimens for the comparison of improved effective crack model and the two-parameter fracture models. It was found that predictions from both the models are in good agreement. From the various sources of experimental results, Karihaloo and Nallathambi (1989a) showed that fracture toughness values obtained from the effective crack model and the two-parameter fracture models and also from the effective crack model and the size-effect model are in good agreement. A similar prediction between the effective crack model and the two-parameter fracture model was also observed from the comparison of fracture parameters using different sources of experimental results (Karihaloo and Nallathambi 1991). It was found that irrespective of the concrete strength, the fracture parameters obtained using the effective crack model ($K_{\mathrm{IC}}^{\mathrm{e}}$ and a_{e}) are practically indistinguishable from the corresponding parameters ($K_{\mathrm{IC}}^{\mathrm{s}}$ and $\mathrm{CTOD_{cs}}$) determined using the two-parameter model.

The size-effect relationships between fictitious crack model, size-effect model, and two-parameter fracture model were developed in the past (Planas and Elices

1990), which predicted almost the same fracture loads for practical size range (100–400 mm) of pre-cracked concrete beam for three-point bending test geometry. It was observed from the size-effect study that fracture loads predicted by the size-effect model and the two-parameter fracture model could diverge about 28 and 3%, respectively, for asymptotically large-size ($D \rightarrow \infty$) beam. Later, based on the similar approach (Planas and Elices 1990), a size-effect study between fictitious crack model and effective crack model was presented by Karihaloo and Nallathambi (1990). In the study, it was shown that the predicted fracture loads from both the models for the practical size range of notched concrete beam in three-point bending test configuration are indistinguishable and in the asymptotic limit (of infinite size), the predictions differed by about 17%.

The numerical results of K_{IC}^{s} in TPFM and G_{FB} in SEM are very similar (Tang et al. 1992, Bazant et al. 1991, Bazant 1994, 2002) and approximately equivalent throughout the whole size range and the second parameter of each model can be obtained by

$$\mathrm{CTOD_{cs}} = \sqrt{\frac{32 G_{FB} c_f}{\pi E'}} \tag{2.36}$$

where E' is $E/(1-\nu^2)$ for plain strain and is E for plain stress case.

Ouyang et al. (1996) established an equivalency between the two-parameter fracture model and the size-effect model based on infinitely large-size specimens. It was found that the relationship between $\mathrm{CTOD_{cs}}$ and c_f theoretically depends on both specimen geometry and initial crack length and both the fracture models can reasonably predict fracture behavior of quasibrittle materials.

Guinea et al. (1997) compared the maximum load size effect for laboratory-size beam specimens, notched or unnotched using cohesive crack model with linear and trapezoidal softening functions. It was found that a linear softening function is accurate enough to predict adequately the observed maximum load size effect for current laboratory beams and the solution of the problem of extracting the softening function from a set of experimental peak loads is nonunique. Additional experimental information, such as the load–displacement curve, is needed to pick out a suitable softening curve.

Planas et al. (1997) proposed a general formula to predict the size effect of pre-cracked specimens with a greater accuracy. The formula is applicable for concrete material being analyzed using cohesive crack model with initial linear softening.

Xu et al. (2003) conducted concrete fracture experiments on both the three-point bending notched beams and the wedge splitting specimens with different relative initial crack length according to the experimental requirements for determining the fracture parameters in the double-K fracture model and the two-parameter fracture model. The comparative results showed that the critical crack length a_c determined using the two different models is hardly different. The values of K_{IC}^{un} and $\mathrm{CTOD_c}$ measured for the double-K fracture model are in good agreement with K_{IC}^{s} and $\mathrm{CTOD_{cs}}$ measured for the two-parameter fracture model.

Hanson and Ingraffea (2003) developed the size-effect, two-parameter, and fictitious crack models numerically to predict crack growth in materials for three-point bending test. The investigation showed that even if the three models predict the same response for infinitely large structures, they do not always predict the same response on the laboratory-size specimens. However, the three models do agree at the laboratory-size specimens for certain ranges of tension-softening parameters. It seemed that the relative size of tension-softening zone must be less than approximately 15% of the ligament length for the two-parameter fracture model to predict similar behavior as of fictitious crack model. Further, it appeared that the total relative size of tension-softening zone is not an indication for the size-effect model to predict the similar response as of the fictitious crack model.

Roesler et al. (2007) plotted the size-effect behavior of experimental results, numerical simulation using cohesive crack model, size-effect model, and two-parameter fracture model for three-point bending test specimens. From the analysis of results, it is found that the size-effect behavior calculated from the size-effect model and two-parameter fracture model resembles closely.

From the fracture tests (Xu and Zhang 2008), it is also clear that the corresponding values of double-K fracture parameters and double-G fracture parameters are equivalent at initial cracking load and unstable peak load.

Cusatis and Schaffert (2009) presented precise numerical simulations based on cohesive crack model for computation of size-effect curves using typical test configurations. The results were analyzed with reference to SEM to investigate the relationship between the size-effect curves and the size-effect law. The practical implications of the study were also discussed in relation to the use of the size-effect curves or the size-effect law for identification of the softening law parameters through the size-effect method.

2.9 Weight Function Approach

The principle of weight function is a powerful method to determine the stress intensity factors (SIFs) of cracked bodies in complex stress fields, viz. thermal or residual stress. In concrete structures, the complex nonlinear stresses develop due to the presence of large FPZ ahead of initial crack tip. In such situation, the use of weight function reduces the need for finite element or boundary element techniques and it provides simple means for calculating stress intensity factors. Bueckner (1970) first introduced the use of weight function to calculate the stress intensity factors. It was shown that for a given cracked body, the weight function depends only on the geometry and it is independent of the applied loads. Subsequently, Rice (1972) further simplified the determination of SIFs using weight function method. For a linear elastic body subjected to any symmetrical loading, it was proved that if the stress intensity factor and corresponding crack face displacement are known as functions of crack length, then the stress intensity factor for any other symmetrical loading acting on the same body may be directly determined. The advantage of using weight function is that once it is obtained for a given cracked geometry and reference loading(s), SIFs of the same geometry for any mode I loading cases may be determined. In case of repeated calculation of SIFs for a given cracked geometry, the use of weight function method will be an alternative and efficient solution method as compared to numerical methods.

2.9.1 Some Existing Weight Functions

Petroski and Achenbach (1978) proposed a general approximate expression of crack surface displacement u_r as given below:

$$u_r(x,a) = \frac{\sigma_o}{E'\sqrt{2}}\left[4F(a/W)\sqrt{a}\sqrt{(a-x)} + G(a/W)\frac{(a-x)^{3/2}}{\sqrt{a}}\right] \tag{2.37}$$

in which

$$\begin{aligned} G(a/W) &= \frac{[I_1(a) - 4F(a/W)I_2(a)\sqrt{a}]\sqrt{a}}{I_3(a)} \\ I_1(a) &= \pi\sqrt{2}\sigma_o \int_0^a [F(a/W)]^2 a\,da \\ I_2(a) &= \int_0^a \sigma(x)\sqrt{(a-x)}dx \\ I_3(a) &= \int_0^a \sigma(x)(a-x)^{3/2}dx \end{aligned} \tag{2.38}$$

where $F(a/W) = K_r/(\sigma_o\sqrt{(\pi a)})$ is the nondimensional geometric correction factor and $G(a/W)$ is an unknown function to be determined using Eq. (2.38). The K_r is the known reference SIF corresponding to externally acting stress σ_o and $\sigma(x)$ is the crack-line stress of an uncracked body due to applied stress σ_o. Complicated geometry even with simple applied stress σ_o will result in complex stress $\sigma(x)$. For such cases, the use of Eqs. (2.37) and (2.38) may be cumbersome. Further, Fett (1988) showed that the inaccurate results of weight function may be obtained for rapidly changing reference stress if expressions of Petroski and Achenbachs (1978) are used. Fett et al. (1987) proposed a general expression of weight function in the following form:

$$m(x,a) = \frac{2}{\sqrt{2\pi(a-x)}}\left[\begin{matrix}1 + M_1(1-x/a) + M_2(1-x/a)^2 \\ +M_3(1-x/a)^3 + \cdots + M_n(1-x/a)^n\end{matrix}\right] \tag{2.39}$$

Sha and Yang (1986) introduced the following form of the general expression of weight function:

$$m(x,a) = \frac{2}{\sqrt{2\pi(a-x)}}\left[\begin{matrix}1 + M_1(1-x/a)^{1/2} + M_2(1-x/a) + \\ M_3(1-x/a)^{3/2} + \cdots + M_n(1-x/a)^{n/2}\end{matrix}\right] \tag{2.40}$$

The number of terms in the expression (2.39) to be considered for a particular geometry is uncertain and it is mentioned elsewhere (Glinka and Shen 1991) that some of the weight functions cannot be accurately expressed even by using seven terms of the expression (2.39).

Glinka and Shen (1991) revealed that using expression (2.40), it is possible to approximate weight functions for mode I cracks with one four-term universal expression. In the four-term universal weight function, the number of unknown parameters are three (M_1, M_2, and M_3) which can be determined directly from three known reference SIFs without using the crack face displacement function. If the three reference SIFs K_{r1}, K_{r2}, and K_{r3} and the corresponding crack-line stress fields σ_{r1}, σ_{r2}, and σ_{r3}, respectively, are known, the simultaneous algebraic equations may be written in the following form. Shen and Glinka (1991) derived the universal form of weight function for different crack geometries subjected to mode I loadings. The results thus obtained were compared with the available analytical solutions and it was observed that the difference in results is less than 2%.

2.9.2 Universal Weight Function for Edge Cracks in Finite Width Plate

Several analytical and numerical solutions are available to compute the approximate weight functions for the edge cracks in finite width plates. These parameters are computed using the analytical weight function for an edge crack in a finite width plate presented by Kaya and Erdogan (1980). Wu (1984) presented approximate weight function for edge cracks in finite plate subjected to a pair of normal forces acting at the crack faces. The general approximate expression (Petroski and Achenbachs 1978) of crack surface displacement was used for the derivation of the weight function. The solution of SIF using weight function and Green's function of Tada et al. (1985) was compared, and it was shown that the difference in the results would be more for deeper cracks. Glinka and Shen (1991) presented the values of parameters M_1, M_2, and M_3 of the four-term universal weight function in a tabular form at discrete values of crack length/width ratios using the analytical weight function for an edge crack in a finite width plate (Kaya and Erdogan 1980). Furthermore, Moftakhar and Glinka (1992) and Wu et al. (2003) have presented the expressions for the parameters of the universal weight functions using the same analytical weight function.

2.9.3 Computation of Stress Intensity Factor and Crack Face Displacement

Generally, numerical integration of weight function is done using specialized method which allows removal of the integrable crack-tip singularity and enables one to obtain the stress intensity factors and crack face displacements in an efficient manner. Moftakhar and Glinka (1992) presented a numerical technique for simple and efficient integration of universal weight function to calculate the stress intensity factors in mode I fracture problem for any nonlinear stress distribution normal to the crack surfaces.

Daniewicz (1994) compared the numerical results of the stress intensity factors and crack face displacements for the edge-cracked half-plane under uniform tension applying both Gauss–Chebyshev and Gauss–Legendre quadratures. It was shown that Gauss–Chebyshev quadrature is superior in terms of both accuracy and computational efficiency.

Kiciak et al. (2003) introduced a method of calculating stress intensity factors for cracks subjected to complex stress fields using the universal weight functions. It was shown that the weight functions enable the determination of stress intensity factors for a variety of geometrical and stress field configurations. Because of singularity problem at integral boundaries, numerical integration of weight functions is difficult and computationally inefficient.

An alternative exact analytical solution for the weight function integral over discrete intervals was introduced by Anderson and Glinka (2006) to obtain the stress intensity factor using constant, linear, and quadratic fit of stress distribution along the crack line. The quadratic method is the most computationally efficient which yields the convergence with a small number of integration increments.

2.10 Scope of the Book

It can be concluded from the extensive literature review that the nonlinear fracture models based on the numerical approach are relatively more cumbersome in computations and the fracture models based on the modified LEFM can be applied to common engineering problems with relatively more computational efficiency to study the crack propagation phenomena of concrete and concrete structures. The double-K fracture model, K_{R}-curve method based on cohesive stress distribution, and the double-G fracture model belonging to the modified LEFM concept can predict the important stages of fracture processes such as crack initiation, stable crack propagation, and unstable fracture unlike the two-parameter fracture model, size-effect model, and effective crack model which can predict the fracture loads at critical condition only. Further, the complete fracture process can be analyzed using the K_{R}-curve method. Existing *analytical methods* for determining the double-K fracture parameters and the K_{R} curve need specialized numerical technique because of singularity problem at the integral boundary. It is also clear that the concrete fracture parameters are influenced by many factors such as softening function of concrete, concrete strength, specimen size (characteristic size, depth D), specimen geometry, geometrical factors, and loading condition. In order to focus the recent development on the behavior of fracture parameters for crack propagation in concrete, the following scopes as shown in Fig. 2.24 are considered in this book which are mainly based on the studies carried out by Kumar (2010) and Kumar and Barai (2008a–c, 2009a–f, 2010a, b):

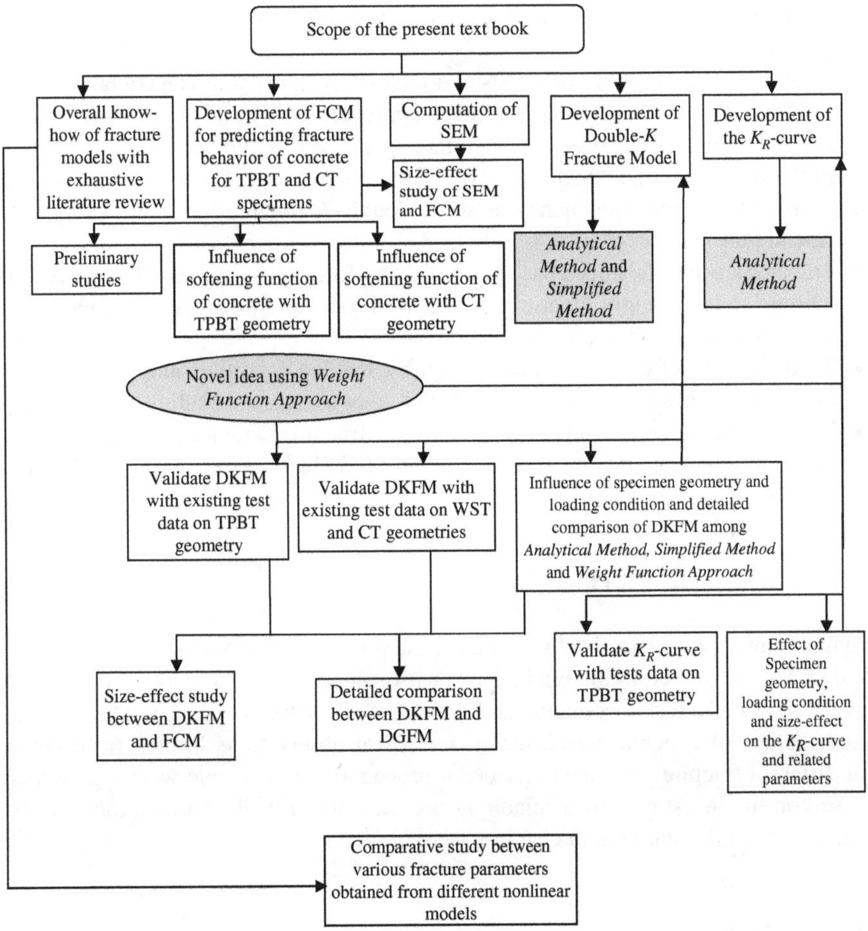

Fig. 2.24 An overview of different topics of the content

- To develop the cohesive crack model for various test configurations (viz. three-point bending, four-point bending tests and compact tension specimen) and carry out a systematic and careful study of the effect of various softening function on the numerical results. Comparison of numerical results obtained from the model with experimental results is available in the literature.
- To introduce closed-form equations using weight function approach for determining the double-*K* fracture parameters and the K_R curve associated with the cohesive stress.
- To validate the weight function approach for determining the double-*K* fracture parameters and carry out in-depth analysis of the results using the experimental test results of various test configurations available in the literature.

- To study size-effect between various concrete fracture models including double-*K* fracture parameters.
- To analyze the comparative results of double-*K* fracture parameters obtained from existing analytical method, simplified approach, and weight function method. Investigation of the effect of specimen geometry, loading condition, relative initial notch length, and softening function on the double-*K* fracture parameters.
- To investigate the equivalency between double-*K* fracture model and double-*G* fracture model.
- To compare the K_R curve based on cohesive force using the existing analytical method and the weight function method as introduced by the authors (Kumar and Barai 2008a–c, 2009a–e).
- To analyze the effect of specimen geometry, loading condition, and size effect on the K_R curve based on cohesive force and other associated fracture parameters.
- To carry out a comparative study between fracture parameters obtained from different concrete fracture models like FCM, TPFM, SEM, ECM, DKFM, and DGFM.

2.11 Closing Remarks

In this chapter, an exhaustive literature review was carried out within the premises of the topics considered in the book. Basic background on various aspects of fracture process, fracture test specimens, and development of the different nonlinear fracture models of concrete was illustrated. Critical observations limited to behavior of different fracture parameters for crack propagation in concrete were highlighted based on the investigations available in the literature. Finally, various topics to be considered in the subsequent chapters were outlined.

References

Alvaredo AM, Torrent RJ (1987) The effect of the shape of the strain-softening on the bearing capacity of concrete beams diagram. Mater Struct 20 (6): 448–454.

Anderson TL (2005) Fracture Mechanics Fundamentals and Applications. CRC Press, Taylor & Francis Group, Boca Raton, FL.

Anderson TL, Glinka G (2006) A closed-form method for integrating weight functions for part-through cracks subject to mode I loading. Eng Fract Mech 73: 2153–2165.

Ansari F (1989). Mechanism of microcrack formation in concrete. ACI Mater J 41: 459–464.

ASTM International Standard E399-06 (2006) Standard Test Method for Linear-Elastic Method Plane-Strain Fracture Toughness K_{IC} of Metallic Materials. Copyright ASTM International, West Conshohocken, PA: 1–32.

Barenblatt GI (1959) The formation of equilibrium cracks during brittle fracture: General ideas and hypotheses, axially symmetric cracks. J Appl Math Mech 23: 622–636.

Barenblatt GI (1962) The mathematical theory of equilibrium cracks in brittle fracture. Adv Appl Mech 7 (1): 55–129.

Barker DB, Hawkins NM, Jeang FL, Cho KZ, Kobayashi AS (1985) Concrete fracture in CLWL specimen. J Eng Mech ASCE 111 (5): 623–638.

Bažant ZP (1984) Size effect in blunt fracture: concrete, rock, metal. J Eng Mech ASCE 110 (4): 518–535.
Bažant ZP (1990) Smeared-tip superposition method for nonlinear and time-dependent fracture. Mech Res Commun 17 (5): 343–351.
Bažant ZP (2002) Concrete fracture models: testing and practice. Eng Fract Mech 69: 165–205.
Bažant ZP, Becq-Giraudon E (2002) Statistical prediction of fracture parameters of concrete and implications for choice of testing standard. Cem Concr Res 32: 529–556.
Bažant ZP, Beissel S (1994) Smeared-tip superposition method for cohesive fracture with rate effect and creep. Int J Fract 65: 277–290.
Bažant ZP, Gettu R, Kazemi MT (1991) Identification of nonlinear fracture properties from size effect tests and structural analysis based on geometry-dependent *R*-curve. Int J Rock Mech Min Sci 28 (1): 43–51.
Bažant ZP, Jirăsek M (1993) R-curve modeling of rate and size effects in quasibrittle fracture. Int J Fract 62: 355–373.
Bažant ZP, Kazemi M (1991) Size dependence of concrete fracture energy determined by RILEM work-of-fracture method. Int J Fract 51: 121–138.
Bažant ZP, Kazemi MT (1990) Determination of fracture energy, process zone length and brittleness number from size effect, with application to rock and concrete. Int J Fract 44: 111–131.
Bažant ZP, Oh BH (1983) Crack band theory for fracture of concrete. Mater Struct 16 (93): 155–177.
Bažant ZP, Zi G (2003) Asymptotic stress intensity factor density profiles for smeared-tip method for cohesive fracture. Int J Fract 119: 145–159.
Bažant ZP, Cedolin L (1984). Approximate linear analysis of concrete fracture by R-curve. J Struct Eng ASCE 110 (6): 1336–1355.
Bažant ZP (1986) Mechanics of distributed cracking. Appl Mech Rev ASME 39 (5): 675–705.
Bažant ZP (1994) Discussion of "Fracture mechanics and size effect of concrete in tension". J Struct Eng ASCE 120 (8): 2555–2558.
Bažant ZP, Kim J-K, Pfeiffer PA (1986) Determination of fracture properties from size effect tests. J Struct Eng ASCE 112 (2): 289–307.
Bažant ZP, Lin F-B (1989) Nonlocal smeared cracking model for concrete fracture. J Struct Eng ASCE 114 (11): 2493–2510.
Bažant ZP, Li YN (1995a) Stability of cohesive crack model: Part I—energy principles. J Appl Mech ASME 62: 959–964.
Bažant ZP, Li YN (1995b) Stability of cohesive crack model: Part II—Eigenvalue analysis of size effect on strength and ductility of structures. J Appl Mech ASME 62: 965–969.
Bažant ZP, Planas J (1998) Fracture and size effect in concrete and other quasibrittle materials, Florida, CRC Press.
Bhargava J, Rehnström A (1975) High speed photography for fracture studies of concrete. Cem Concr Res 5: 239–248.
Bocca P, Carpinteri A, Valente S (1991) Mixed mode fracture of concrete. Int J Solid Struct 27 (9): 1139–1153.
Brown JH (1972) Measuring the fracture toughness of cement paste and mortar. Magn Concr Res 24 (81): 185–196.
Brown JH, Pomeroy CD (1973) Fracture toughness of cement paste and mortars. Cem Concr Res 3: 475–480.
Brühwiler E, Wittmann FH (1990) The wedge splitting test: A method of performing stable fracture mechanics tests. Eng Fract Mech 35: 117–125.
Bueckner HF (1970) A novel principle for the computation of stress intensity factors. Z Angew Math Mech 50: 529–546.
Carpinteri A (1982) Application of fracture mechanics to concrete structures. J Struct Div ASCE 108 (ST4): 833–847.

Carpinteri A (1989a) Decrease of apparent tensile and bending strength with specimen size: two different explanations based on fracture mechanics. Int J Solid Struct 25 (4): 407–429.

Carpinteri A (1989b) Cusp catastrophe interpretation of fracture instability. J Mech Phys Solid 37 (5): 567–582.

Carpinteri A (1989c) Post-peak and post-bifurcation analysis of cohesive crack propagation. Eng Fract Mech 32: 265–278.

Carpinteri A (1990) A catastrophe theory approach to fracture mechanics. Int J Fract 44: 57–69.

Carpinteri A, Colombo G (1989) Numerical analysis of catastrophic softening behaviour (snap-back instability). Comput Struct 31 (4): 607–636.

Carpinteri A, Cornetti P, Barpi F, Valente S (2003) Cohesive crack model description of ductile to brittle size-scale transition: dimensional analysis vs. renormalization group theory. Eng Fract Mech 70: 1809–1839.

Carpinteri A, Cornetti P, Puzzi S (2006) Scaling laws and multiscale approach in the mechanics of heterogeneous and disordered materials. Appl Mech Rev ASME 59: 283–305.

CEB-Comite Euro-International du Beton-EB-FIP Model Code 1990 (1993) Bulletin D'Information No. 2123/214, Lausanne.

Cendón DA, Gálvez JC, Elices M, Planas J (2000) Modelling the fracture of concrete under mixed loading. Int J Fract 103: 293–310.

Cho KZ, Kobayashi AS, Hawkins NM, Barker DB, Jeang FL (1984) Fracture process zone of concrete cracks. J Eng Mech ASCE 110 (8): 1174–1184.

Cook DJ, Crookham GD (1978) Fracture toughness measurements of polymer concretes. Magn Concr Res 30 (105): 205–214.

Cusatis G, Schauffert EA (2009) Cohesive crack analysis of size effect. Eng Frac Mech 76: 2163–2173.

Daniewicz SR (1994) Accurate and efficient numerical integration of weight function using Gauss–Chebyshev quadrature. Eng Fract Mech 48: 541–544.

de Borst R (2003). Numerical aspects of cohesive-zone models. Eng Fract Mech 70: 1743–1757.

de Borst R, Remmers JJC, Needleman A, Abellan MA (2004) Discrete vs smeared crack models for concrete fracture: bridging the gap. Int J Numer Anal Meth Geomech 28: 583–607.

Derucher KN (1978) Application of the scanning electron microscope to fracture studies of concrete. Build Environ 13: 135–141.

Dhir RK, Sangha RM (1974) Development and propagation of microcracks in plain concrete. Mater Struct 7(1): 17–23.

Dugdale DS (1960) Yielding of steel sheets containing slits. J Mech Phys Solid 8 (2): 100–104.

Elices M, Guinea GV, Gómez J, Planas J (2002) The cohesive cone model: Advantages, limitations and challenges. Eng Fract Mech 69: 137–163.

Elices M, Guinea GV, Planas J (1992) Measurement of the fracture energy using three-point bend tests: Part 3 – Influence of cutting the P–δ tail. Mater Struct 25 (6): 327–334.

Elices M, Planas J (1993) The equivalent elastic crack: 1. Load-Y equivalences. Int J Fract 61: 159–172.

Elices M, Rocco C, Roselló C (2009) Cohesive crack modeling of a simple concrete: Experimental and numerical results. Eng Fract Mech 76: 1398–1410.

Elices M, Planas J (1996). Fracture mechanics parameters of concrete an overview. Adv Cem Based Mater 4: 116–127.

Evans AG, Clifton JR, Anderson E (1976) The fracture mechanics of mortars. Cem Concr Res 6: 535–548.

Evans RH, Marathe MS (1968) Microcracking and stress–strain curves for concrete in tension. Mater Struct 1(1): 61–64.

Fett T (1988) Limitations of the Petroski–Achenbach procedure demonstrated for a simple load case. Eng Fract Mech 29: 713–716.

Fett T, Mattheck C, Munz D (1987) On calculation of crack opening displacement from the stress intensity factor. Eng Fract Mech 27: 697–715.

Gálvez JC, Červenka J, Cendón DA, Saouma V (2002) A discrete crack approach to normal/shear cracking of concrete. Cem Concr Res 32: 1567–1585.

Gálvez JC, Elices M, Guinea GV, Planas J (1998) Mixed mode fracture of concrete under proportional and nonproportional loading. Int J Fract 94: 267–284.

Gdoutos EE (2005) Fracture Mechanics an Introduction, Second Edition, Dordrecht, The Netherlands, Springer.

Gerstle H, Xie M (1992) FEM modeling of fictitious crack propagation in concrete. J Eng Mech ASCE 118 (2): 416–434.

Gettu R, Bažant ZP, Karr Martha E (1990) Fracture properties and brittleness of high strength concrete. ACI Mater J 87 (6): 608–618.

Gjϕrv OE, Sϕrensen SI, Arnesen A (1977) Notch sensitivity and fracture toughness of concrete. Cem Concr Res 7: 333–344.

Glinka G, Shen G (1991) Universal features of weight functions for cracks in Mode I. Eng Fract Mech 40: 1135–1146.

Glucklich J. (1963). Fracture of plain concrete. J Eng Mech Div ASCE 89 (EM6): 127– 137.

Gopalaratnam VS, Shah SP (1985) Softening response of plain concrete in direct tension. J Am Concr Inst 82: 310–323.

Gopalaratnam VS, Ye BS (1991) Numerical characterization of the nonlinear fracture process in concrete. Eng Fract Mech 40: 991–1006.

Griffith AA (1921) The phenomena of rupture and flow in solids. Philos Trans Roy Soc A221: 163–197.

Guinea GV (1995) Modelling the fracture of concrete: the cohesive crack. Mater Struct 28 (4): 187–194.

Guinea GV, Elices M, Planas J (1997) On the initial shape of the softening function of cohesive materials. Int J Fract 87: 139–149.

Guinea GV, Planas J, Elices M (1992) Measurement of the fracture energy using three-point bend tests: Part 1 – Influence of experimental procedures. Mater Struct 25 (4), 212–218.

Guinea GV, Planas J, Elices M (1994) A general bilinear fit for the softening curve of concrete. Mater Struct 27 (2): 99–105.

Hadjab HS, Thimus J-Fr, Chabaat M (2007) The use of acoustic emission to investigate fracture process zone in notched concrete beams. Curr Sci 93 (5): 648–653.

Hanson JH, Ingraffea AR (2003) Using numerical simulations to compare the fracture toughness values for concrete from the size-effect, two-parameter and fictitious crack models. Eng Fract Mech 70: 1015–1027.

Heilmann HG, Hilsdorf HH, Finsterwalder K (1969) Festigkeit und Verformung von Beton unter Zugspannungen. Deutscher Ausschuss für Stahlbeton, Heft 203, Berlin, W. Ernst & Sohn.

Hillemeier B, Hilsdorf HK (1977) Fracture mechanics studies on cement compound. Cem Concr Res 7: 523–536.

Hillerborg A (1985a) The theoretical basis of a method to determine the fracture energy G_F of concrete. Mater Struct 18 (4): 291–296.

Hillerborg A (1985b) Results of three comparative test series for determining the fracture energy G_F of concrete. Mater Struct 18 (5): 407–413.

Hillerborg A, Modeer M, Petersson PE (1976) Analysis of crack formation and crack growth in concrete by means of fracture mechanics and finite elements. Cem Concr Res 6: 773–782.

Hilsdorf HK, Brameshuber W (1984) Size effects in the experimental determination of fracture mechanics parameters. Application of Fracture Mechanics to Cementitious Composites (Edited by S.P. Shah), NATOARW, Northwestern University, USA: 361–397.

Hu XZ, Wittmann FH (1990) Experimental method to determine extension of fracture-process zone. J Mater Civ Eng ASCE 2 (1): 459–464.

Hughes BP, Chapman GP (1966) The complete stress–strain curve for concrete in direct tension. Bull RILEM 30: 95–97.

Inglis CE (1913) Stresses in a plate due to the presence of cracks and sharp corners. Trans Inst Naval Archit 55: 219–241.

Ingraffea AR, Gerstle WH (1984) Nonlinear fracture models for discrete crack propagation. Proceedings of the NATO Advanced Workshop on Application of Fracture Mechanics to Cementitious Composites (Edited by SP Shah), M. Nijhoff, Hingham, MA: 171–209.

Ingraffea AR, Gerstle WH, Gergely P, Saouma V (1984) Fracture mechanics of bond in reinforced concrete. J Struct Eng ASCE 110 (4): 871–890

Irwin GR (1955) Fracturing of Metals, ASM, Cleveland, OH: 147–166.

Irwin GR (1957) Analysis of stresses and strains near the end of a crack traversing a plate. J Appl Mech Trans ASME 24: 361–364.

Irwin GR (1960) Crack extension force for a part-through crack in a plate. ASTM Bullet 243: 29–40.

Issa MA, Issa MA, Islam MS, Chudnovsky A (2000a). Size effects in concrete fracture: Part I, experimental setup and observations. Int J Fract 102: 1–24.

Issa MA, Hammad AM, Chudnovsky A (1993) Correlation between crack tortuosity and fracture toughness in cementitious material. Int J Fract 69: 97–105.

Issa MA, Issa MA, Islam MS, Chudnovsky A (2000b) Size effects in concrete fracture – Part II: Analysis of test results. Int J Fract 102: 25–42.

Issa MA, Issa MA, Islam MS, Chudnovsky A (2003) Fractal dimension—a measure of fracture roughness and toughness of concrete. Eng Fract Mech 70: 125–137.

Jenq YS, Shah SP (1985a) Two parameter fracture model for concrete. J Eng Mech ASCE 111 (10): 1227–1241.

Jenq YS, Shah SP (1985b) A fracture toughness criterion for concrete. Eng Fract Mech 21: 1055–1069.

Jenq YS, Shah SP (1988a) Geometrical effects on mode I fracture parameters. Report to RILEM Committee 89-FMT.

Jenq YS, Shah SP (1988b) On concrete fracture testing methods. In Fracture Toughness and Fracture Energy: Test Method for Concrete and Rock. (Edited by H Mihashi, H Takahashi, FH Wittmann), Balkema, Rotterdam: 443–463.

Jiraśek M, Zimmermann T (2001a) Embedded crack model: Part I. Basic formulation. Int J Numer Meth Eng 50: 1269–1290.

Jiraśek M, Zimmermann T (2001b) Embedded crack model. Part II: Combination with smeared cracks. Int J Numer Meth Eng 50: 1291–1305.

Kaplan MF (1961) Crack propagation and the fracture of concrete. J Am Concr Inst 58 (5): 591–610.

Karihaloo BL (1987) Do plain and fiber-reinforced concretes have an *R*-curve behaviour? Fracture of Concrete and Rock (Edited by SP Shah, SE Swartz), Springer, Houston, Texas: 96–105.

Karihaloo BL (1995) Fracture Mechanics and Structural Concrete, Concrete Design and Construction Series. Longman Scientific & Technical, Harlow.

Karihaloo BL, Nallathambi P (1989a) An improved effective crack model for the determination of fracture toughness of concrete. Cem Concr Res 19: 603–610.

Karihaloo BL, Nallathambi P (1989b) Fracture toughness of plain concrete from three-point bend specimens. Mater Struct 22 (3): 185–193.

Karihaloo BL, Nallathambi P (1990) Size-effect prediction from effective crack model for plain concrete. Mater Struct 23 (3): 178–185.

Karihaloo BL, Nallathambi P (1991) Notched beam test: Mode I fracture toughness. Fracture Mechanics Test Methods for Concrete. Report of RILEM Technical Committee 89-FMT (Edited by SP Shah, A Carpinteri), Chapman & Hall, London: 1–86.

Kaya AC, Erdogan F (1980) Stress intensity factors and COD in an orthotropic strip. Int J Fract 16: 171–190.

Kesler CE, Naus DJ, Lott JL (1972) Fracture mechanics: its applicability to concrete. Proceedings of the International Conference on Mechanical Behavior of Materials, The Society of Material Science, Kyoto, Japan 4, pp. 113–124.

Kiciak A, Glinka G, Burns DJ (2003) Calculation of stress intensity factors and crack opening displacements for cracks subjected to complex stress fields. J Press Vessel Technol Trans ASME 125: 260–266.

Kim JK, Lee Y, Yi ST (2004) Fracture characteristics of concrete at early ages. Cem Concr Res 34: 507–519.

Krafft JM, Sullivan AM, Boyle RW (1961) Effect of dimensions on fast fracture instability of notched sheets. Proceedings of the Crack Propagation Symposium, College of Aeronautics, Cranfield, England, Vol. 1, pp. 8–26.

Kumar S (2010) Behaviour of fracture parameters for crack propagation in concrete. Ph.D. Thesis submitted to Department of Civil Engineering, Indian Institute of Kharagpur, India.

Kumar S, Barai SV (2008a) Influence of specimen geometry and size-effect on the K_R-curve based on the cohesive stress in concrete. Int J Fract 152: 127–148.

Kumar S, Barai SV (2008b) Influence of specimen geometry on determination of double-K fracture parameters of concrete: A comparative study. Int J Fract 149: 47–66.

Kumar S, Barai SV (2008c) Cohesive crack model for the study of nonlinear fracture behaviour of concrete. J Inst Eng (India), CV 89 (Nov.):7–15.

Kumar S, Barai SV (2009a) Determining double-K fracture parameters of concrete for compact tension and wedge splitting tests using weight function. Eng Fract Mech 76: 935–948.

Kumar S, Barai SV (2009b) Effect of softening function on the cohesive crack fracture parameters of concrete CT specimen. Sadhana-Acad Proc Eng Sci 36(6): 987–1015.

Kumar S, Barai SV (2009c) Equivalence between stress intensity factor and energy approach based fracture parameters of concrete. Eng Fract Mech 76: 1357–1372.

Kumar S, Barai SV (2009d) Size-effect of fracture parameters in concrete: a comparative study. Comput Concr An Int J (under review).

Kumar S, Barai SV (2009e) Weight function approach for determining crack extension resistance based on the cohesive stress distribution in concrete. Eng Fract Mech 76: 1131–1148.

Kumar S, Barai SV (2009f) Influence of loading condition and size-effect on the K_R-curve based on the cohesive stress in concrete. Int J Fract 156:103–110.

Kumar S, Barai SV (2010a) Determining the double-K fracture parameters for three-point bending notched concrete beams using weight function. Fatigue Fract Eng Mater Struct DOI: 10.1111/j.1460-2695.2010.01477.x.

Kumar S, Barai SV (2010b) Size-effect prediction from the double-K fracture model for notched concrete beam. Int J Damage Mech 9: 473–497.

Kwon SH, Zhao Z, Shah SP (2008) Effect of specimen size on fracture energy and softening curve of concrete: Part II. Inverse analysis and softening curve. Cem Concr Res 38: 1061–1069.

Lee NK, Mayfield B, Snell C (1981) Detecting the progress of internal cracks in concrete by using embedded graphite rods. Magn Concr Res 33(116): 180–183.

Li VC, Liang E (1986) Fracture processes in concrete and fiber reinforced cementitious composites. J Eng Mech ASCE 112(6): 567–586.

Li YN, Bažant ZP (1994) Eigenvalue analysis of size effect for cohesive crack model. Int J Fract 66: 213–226.

Li YN, Liang RY (1992) Stability theory of cohesive crack model. J Eng Mech ASCE 118 (3): 587–603.

Li YN, Liang RY (1993) The theory of the boundary eigenvalue problem in the cohesive crack model and its application. J Mech Phys Solid 41(2): 331–350.

Li YN, Liang RY (1994) Peak load determination in linear fictitious crack model. J Eng Mech ASCE 120 (2): 232–249.

Liang RY, Li YN (1991b) Study of size effect in concrete using fictitious crack model. J Eng Mech ASCE 117 (7): 1931–1651.

Liang RYK, Li Y-N (1991a) Simulations of nonlinear fracture process zone in cementitious material—a boundary element approach. Comput Mech 7: 413–427.

Linsbauer HN, Tschegg EK (1986) Fracture energy determination of concrete with cube-shaped specimens. Zem Beton 31:38–40.

Mai YW (1984) Fracture measurements of cementitious composites. Application of Fracture Mechanics to Cementitious Composites (Edited by S.P. Shah), NATO-ARW, Northwestern University, USA: 399–429.
Maji A, Ouyang C, Shah SP (1990) Fracture mechanisms of concrete based on acoustic emission. J Mater Res 5(1): 206–217.
Maji A, Shah SP (1988) Process zone and acoustic-emission measurements in concrete. Exp Mech 28(1): 27–33.
Mariani S, Perego U (2003) Extended finite element method for quasibrittle fracture. Int J Numer Meth Eng 58: 103–126.
Mergheim J, Kuhl E, Steinmann P (2005) A finite element method for the computational modelling of cohesive cracks. Int J Numer Meth Eng 63: 276–289.
Meschke G, Dumstorff P (2007) Energy-based modeling of cohesive and cohesionless cracks via X-FEM. Comput Meth Appl Mech Eng 196: 2338–2357.
Mindess S, Diamond S (1980) A preliminary SEM study of crack propagation in mortar. Cem Concr Res 10: 509–519.
Mindess S, Diamond S (1982) A device for direct observation of cracking of cement paste or mortar under compressive loading within a scanning electron microscope. Cem Concr Res 12: 569–576.
Mindess S, Nadeau JS (1976) Effect of notch width on K_{IC} for mortar and concrete. Cem Concr Res 6: 529–534.
Moftakhar AA, Glinka G (1992) Calculation of stress intensity factors by efficient integration of weight function. Eng Fract Mech 43: 749–756.
Moës N, Belytschko T (2002) Extended finite element method for cohesive crack growth. Eng Fract Mech 69: 813–833.
Mulmule SV, Dempsey JP (1997). Stress-separation curves for saline ice using fictitious crack model. J Eng Mech ASCE 123 (8): 870–877.
Nallathambi P, Karihaloo BL (1986) Determination of specimen-size independent fracture toughness of plain concrete. Magn Concr Res 38 (135): 67–76.
Nallathambi P, Karihaloo BL, Heaton BS (1984) Effect of specimen and crack size, water/cement ratio and coarse aggregate texture upon fracture toughness of concrete. Magn Concr Res 36 (129): 227–236.
Naus D, Batson GB, Lott JL (1974) Fracture mechanics of concrete. Fracture Mechanics of Ceramics (Edited by Bradt RC, Hasselman DPH, Lange FF), Plenum, New York, NY, Vol. 2: 469–481.
Naus DJ, Lott JL (1969) Fracture toughness of Portland cement concretes. Am Concr Inst J Title No. 66–39: 481–489.
Ngo D, Scordelis AC (1967) Finite element analysis of reinforced concrete beams. J Am Concr Inst 64: 152–163.
Oliver J (1995) Continuum modelling of strong discontinuities in solid mechanics using damage models. Comput Mech 17: 49–61.
Orowan E (1955) Energy criteria of fracture. Weld J 34: 1575–1605.
Ostuka K, Date H (2000) Fracture process zone in concrete tension specimen. Eng Fract Mech 65: 111–131.
Ouyang C, Barzin Mobasher B, Shah SP (1990) An R-curve approach for fracture of quasi-brittle materials. Eng Fract Mech 37: 901–913.
Ouyang C, Landis E, Shah SP (1991) Damage assessment in concrete using quantitative acoustic emission. J Eng Mech ASCE 117(11): 2681–2698.
Ouyang C, Tang T, Shah SP (1996) Relationship between fracture parameters from two parameter fracture model and from size effect model. Mater Struct 29(2): 79–86.
Park K, Paulino GH, Roesler JR (2008) Determination of the kink point in the bilinear softening model for concrete. Eng Fract Mech 7: 3806–3818.
Petersson PE (1981) Crack growth and development of fracture zone in plain concrete and similar materials. Report No. TVBM-100, Lund Institute of Technology.
Petroski HJ, Achenbach JD (1978) Computation of weight the function from a stress intensity factor. Eng Fract Mech 10: 257–266.

Planas J, Elices M (1990) Fracture criteria for concrete: mathematical validations and experimental validation. Eng Fract Mech 35: 87–94.

Planas J, Elices M (1991) Nonlinear fracture of cohesive material. Int J Fract 51: 139–157.

Planas J, Elices M (1992) Asymptotic analysis of a cohesive crack: 1. Theoretical background. Int J Fract 55: 153–177.

Planas J, Elices M, Guinea GV (1992) Measurement of the fracture energy using three-point bend tests: Part 2 – influence of bulk energy dissipation. Mater Struct 25 (5): 305–312.

Planas J, Elices M, Guinea GV, Gómez FJ, Cendón DA, Arbilla I (2003) Generalizations and specializations of cohesive crack models. Eng Fract Mech 70: 1759–1776.

Planas J, Guinea GV, Elices M (1997) Generalized size effect equation for quasibrittle materials. Fatigue Fract Engng Mater Struct 20(5): 671–687.

Planas J, Elices M. and Ruiz G. (1993b). The equivalent elastic crack: 2. X–Y equivalences and asymptotic analysis. Int J Fract 61: 231–246.

Plans J, Elices M, Guinea GV (1993a) Cohesive cracks versus nonlocal models: closing the gap. Int J Fract 63: 173–187.

Prasad MVKV, Krishnamoorthy CS (2002) Computational model for discrete crack growth in plain and reinforced concrete. Comput Meth Appl Mech Eng 191: 2699–2725.

Radjy F, Hansen TC (1973) Fracture of hardened cement paste and concrete. Cem Concr Res 3: 343–361

Raghu Prasad BK, Renuka Devi MV (2007) Extension of FCM to plain concrete beams with vertical tortuous cracks. Eng Fract Mech 74: 2758–2769.

Rashid YR (1968) Analysis prestressed concrete pressure vessels. Nucl Eng Des 7(4): 334–355.

Refai TME, Swartz SE (1987) Fracture behavior of concrete beams in three-point bending considering the influence of size effects. Report No. 190, Engineering Experiment Station, Kansas State University.

Reinhardt HW, Cornelissen HAW, Hordijk DA (1986) Tensile tests and failure analysis of concrete. J Struct Eng ASCE 112 (11): 2462–2477.

Reinhardt HW, Xu S (1999) Crack extension resistance based on the cohesive force in concrete. Eng Fract Mech 64: 563–587.

Rice JR (1968a) A path independent integral and the approximate analysis of strain concentration by notches and cracks. J Appl Mech ASME 35: 379–386.

Rice JR (1968b) Mathematical analysis in the mechanics of fracture. In Fracture; An Advanced Treatise (Edited by Liebowitz H), Academic, New York, NY, Vol. 2: 191–311.

Rice JR (1972) Some remarks on elastic crack-tip stress fields. Int J Solid Struct 8:751–758.

RILEM Draft Recommendation (TC50-FMC) (1985) Determination of fracture energy of mortar and concrete by means of three-point bend test on notched beams. Mater Struct 18 (4): 287–290.

RILEM Draft Recommendations (TC89-FMT) (1990a) Determination of fracture parameters ($K_{Ic}{}^{s}$ and $CTOD_c$) of plain concrete using three-point bend tests, proposed RILEM draft recommendations. Mater Struct 23 (138): 457–460.

RILEM Draft Recommendations (TC89-FMT) (1990b) Size-effect method for determining fracture energy and process zone size of concrete. Mater Struct 23 (138): 461–465.

Roesler J, Paulino GH, Park K, Gaedicke C (2007) Concrete fracture prediction using bilinear softening. Cem Concr Compos 29: 300–312.

Sakata Y, Ohtsu M (1995) Crack evaluation in concrete members based on ultrasonic spectroscopy. ACI Mater J 92(6): 686–698.

Saleh AL, Aliabad MH (1995) Crack growth analysis in concrete using boundary element method. Eng Fract Mech 51(4): 533–545.

Sha GT, Yang CT (1986) Weight functions of radial cracks emanating from a circular hole in a plate. In Fracture Mechanics: Seventh Volume (Edited by JH Underwood et al.), ASTM STP, Philadelphia, 905: 573–600.

Shah SP, McGarry FJ (1971) Griffith fracture criterion and concrete. J Eng Mech ASCE 97(EM6): 1663–1676.

Shah SP, Ouyang C (1994) Fracture mechanics for failure of concrete. Annu Rev Mater Sci 24: 193–320.

Shen G, Glinka G (1991) Determination of weight functions from reference stress intensity factors. Theor Appl Fract Mech 15: 237–245.

Simo JC, Oliver J, Armero F (1993) An analysis of strong discontinuities induced by strain-softening in rate-independent inelastic solids. Comput Mech 12: 277–296.

Strange PC, Bryant H (1979) Experimental test on concrete fracture. J Eng Mech ASCE 105(EM2): 337–342.

Swartz SE, Go CG (1984) Validity of compliance calibration to cracked concrete beams in bending. Exp Mech 24 (2): 129–134.

Tada H, Paris PC, Irwin G (1985) The Stress Analysis of Cracks Handbook. Del Research Corporation, Hellertown, Pennsylvania, PA.

Tang T, Shah SP, Ouyang C (1992) Fracture mechanics and size effect of concrete in tension. J Struct Eng ASCE 118 (11): 3169–3185.

Walsh PF (1972) Fracture of plain concrete. Indian Concr J 46 (11): 469–470, 476.

Walsh PF (1976) Crack initiation in plain concrete. Magn Concr Res 28: 37–41.

Wecharatana M, Shah SP (1982) Slow crack growth in cement composites. J Struct Div ASCE 108(ST6): 1400–1413.

Wecharatana M, Shah SP (1983) Predictions of nonlinear fracture process zone in concrete. J Eng Mech ASCE 109(5): 1231–1246.

Wells AA (1962) Unstable crack propagation in metals: damage and fast fracture. Proceedings of the Crack Propagation Symposium Cranfield Vol. 1: 210–230.

Wells GN, Sluys LJ (2001) A new method for modelling cohesive cracks using finite elements. Int J Numer Meth Eng 50: 2667–2682.

Westergaard HM (1939) Bearing pressures and cracks. J Appl Mech Trans ASME 60: A49–A53.

Wittmann FH, Rokugo K, Bruhwiller E, Mihashi H, Simopnin P (1988) Fracture energy and strain softening of concrete as determined by compact tension specimens. Mater Struct 21(1):21–32.

Wu XR (1984) Approximate weight functions for center and edge cracks in finite bodies. Eng Fract Mech 20 (1): 35–49.

Wu Z, Jakubczak H, Glinka G, Molski K, Nilsson L (2003). Determination of stress intensity factors for cracks in complex stress fields. Arch Mech Eng L(1): s41–s67.

Xie M, Gerstle WH (1995) Energy-based cohesive crack propagation modeling. J Eng Mech ASCE 121(12): 1349–1358.

Xu S, (1999) Determination of parameters in the bilinear, Reinhardt's non-linear and exponentially non-linear softening curves and their physical meanings. Werkstoffe und Werkstoffprüfung im Bauwesen, Hamburg, Libri BOD: 410–424.

Xu S, Reinhardt HW (1998) Crack extension resistance and fracture properties of quasi-brittle materials like concrete based on the complete process of fracture. Int J Fract 92: 71–99.

Xu S, Reinhardt HW (1999a) Determination of double-*K* criterion for crack propagation in quasi-brittle materials, Part I: Experimental investigation of crack propagation. Int J Fract 98: 111–149.

Xu S, Reinhardt HW (1999b) Determination of double-*K* criterion for crack propagation in quasi-brittle materials, Part II: analytical evaluating and practical measuring methods for three-point bending notched beams. Int J Fract 98: 151–177.

Xu S, Reinhardt HW (1999c) Determination of double-*K* criterion for crack propagation in quasi-brittle materials, Part III: compact tension specimens and wedge splitting specimens. Int J Fract 98: 179–193.

Xu S, Reinhardt HW (2000) A simplified method for determining double-*K* fracture meter parameters for three-point bending tests. Int J Fract 104: 181–209.

Xu S, Reinhardt HW, Wu Z, Zhao Y (2003) Comparison between the double-*K* fracture model and the two parameter fracture model. Otto-Graf J 14: 131–158.

Xu S, Zhang X (2008) Determination of fracture parameters for crack propagation in concrete using an energy approach. Eng Fract Mech 75: 4292–4308.

Xu S, Zhao Y, Wu Z (2006) Study on the average fracture energy for crack propagation in concrete. J Mater Civ Eng ASCE 18(6): 817–824.

Xu S, Zhu Y (2009) Experimental determination of fracture parameters for crack propagation in hardening cement paste and mortar. Int J Fract 157:33–43.

Zhang X, Xu S (2007) Fracture resistance on aggregate bridging crack in concrete. Front Archit Civ Eng China 1(1): 63–70.

Zhang X, Xu S, Zheng S (2007) Experimental measurement of double-K fracture parameters of concrete with small-size aggregates. Frontiers Archit Civ Eng China 1(4): 448–457.

Zhao Y, Xu S (2002) The influence of span/depth ratio on the double-K fracture parameters of concrete. J China Three Georges Univ (Nat Sci) 24 (1): 35–41.

Zhao Y, Xu S (2004) An analytical and computational study on energy dissipation along fracture process zone in concrete. Comput Concr 1(1): 47–60.

Zhao Y, Xu S, Wu Z (2007) Variation of fracture energy dissipation along evolving fracture process zones in concrete. J Mater Civ Eng ASCE 19(8): 625–633.

Zhao Z, Kwon SH, Shah SP (2008) Effect of specimen size on fracture energy and softening curve of concrete: Part I. Experiments and fracture energy. Cem Concr Res 38: 1049–1060.

Zi G, Bažant ZP (2003) Eigenvalue method for computing size effect of cohesive cracks with residual stress, with application to kink-bands in composites. Int J Eng Sci 41: 1519–1534.

Zi G, Belytschko T (2003) New crack-tip elements for X-FEM and applications to cohesive cracks. Int J Numer Meth Eng 57: 2221–2240.

Chapter 3
Fracture Behavior of Concrete using Cohesive Crack and Size-Effect Models

3.1 Introduction

This chapter gives an idea of the nonlinear fracture behavior of concrete using cohesive crack and size-effect models. Initially, the formulation and development of cohesive crack model for three-point bending test and compact tension specimen are presented. Then a systematic study on the several cohesive crack fracture parameters for the TPBT and CT specimen with various softening functions is carried out. Toward the end, with reference to cohesive crack model, the size-effect study from size-effect law is investigated using three-point bending test. The contribution in this chapter is mainly based on work done by Kumar (2010) and Kumar and Barai (2008, 2009).

3.2 Cohesive Crack Model for Three-Point Bending Test

3.2.1 Formulation Based on Energy Principle

Assuming that under condition of proportional and monotonic loading in mode I crack propagation, no unloading occurs in the fracture process zone and the process zone boundary can be defined by only one variable, i.e., the crack length a. The potential energy Π of a cracked body as shown in Fig. 3.1 is written as (Li and Liang 1994)

$$\Pi\,(u_i, P, a) = \int_V W(\varepsilon_{ij})\mathrm{d}V - \int_V F_i u_i\,\mathrm{d}V - P\int_{A_\mathrm{T}} b_i u_i\,\mathrm{d}A + \int_{A_\mathrm{P}} \phi(w)\mathrm{d}A \tag{3.1}$$

where W is the strain energy density function defined in the body, ε_{ij} is the strain field, F_i is the body force, u_i is an admissible displacement field for the system, b_i is the boundary force distribution defined in the part of boundary A_T. The potential ϕ can be defined for any given cohesive law $\sigma = f(w)$ as

S. Kumar, S.V. Barai, *Concrete Fracture Models and Applications*,
DOI 10.1007/978-3-642-16764-5_3, © Springer-Verlag Berlin Heidelberg 2011

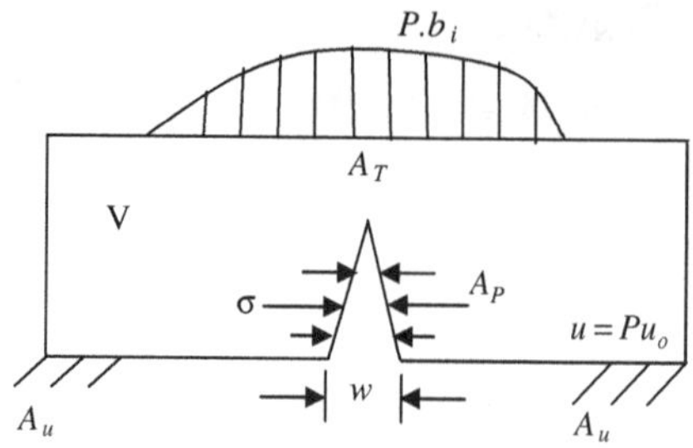

Fig. 3.1 Definition of potential energy of a cracked body

$$\phi(w) = \int_0^w f(v)\mathrm{d}v \tag{3.2}$$

The positive sign of the surface potential is due to fact that the positive direction of cohesive forces is in the opposite direction to that of the crack opening displacement. The displacement equilibrium of the system is obtained by equating to zero the first-order variation of the potential Π with respect to displacement:

$$\begin{aligned}0 = \Pi_u.\delta u_i = &\int_V \sigma_{ij}\delta\varepsilon_{ij}\,\mathrm{d}V - \int_V F_i\delta u_i\,\mathrm{d}V - P\int_{A_T} b_i\delta u_i\,\mathrm{d}A\\ &+ \int_{A_P} f(w)\delta w\,\mathrm{d}A\end{aligned} \tag{3.3}$$

All possible equilibrium states of the system cannot be determined using Eq. (3.3). Among all the displacement equilibrium states that satisfy Eq. (3.3) for different combinations of given P and a, there is one combination which satisfies the following crack equilibrium equation:

$$\begin{aligned}0 = \Pi_a = \frac{\partial \Pi}{\partial a} = &\frac{\partial}{\partial a}\left(\int_V W(\varepsilon_{ij})\mathrm{d}V - \int_V F_i u_i\,\mathrm{d}V - P\int_{A_T} b_i u_i\,\mathrm{d}A\right)\\ &+ \frac{\partial}{\partial a}\int_{A_P}\phi(w)\mathrm{d}A,\ \forall\delta a > 0\end{aligned} \tag{3.4}$$

For linear elastic body and constant body forces, the terms in the parenthesis on the right-hand side of Eq. (3.4) can be recognized as the energy release rate, whereas the last term is the energy needed to create a unit area of process zone, which is deformation dependent and not as constant material property. Hence, Eq. (3.4) is equivalently written in terms of stress intensity factors as

$$(Pk_{\mathrm{P}} + K_\sigma)^2 = 0 \text{ or } Pk_{\mathrm{P}} + K_\sigma = 0 \tag{3.5}$$

where k_{P} is the stress intensity factor at the process zone tip due to unit applied load, $P = 1$, and K_σ is the stress intensity factor at the process zone tip due to cohesive stress in the process zone. The basic feature of cohesive crack model is represented by Eq. (3.5). The equation states that the algebraic summation of stress intensity factors at the crack tip due to the external loading, constant body force, boundary force, if any and the cohesive stress distribution in the process zone is zero. This statement eliminates the stress singularity at the crack tip of the structure and a finite stress acts near the crack tip.

3.2.2 Basic Assumptions

The following assumptions are considered in the development of cohesive crack model (Petersson 1981, Carpinteri 1989a–c, Carpinteri and Colombo 1989, Carpinteri et al. 2006):

1. The bulk of material behaves in a linear elastic and isotropic manner.
2. The cohesive process zone begins to develop when the maximum principal stress becomes equal to the tensile strength.
3. The material is in partially damaged condition and is still able to transfer the stress known as cohesive stress σ after formation of cohesive fracture zone. The cohesive stress depends on the crack opening displacement w. For mode I opening, σ acts normal to the crack faces as shown in Fig. 3.2 and is related to w as below:

$$\sigma = f(w) \tag{3.6}$$

Also from the definition of softening function of concrete

$$f_t = f(0) \tag{3.7}$$

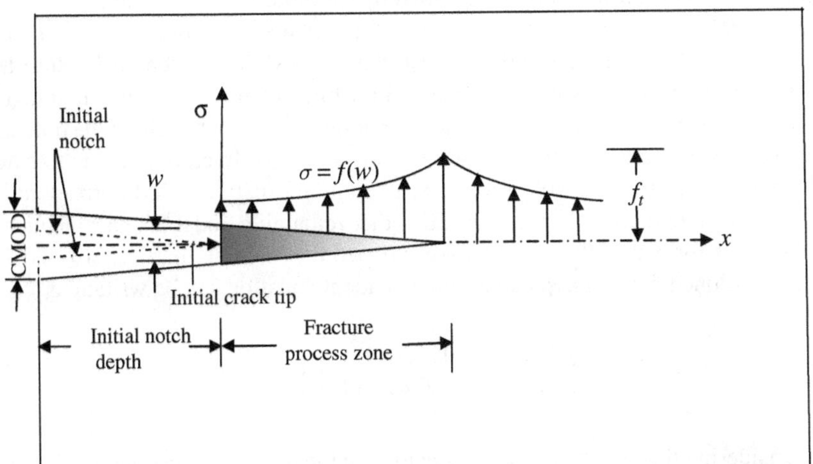

Fig. 3.2 Stress distribution and cohesive crack growth in mode I opening for concrete

and total fracture energy G_F is mathematically written as

$$G_F = \int_0^{w_c} f(w)dw \tag{3.8}$$

The structural behavior can be analyzed using the parameter called *characteristic length* l_{ch}:

$$l_{ch} = \frac{EG_F}{f_t^2} \tag{3.9}$$

The characteristic length is an inverse measure of the brittleness of the material.

3.2.3 Finite Element Discretization

Two material lengths (i) the fracture process zone and (ii) the width of the FPZ have always been a matter of research in the fracture modeling of concrete. Experimental results (Petersson 1981) show that the size of FPZ is significant and comparable, whereas its width is normally small (in the order of maximum aggregate size) with respect to the characteristic dimension of a structure. Depending upon a method of analysis, these material lengths are considered or ignored during numerical simulation. The fictitious crack model or cohesive crack model assumes the FPZ collapsed into a line or a surface in two- or three-dimensional analysis, respectively. When a single macroscopic crack opens in a fixed direction under a pure mode I loading, the cohesive crack can be modeled easily by employing procedures proposed by Petersson (1981) and was further modified by Carpinteri (1989a–c) and Planas and Elices (1991). In this method the number of fracture nodes along the potential crack line is kept as fixed in the standard linear elastic FEM. The potential fracture line for the TPBT specimen is shown in Fig. 3.3 for finite element implementation. The crack opening nodes are taken from 1 at the bottom to n at the top along the potential fracture line. The crack opening displacements along the fracture line due to external force of the structure are grouped in the column matrix w and corresponding nodal forces into a column matrix F. The other remaining nodal displacements and nodal forces for rest of the structure are grouped into column matrices w_R and F_R. Then partitioned form of associated finite element equation can be written as

$$\begin{Bmatrix} F \\ F_R \end{Bmatrix} = -\begin{bmatrix} K_{CC} & K_{CR} \\ K_{RC} & K_{RR} \end{bmatrix} \begin{Bmatrix} w \\ w_R \end{Bmatrix} \tag{3.10}$$

Considering that no other forces are acting on the rest of the specimen, $[F_R] = 0$, and eliminating out all the displacement components of w_R, the matrix equilibrium equation is obtained as

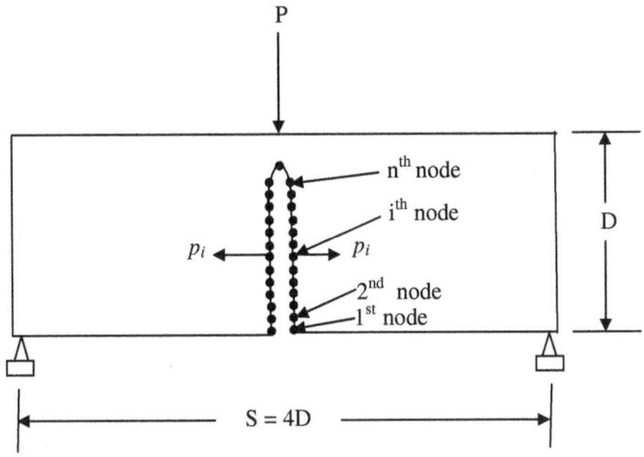

Fig. 3.3 Finite element pair of nodes along fracture line

$$\{F\} = [K_{\mathrm{F}}]\{w\} \tag{3.11}$$

in which

$$[K_{\mathrm{F}}] = [K_{\mathrm{CC}}] - [K_{\mathrm{CR}}][K_{\mathrm{RR}}]^{-1}[K_{\mathrm{RC}}] \tag{3.12}$$

The crack opening for n nodes along the fracture zone is further expressed in the following form for convenience:

$$\{w\} = [K]\{p\} + \{C\}P + \{p_{\mathrm{g}}\} \tag{3.13}$$

where $\{w\}$ is the vector of crack opening displacement at node i, $[K]$ is the symmetric matrix and the value of K_{ij} is the crack opening displacement at node i by a unit opening nodal force applied at node j, $\{p\}$ is the vector of nodal forces, $\{C\}$ is the vector of crack opening displacement at node i when $P =$ unit value, and $\{p_{\mathrm{g}}\}$ is the vector of crack opening displacement at node i due to self-weight of specimen.

In the present study, the enhanced influence method (Planas and Elices 1991) is used to solve Eq. (3.13). Assuming that the total number of fracture nodes are n and initial crack tip lies at kth node, according to enhanced influenced method, Eq. (3.13) may be partitioned between the notches from nodes $i = 1, 2, 3, \ldots, (k-1)$ in the notch portion and from $i = k, (k+1), (k+2), \ldots, n$ in the ligament portion. Hence, the partitioned equation (3.13) may be written in the following form:

$$\begin{Bmatrix} w_{\mathrm{N}} \\ w_{\mathrm{L}} \end{Bmatrix} = \begin{bmatrix} K_{\mathrm{NN}} K_{\mathrm{NL}} \\ K_{\mathrm{LN}} K_{\mathrm{LL}} \end{bmatrix} \begin{Bmatrix} p_{\mathrm{N}} \\ p_{\mathrm{L}} \end{Bmatrix} + \begin{Bmatrix} C_{\mathrm{N}} \\ C_{\mathrm{L}} \end{Bmatrix} P + \begin{Bmatrix} p_{\mathrm{gN}} \\ p_{\mathrm{gL}} \end{Bmatrix} \tag{3.14}$$

The subscripts N and L denote notch portion for $i = 1, 2, \ldots, (k-1)$ and ligament zone for $i = k, (k+1), (k+2), \ldots, n$. Since width of crack opening in ligament zone is zero and initial notch portion is traction-free zone, it may be written as

$$\left.\begin{aligned} \{p_{\mathrm{N}}\} &= \{0\},\ \text{for } i = 1,\ 2,\ 3,\ \ldots,\ (k-1) \\ \{w_{\mathrm{L}}\} &= \{0\},\text{for } i = k, (k+1), (k+2), \ldots, n \end{aligned}\right\} \tag{3.15}$$

Equations (3.14) and (3.15) yield

$$\{p_{\mathrm{L}}\} = [M_{\mathrm{LL}}]\{w_{\mathrm{L}}\} - [M_{\mathrm{LL}}]\{C_{\mathrm{L}}\}P - [M_{\mathrm{LL}}]\{p_{g_{\mathrm{L}}}\} \tag{3.16}$$

where

$$M_{\mathrm{LL}} = K_{\mathrm{LL}}^{-1}, \{T_{\mathrm{L}}\} = [M_{\mathrm{LL}}]\{C_{\mathrm{L}}\}, \{T_{\mathrm{g}}\} = [M_{\mathrm{LL}}]\{p_{\mathrm{gL}}\}$$

$$\{p_{\mathrm{L}}\} = [M_{\mathrm{LL}}]\{w_{\mathrm{L}}\} - \{T_{\mathrm{L}}\}P - \{T_{\mathrm{g}}\} \tag{3.17}$$

Further, the ligament portion may be partitioned between cohesive (damaged) zone and uncracked (or undamaged) zones along the fracture line. Let cohesive zone form between the nodes $j = k,\ (k+1), \ldots, l$ and undamaged part is $j = (l+1),\ (l+2), \ldots, n$. Hence after further partitioning of Eq. (3.17), it may be written as

$$\begin{Bmatrix} p_{\mathrm{LC}} \\ p_{\mathrm{LU}} \end{Bmatrix} = \begin{bmatrix} M_{\mathrm{LLCC}} & M_{\mathrm{LLCU}} \\ M_{\mathrm{LLUC}} & M_{\mathrm{LLUU}} \end{bmatrix} \begin{Bmatrix} w_{\mathrm{LC}} \\ w_{\mathrm{LU}} \end{Bmatrix} - \begin{Bmatrix} T_{\mathrm{LC}} \\ T_{\mathrm{LU}} \end{Bmatrix} P - \begin{Bmatrix} T_{\mathrm{gC}} \\ T_{\mathrm{gU}} \end{Bmatrix} \tag{3.18}$$

where subscripts C and U denote cohesive zone for $j = k,\ (k+1), \ldots, l$ and undamaged zone for $j = (l+1), (l+2), \ldots, n$. Since width of crack opening displacements in undamaged zone and also at the last node of cohesive zone is zero, mathematically it is expressed as

$$\{w_{\mathrm{LU}}\} = \{0\},\ \{w_{\mathrm{LC}}\}_{j=l} = 0 \tag{3.19}$$

Equations (3.18) and (3.19) yield

$$\{p_{\mathrm{LC}}\} = [M_{\mathrm{LLCC}}]\{w_{\mathrm{LC}}\} - \{T_{\mathrm{LC}}\}P - \{T_{\mathrm{gC}}\} \tag{3.20}$$

For $j = l$th node, the applied load P may be expressed as

$$P = \frac{\sum_{j=k}^{l-1} \{M_{\mathrm{LLCC}}\}_{l,j}\{w_{\mathrm{LC}}\}_j - \{p_{\mathrm{LC}}\}_{j=l} - \{T_{\mathrm{gC}}\}_{j=l}}{\{T_{\mathrm{LC}}\}_{j=l}} \tag{3.21}$$

For the nodes $j = k,\ (k+1),\ (k+2), \ldots,\ (l-1)$ and using Equations (3.20) and (3.21), it can be formed as many number of nonlinear simultaneous equations as the cracks are opened in the cohesive zone. The functional form of the nonlinear simultaneous equations may be expressed as

$$\{Z\} = [M_{\mathrm{LLCC}}]\{w_{\mathrm{LC}}\} - \{p_{\mathrm{LC}}\} - \{T_{\mathrm{LC}}\}\frac{\sum\limits_{j=k}^{l-1}\{M_{\mathrm{LLCC}}\}_{l,j}\{w_{\mathrm{LC}}\}_j - \{p_{\mathrm{LC}}\}_{j=l} - \left\{T_{\mathrm{gC}}\right\}_{j=l}}{\{T_{\mathrm{LC}}\}_{j=l}} - \left\{T_{\mathrm{gC}}\right\} \tag{3.22}$$

If proper sign convention is used, it can be written as $\{p_{\mathrm{LC}}\} = -F_{\mathrm{uc}}f(w_{\mathrm{LC}})$, where for regular discretization of uniform mesh size h, $F_{\mathrm{uc}} = Bhf_{\mathrm{t}}$ for all values of i except $i = k$ and $i = n$; and $F_{\mathrm{uc}} = Bhf_{\mathrm{t}}/2$ for $i = k$ and $i = n$. The solution of Eq. (3.22) is obtained using Newton–Raphson method. The function Z and its Jacobian matrix for unknown crack opening displacements using different softening functions are programmed in separate functions. Thus, this method becomes powerful and as many types of softening laws as possible can be incorporated into a program. After the solution of Eq. (3.22) for $\{w_{\mathrm{LC}}\}$, the values of P and $\{p_{\mathrm{LC}}\}$ can be determined using Equations (3.21) and (3.20), respectively. Going back to Equations (3.19), (3.18), (3.17), (3.15), (3.14), and (3.13), it is possible to determine all unknown values of $\{w\}$ and $\{p\}$.

3.2.4 Beam Deflection

Also, for the known values of $\{p\}$ and P, the central deflection of beam δ is determined as

$$\delta = D_{\mathrm{L}}P + \left\{D_{\mathrm{p}}\right\}^{\mathrm{T}}\{p\} + D_{\mathrm{g}} \tag{3.23}$$

where D_{L} is the load point deflection when external load P is unity, $\left\{D_{\mathrm{p}}\right\}$ is the vector of load point deflection when unit load vector $\{p\} = \{1\}$, D_{g} is the load point deflection due to self-weight of the specimen. $[K], \{C\}$, $\left\{p_{\mathrm{g}}\right\}$, $\left\{D_{\mathrm{p}}\right\}$, and D_{g} in Eqs. (3.13) and (3.23) are the known quantities and can be computed using standard linear elastic finite element method. In the present study, due to symmetry, half of the beam is discretized using four-node isoparametric quadrilateral elements as shown in Fig. 3.4. A typical four-node isoparametric plane element is also shown in Fig. 3.5. For the sake of simplicity in element connectivity and nodal coordinate, the beam is divided into three bands of length: $D/4$, $3D/4$, and D, respectively, along its longitudinal direction. For this particular case, the number of meshes in each band is taken to be 16, 5, and 5, respectively. However, these numbers can be chosen arbitrarily to any value depending upon the memory capacity of the computer. The number of divisions along depth of the beam is fixed as 60 for this particular discretization. Six number of nodes from the top is restrained against movement in horizontal direction to allow the development of cohesive zone along potential fracture line.

For the sake of simplicity in element connectivity and nodal coordinate, the beam is divided into three bands of length: $D/4$, $3D/4$, and D, respectively, along its longitudinal direction. The number of division in the band $D/4$ is so selected such that

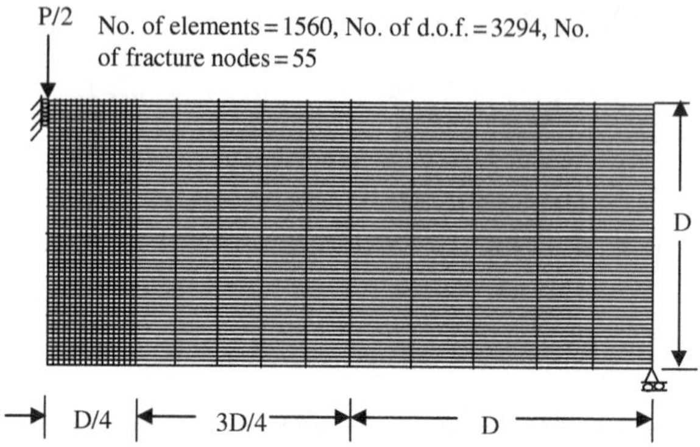

Fig. 3.4 Finite element mesh for half of the beam with 55 fracture nodes

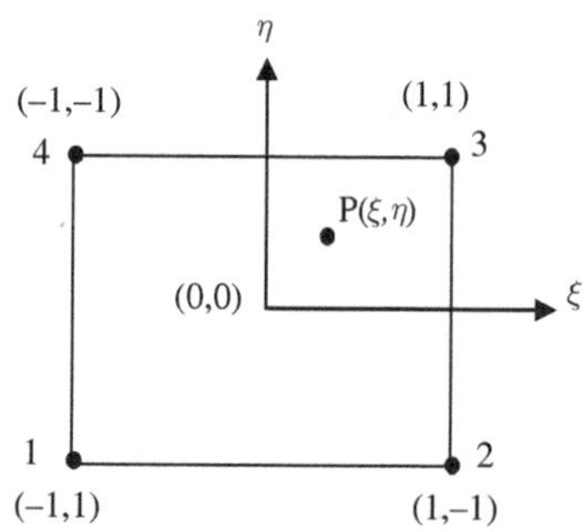

Fig. 3.5 Typical four-node isoparametric plane element in (ξ,η) coordinates

the aspect ratio of the FE mesh becomes 1 or almost 1. In the other two bands $3D/4$ and D along the length of the beam, the number of divisions was kept arbitrary to be low. Numerical experiments were carried out by varying the aspect ratios in the bands of $3D/4$ and D while keeping the aspect ratio of the FE mesh in the band $D/4$ as 1 for relatively coarser FE mesh size. It was found that the influence coefficients along the crack line did not change appreciably in these cases. Further, a convergence study was also carried out with a benchmark problem (Carpinteri and Colombo 1989) of TPBT specimen for which $B = 150$ mm, $D = 150$ mm, and $S = 600$ mm were taken into consideration. The material properties were taken as $\nu = 0.10$, $E = 36.5$ Gpa, $f_t = 3.19$ MPa, and $G_F = 50$ N/m with linear softening function. In this case, the number of divisions along the dimension D was taken as 40. The numbers of divisions of 10, 15, and 5 were considered in the bands $D/4$, $3D/4$, and D, respectively. The number of nodes along fracture line was taken to be 37. The P–δ curves for the a_0/D ratios 0 and 0.5 were simulated from the present model and the results were compared (Kumar, 2010) with the solution obtained by Carpinteri and Colombo (1989). From the comparison, it was observed that the numerical results obtained in the present calculation were very close to the reported results (Carpinteri and Colombo 1989).

In the subsequent numerical simulation, the FE mesh has been further refined and the number of meshes in each band is taken to be 16, 5, and 5, respectively. The number of divisions along the depth of the beam is fixed as 60 for this particular discretization. Six numbers of nodes from the top is restrained against movement in horizontal direction to allow the development of cohesive zone along potential fracture line.

3.2.5 Model Implementation

The finite element discretization as shown in Fig. 3.4 is adopted in which the number of elements is 1560, the number of degree of freedom (d.o.f.) is 3294, and the number of fracture nodes along the potential fracture line is 55. First, the linear elastic finite element program is executed to obtain the influence coefficients such as $[K], \{C\}, \{p_g\}, \{D_p\}, D_L$, and D_g. These coefficients are determined once and for all and used for cohesive crack propagation program. Initially, since the structure is linear elastic, the tensile stress at the kth node will exceed the tensile strength of concrete and this node is opened first. Thereafter, cohesive stress starts to act across the opened node pair according to a particular softening law. At this point the basic unknown quantities, viz. $\{w_{LC}\}_{j=k,(k+1),\ldots,(l-1)}$ and P, are determined using Eqs. (3.21) and (3.22). Subsequently, the values of $\{w\}$ and $\{p\}$ are computed and then the deflection of the beam is obtained using Eq. (3.23). In this step, the first point of P-CMOD and load–deflection (P–δ) curves are thus determined. In the second stage of calculation, the external load is increased and at some particular value of the load, the tensile stress at node $(k + 1)$ is greater than the tensile strength of concrete. At this point, $(k+1)$th node pair also gets opened and cohesive stress starts to act at this node. Following similar process as mentioned above, the second point of P-CMOD and P–δ curves are obtained. This process is continued till the node pair (l–1) in cohesive zone is opened and the program gets terminated when $l = n$. The basic unknowns in every step are as many as the numbers of nodes opened in the cohesive zone. At each step of computation the value of COD is checked and when it becomes greater than or equal to w_c, the corresponding value of nodal force goes down to zero. A flowchart as shown in Fig. A.1 for the development of this model is presented in Appendix.

3.3 Softening Function of Concrete

Some widely used softening functions such as linear (Hillerborg et al. 1976), bilinear (Petersson 1981, Wittmann et al. 1988, CEP-FIP Model Code-1990 1993), exponential (Gopalaratnam and Shah 1985, Karihaloo 1995), nonlinear (Reinhardt et al. 1986), and quasi-exponential (Planas and Elices 1990) available in the literature are mathematically expressed below.

3.3.1 Linear Softening Function

$$\sigma = f_t\left(1 - \frac{w}{w_c}\right), \quad \text{for } 0 \le w \le w_c$$
$$\sigma = 0, \quad \text{for } w \ge w_c \tag{3.24}$$

and

$$w_c = \frac{2G_F}{f_t} \tag{3.25}$$

3.3.2 Bilinear Softening

The bilinear softening as shown in Fig. 3.6 is completely expressed as

$$\sigma = f_t - (f_t - \sigma_s)\frac{w}{w_s}, \text{ for } 0 \le w \le w,$$
$$\sigma = \sigma_s \frac{w_c - w}{w_c - w_s}, \text{ for } w_s \le w \le w_c \tag{3.26}$$
$$\sigma = 0, \text{ for } w \ge w_c$$

The area under the softening curve is

$$G_F = \frac{(\alpha_s + \beta_s)}{2} f_t w_c \tag{3.27}$$

The values of kink point in nondimensional form are $\alpha_s = \sigma_s/f_t$ and $\beta_s = w_s/w_c$, where σ_s and w_s are the ordinate and abscissa at the point of slope change of bilinear

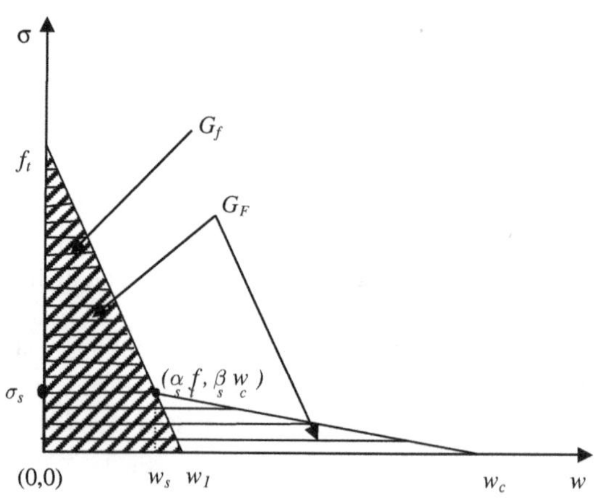

Fig. 3.6 Generalized bilinear softening function

softening curve, respectively. Petersson (1981) used the values of $\alpha_s = 1/3$ and $\beta_s = 2/9$ and hence Eq. (3.27) is expressed as

$$w_c = \frac{3.6G_F}{f_t} \tag{3.28}$$

The values of $\alpha_s = 1/4$ and $\beta_s = 3/20$ in bilinear strain-softening diagram were found suitable to simulate the P-COD behavior of CT specimens (Wittmann et al. 1988). In this case, the following relation can be obtained:

$$w_c = \frac{5G_F}{f_t} \tag{3.29}$$

CEB-FIP Model Code-1990 (1993) introduced various empirical relations to obtain the different parameters of bilinear softening curve. According to the code, the values of w_c and the constant k_d depend on the maximum size of aggregates d_a as given below:

d_a (mm)	w_c (mm)	k_d
8	0.12	4
16	0.15	6
32	0.25	10

For the maximum size of aggregate 19 mm (say), the values of w_c and k_d are linearly interpolated as $w_c = 0.1688$ mm and $k_d = 6.75$. The value of $\alpha_s = 0.15$ is kept fixed as prescribed in the CEB-FIP Model Code, whereas the value of β_s for a given value of G_F (in N/m) may be obtained using the empirical relations given in the code:

$$w_s = \frac{G_F - 22w_c\left(\frac{G_F}{k_d}\right)^{0.95}}{150\left(\frac{G_F}{k_d}\right)^{0.95}}\text{mm}, \quad \beta_s = \frac{w_s}{w_c} \tag{3.30}$$

The following relations are obtained using Eqs. (3.27) and (3.30):

$$\beta_s = 0.159, \quad w_c = \frac{6.475G_F}{f_t} \tag{3.31}$$

Characteristic parameters of bilinear softening are also determined with the help of the following relations proposed by Xu and Reinhardt (1999) and Xu and Zhang (2008). This softening function may be termed as the modified bilinear softening function and in this case, the G_F may be computed according to relations presented in the CEP-FIP Model Code (1990). The other parameters of the modified bilinear softening are

$$\alpha_s = \frac{\sigma_s}{f_t} = \frac{2 - f_t.\mathrm{CTOD_c}/G_F}{\alpha_F}$$
$$\beta_s = \frac{w_s}{\mathrm{CTOD_c}} = 1 \tag{3.32}$$

where $\alpha_F = 9 - d_a/8$ and $w_c = \alpha_F.G_F/f_t$.

3.3.3 Exponential Softening

Exponential softening (Gopalaratnam and Shah 1985) was originally proposed in the following form:

$$\sigma = f_t \exp(-kw^{\lambda}) \tag{3.33}$$

where λ and k are constant values: $\lambda = 1.01$ and $k = 0.063$ and w in micrometers. However, in the present study the following exponential relationship (Karihaloo 1995) is used:

$$\sigma = f_t \exp\left(-\mu \frac{w}{w_c}\right) \tag{3.34}$$

where μ is the material constant taken as $\mu = 4.6052$ for $\sigma = 0.01 f_t$ and $w = w_c$, and w_c is computed as below:

$$w_c = \frac{4.6517 G_F}{f_t} \tag{3.35}$$

3.3.4 Nonlinear Softening

Nonlinear softening relation (Reinhardt et al. 1986) was originally proposed in the following form:

$$\sigma = f_t \left\{ \left[1 + \left(\frac{c_1 w}{w_c} \right)^3 \right] \exp\left(\frac{-c_2 w}{w_c} \right) - \frac{w}{w_c} \left(1 + c_1^3 \right) \exp\left(-c_2\right) \right\} \tag{3.36}$$

where, c_1, c_2, and w_c are the material constants. For normal concrete, the three parameters in Eq. (3.36) are considered as $c_1 = 3$, $c_2 = 6.93$, $w_c = 160\,\mathrm{mm}$. Also

$$G_F = w_c f_t \left\{ \begin{array}{l} \frac{1}{c_2} \left[1 + 6 \left(\frac{c_1}{c_2} \right)^3 \right] - \left[1 + c_1^3 \left(1 + \frac{3}{c_2} + \frac{6}{c_2^2} + \frac{6}{c_2^3} \right) \right] \\ \times \frac{\exp(-c_2)}{c_2} - \left(\frac{1 + c_1^3}{2} \right) \exp(-c_2) \end{array} \right\} \tag{3.37}$$

w_c is computed using the above values of c_1 and c_2 and for a given value of G_F in Eq. (3.37) as

$$w_c = \frac{5.136 G_F}{f_t} \tag{3.38}$$

3.3.5 *Quasi-exponential Softening*

Quasi-exponential softening function (Planas and Elices 1990) is characterized by the following expression:

$$\begin{aligned} &\sigma(w) = f_t \left\{ (1 + c_1) \exp\left(\frac{-c_2 w f_t}{G_{FC}} \right) - c_1 \right\}, \text{ for } 0 < w \leq 5G_{FC}/f_t \\ &\sigma(w) = 0, \qquad \text{for } 5G_{FC}/f_t \leq w, \\ &c_1 = 0.0082896, \ c_2 = 0.96020 \end{aligned} \tag{3.39}$$

where c_1 and c_2 are the material constants.

3.4 Numerical Study Using TPBT Specimen

3.4.1 *Experimental Results and Numerical Computation*

Complete computer programs for calculating the influence coefficients and studying crack propagation are developed in MATLAB. For executing the finite element program and crack propagation study for TPBT configuration of notched plain concrete beam, the actual tests data recently published by Roesler et al. (2007) are considered. The test beam series of B250-80a-c and B150-80b-c for specimen depth 250 and 150 mm, respectively and test beam series of B63-80b-c and CB63-80a-c for specimen depth 63 mm are considered for comparison with the numerical results. The beam designation starting with "B" and "CB" is the cast and cut beams, respectively. These beams have constant beam thickness B of 80 mm and a_o/D ratio of 1/3 for all the specimens. Average values of splitting tensile strength $f_t = 4.15$ MPa and modulus of elasticity $E = 32$ GPa were obtained from tests of the standard concrete cylinders (150 mm × 300 mm). Salient features of test result are reported in Table 3.1. The average values of $G_F = 167$, 164, and 119 N/m for specimens depth of 250, 150, and 63 mm, respectively, as obtained from the tests are used as material property in the model. The details of the experimental results can be seen in the work of Roesler et al. (2007). In the present study, splitting tensile strength instead of direct tensile strength is approximated as material property of concrete and the value of υ for normal concrete is assumed to be 0.20. The linear softening law, Petersson's (1981) bilinear, CEB-FIP Model Code-1990's (1993) bilinear,

Table 3.1 Test results of three-point bending geometry for notched concrete beam

D (mm)	Beam No.	S (mm)	a_o (mm)	$P_{u,exp}$ (kN)	σ_{Nu} (MPa)	Mean $P_{u,exp}$ (kN)	G_F (N/m)	Mean G_F(N/m)
	B250-80a			6.926	2.08		193	
250	B250-80b	1000	83	6.303	1.89	6.6983	139	167
	B250-80c			6.866	2.06		169	
	B150-80a			–			–	
150	B150-80b	600	50	4.089	2.04	4.1235	170	164.5 ≈
	B150-80c			4.158	2.08		159	164 Roesler et al. (2007)
	B63-80a			–			–	
	B63-80b			2.264	2.70		106	
63	B63-80c	250	21	2.054	2.45	2.5224	–	119
	CB63-80a			2.725	3.25		123	
	CB63-80b			2.749	3.28		124	
	CB63-80b			2.820	3.36		124	

Roesler et al. (2007)
Note: $P_{u,exp}$ denotes the peak load obtained from experiment

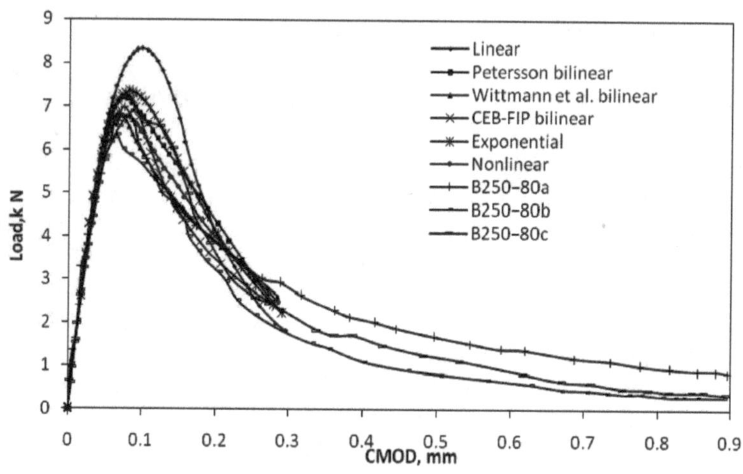

Fig. 3.7 P-CMOD curves for specimen depth 250 mm

Wittmann et al.'s (1988) bilinear, exponential (Karihaloo 1995), and nonlinear softening functions (Reinhardt et al. 1986) are incorporated into the computer program. The P-CMOD curves obtained from tests as well as numerical result using various softening functions for D = 250, 150, and 63 mm are shown in Figs. 3.7, 3.8, and 3.9 respectively.

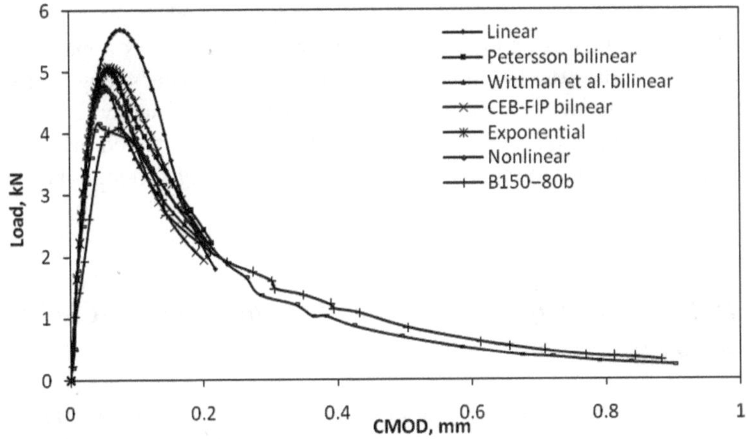

Fig. 3.8 P-CMOD curves for specimen depth 150 mm

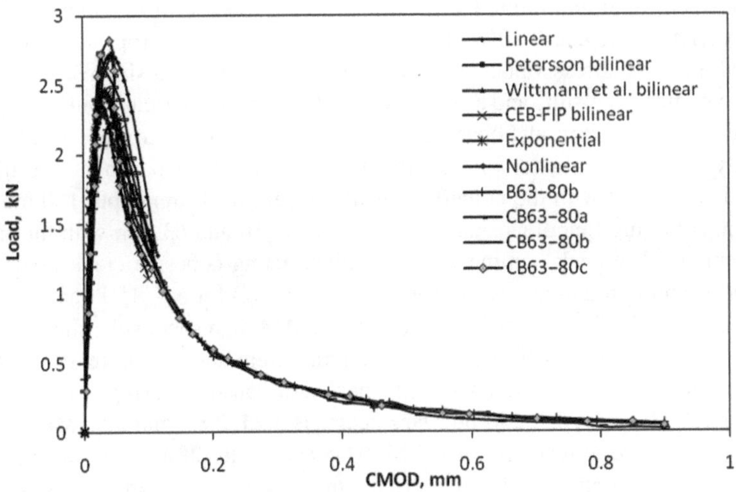

Fig. 3.9 P-CMOD curves for specimen depth of 63 mm

3.4.2 Comparison with Numerical Results Using Linear Softening

The peak load determined using the numerical model with linear softening function is 8319.6 N, whereas the average load obtained in the experiment is 6698.3 N for specimen depth of 250 mm. There is an overestimation in the peak load by 24.20% for this particular example.

Similarly, the overestimation in the peak loads using cohesive crack model with linear softening is 37.99 and 7.98% for specimen depths 150 and 63 mm, respectively. The overestimation in predicting the peak load may be attributed to two important reasons: (i) the use of straight line softening which is the simplest form of the softening relations and (ii) the splitting tensile strength as a material property, which is roughly 10% higher than the mean uniaxial tensile strength. However, the numerical results of cohesive crack model cannot be adequately judged from a single set of experimental results. The post-peak response obtained using numerical simulation is similar to the experimental results showing softening behavior of concrete.

3.4.3 Influence of Softening Function on the Global P-CMOD Response

The peak nominal stress obtained from the test results (Roesler et al. 2007) is presented in Table 3.1. As a particular case, the range of peak nominal stress for beam depths 63, 150, and 250 mm is 2.45–3.36, 2.04–2.08, and 1.89–2.08 MPa, respectively, showing a wide scatter in results for some cases, and the percentage difference between minimum and maximum peak stresses for the above specimen sizes is 27.1, 1.92, and 9.13%, respectively. It may be inappropriate to compare the peak nominal stress values obtained from test results of a single concrete mix with the numerical results and also in present study, the numerical results are gained using cohesive crack model considering the splitting tensile strength as a material property instead of uniaxial tensile strength. However, the experimental results are in good agreement with the numerical results except for beam depth 150 mm with different softening functions and for beam depths 250 and 63 mm with linear softening curves. The peak nominal stress obtained using cohesive crack model with various softening relations (except for linear softening) for 63-, 150-, and 250-mm-size beams is 2.73–2.93, 2.35–2.53, and 2.03–2.20 MPa, respectively, showing a less scatter in results, and the difference between minimum and maximum peak stresses for all sizes of beams is less than 8%. The peak nominal stress using linear softening function for 63-, 150-, and 250-mm-size beams is 3.24, 2.85, and 2.50 MPa, respectively. The post-peak response of P-CMOD curves with different softening curves (except linear softening) is slightly varied. On an average, the peak loads predicted using linear softening is approximately 10% larger than those obtained using other softening functions (bilinear curves, exponential, nonlinear).

3.4.4 Influence of Kink Point in the Bilinear Softening on the Global P-CMOD Response

Bazant (2002) proposed three-parameter (G_f ,G_F, and f_t) bilinear fit which completely characterizes the bilinear softening of concrete as given below:

$$w_c = \frac{2}{\alpha f_t}\left[G_F - (1-\alpha)\,G_f\right] \tag{3.40}$$

where

$$G_{\mathrm{f}} = \frac{f_{\mathrm{f}} w_1}{2} \tag{3.41}$$

in which w_1 is the horizontal intercept of the initial tangent of softening. Bazant (2002) assumed $\alpha = 1/4$ as proposed by Wittmann et al. (1988) and a fixed relationship as

$$G_{\mathrm{F}} = 2.5 G_{\mathrm{f}} \tag{3.42}$$

Equation (3.42) is derived by Bažant and Becq-Giraudon (2002) on the basis of 238 different test results. In the present calculation, assuming that the relationship (Eq. (3.42)) may be influenced by concrete mixes, the value of w_1 may be computed from the size effect of the experimental results according to the proposed method mentioned in Guinea et al. (1994) for which the size-effect curve using linear softening may be written as

$$\frac{\sigma_{\mathrm{Nu}}}{f_{\mathrm{t}}} = \psi \left(\frac{2Df_{\mathrm{t}}}{Ew_1} \right) \tag{3.43}$$

where σ_{Nu} is the peak nominal stress for three-point bending test obtained as

$$\sigma_{\mathrm{Nu}} = \frac{3}{2} \left(\frac{P_{\mathrm{u}} S}{BD^2} \right) \tag{3.44}$$

The function ψ in Eq. (3.43) is unique and may be defined once and for all with arbitrary values of f_{t}, E, and w_1. In this study the average of G_{F} for the values of 167, 164, and 119 N/m = 150 N/m is used to generate the curve using cohesive crack method with linear softening. This curve is known as *master curve* having abscissa x_{M} and ordinate y (Fig. 3.10) as given below:

$$x_{\mathrm{M}} = \log \left(\frac{2Df_{\mathrm{t}}}{Ew_1} \right), \quad y = \frac{\sigma_{\mathrm{Nu}}}{f_{\mathrm{t}}} \tag{3.45}$$

To get a particular value of w_1 so that peak load obtained from the test results matches with the theoretical computation, all the nine points of experimental results are plotted on the same figure. It is interesting to note that excepting three points which belong to the test results of cut beam series CB63-80a-c, the experimental results show some definite relationship with the *master curve*. A straight line fit is then obtained using these experimental results which is termed as *shifted curve*. From Fig. 3.10, the magnitude of the shift Δx between the *master curve* and the *shifted curve* is expressed as

$$\Delta x = x_{\mathrm{M}} - x = \log \left(\frac{2G_{\mathrm{F}}}{w_1 f_{\mathrm{t}}} \right) \tag{3.46}$$

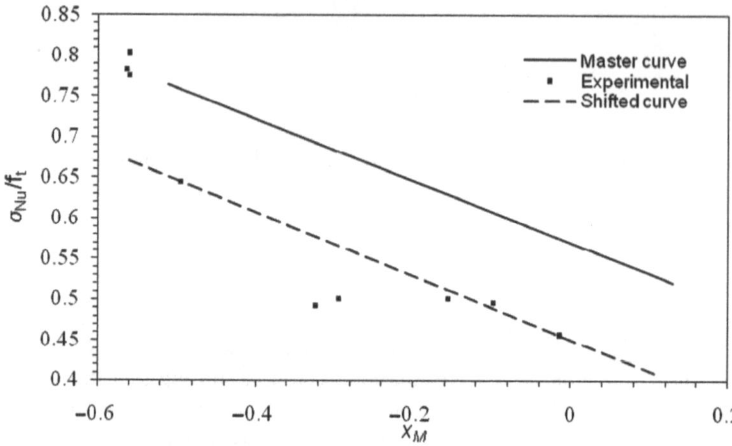

Fig. 3.10 Peak-load size effect for linear softening

The value of Δx is read from the plot and w_1 is determined as

$$w_1 = \frac{2G_F}{f_t} 10^{-\Delta x} \tag{3.47}$$

For this particular test series, the value of w_1 is obtained as

$$w_1 = \frac{1.0024G_F}{f_t} \tag{3.48}$$

Once Eq. (3.48) is established, the relation between G_F and G_f can be obtained within limited test series as

$$G_F \approx 2G_f \tag{3.49}$$

For the study of the influence of kink point on the global P-CMOD response, the cohesive crack model with three-parameter bilinear softening curve with varying values of kink points is simulated. Five different values of α_s as 1/3, 1/4, 1/5, 0.15, and 0.10 and corresponding obtained values of β_s as 0.1673, 0.1506, 0.1339, 0.1113, and 0.0822 are arbitrarily selected. Thus, the numerical results obtained for specimen sizes of 250 mm with G_F 167 N/m, 150 mm with G_F 164 N/m, and 63 mm with G_F 119 N/m are compared with the resulting P-CMOD responses at different kink points for each size of specimen as plotted in Figs. 3.11, 3.12, and 3.13, respectively. It is observed that changing the values of α_s does not affect the values of peak load, whereas the softening branch follows different paths with different values of α_s after the peak. This divergence starts at a decrease of approximately 7.74, 4.16, and 1.94% of peak load in the softening branch for beam depths 63, 150, and 250 mm, respectively. It is also observed from P-CMOD curves that the softening branch

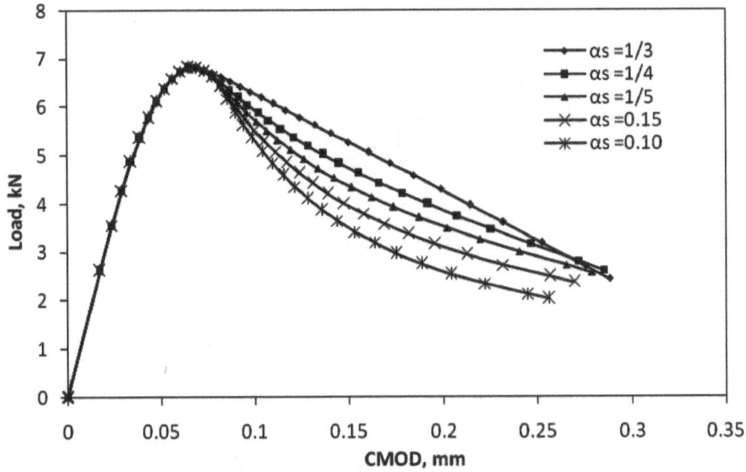

Fig. 3.11 Effect of kink point on P-CMOD curve using three-parameter bilinear curve for specimen depth 250 mm

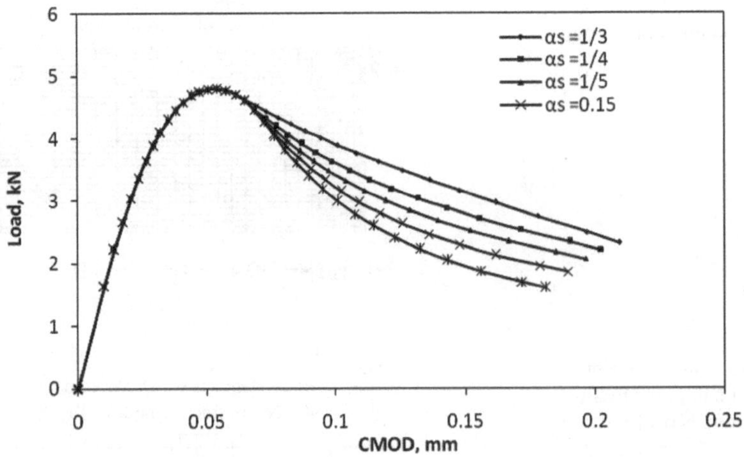

Fig. 3.12 Effect of kink point on P-CMOD curve using three-parameters bilinear curve for specimen depth 150 mm

becomes deeper and deeper (means the values of CMOD decrease at a particular load) as the value of α_s decreases from 1/3 to 0.1.

3.4.5 Effect of Finite Element Mesh Size on Bearing Capacity of the Beam

Two examples for beam depths of 250 and 150 mm are considered to notice the mesh size dependency on the peak load obtained using the numerical results with linear

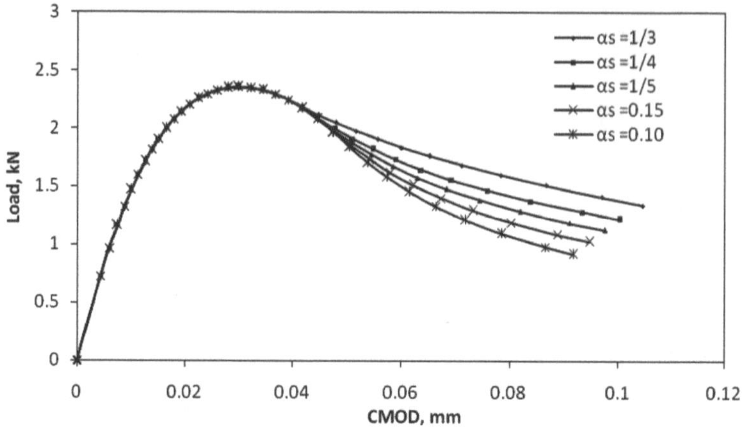

Fig. 3.13 Effect of kink point on P-CMOD curve using three-parameter bilinear curve for specimen depth 63 mm

Fig. 3.14 Finite element mesh for half of the beam with 27 numbers of fracture nodes

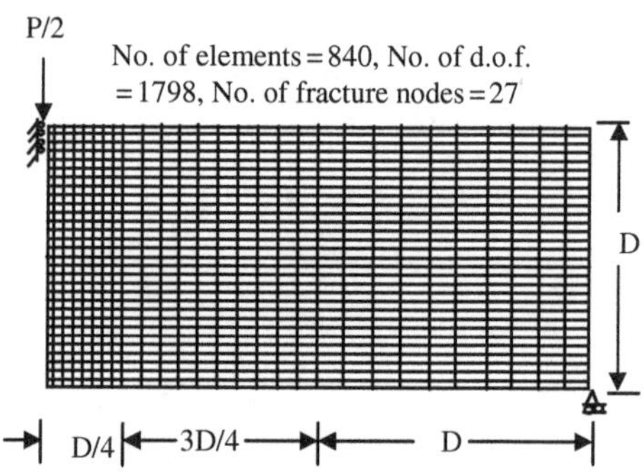

Fig. 3.15 Finite element mesh for half of the beam with 14 numbers of fracture nodes

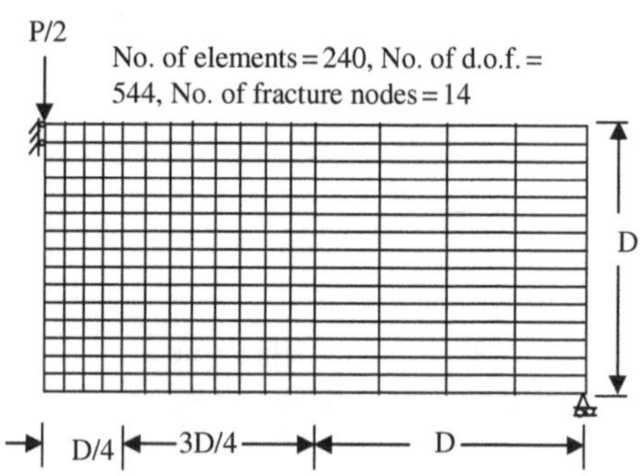

softening function. Three different finite element mesh sizes as shown in Figs. 3.4, 3.14, and 3.15 are considered in which the numbers of fracture nodes are taken to be 55, 27, and 14 in the finite element mesh, respectively.

The fracture energy and the tensile strength for the calculation are taken as 150 N/m and $f_t = 4.15$ MPa, respectively. The beam thickness for both the specimens is the same as 80 mm. For fracture nodes 55, 27, and 14, the peak loads determined from the numerical result are 8.0849, 8.0576, and 8.0545 kN, respectively, for $D = 250$ mm, whereas these are 5.5721, 5.576, and 5.5715 kN, respectively, for $D = 150$ mm. It is clear that even if the mesh size is relatively coarser, the model is able to predict almost the same values of peak load for both sizes of the specimen. For the coarser mesh size, the distances between the closing forces are relatively large and it becomes larger when the size of specimen is increased. However, as mentioned elsewhere (Petersson 1981), it seems unsuitable to use larger distances between the closing forces than $0.2l_{ch}$, which ranges from 40 to 80 mm for concrete.

3.4.6 Effect of Size Scale on the Type of Failure

For this study the dimensions of the beams are taken as $B = 80$ mm and $S = 4D$ mm and the material properties are $G_F = 150$ N/m and $f_t = 4.15$ MPa with linear softening function. Number of fracture nodes is taken to be 55 for each five specimens having $D = 63$, 150, 250, 450, and 650 mm. The load–deflection responses are obtained using cohesive crack model and are plotted in Figs. 3.16, 3.17, 3.18, 3.19, and 3.20. When the depth of the beam is up to 150 mm (Figs. 3.16 and 3.17), almost similar trend of global load–deflection behavior is observed.

It is clear that the load-carrying capacity decreases and the ductile behavior of the beam increases as a_o/D ratio increases. The slope of softening portion is the maximum when the beam is uncracked. This slope is approximately infinite for beam of depth 250 mm when the beam is uncracked (Fig. 3.18). For the beam of depth equal

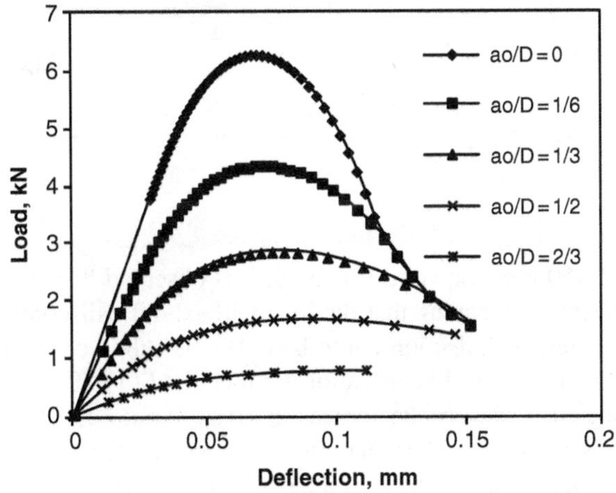

Fig. 3.16 Load vs. deflection curve for $D = 63$ mm

Fig. 3.17 Load vs. deflection curve for $D = 150\,\mathrm{mm}$

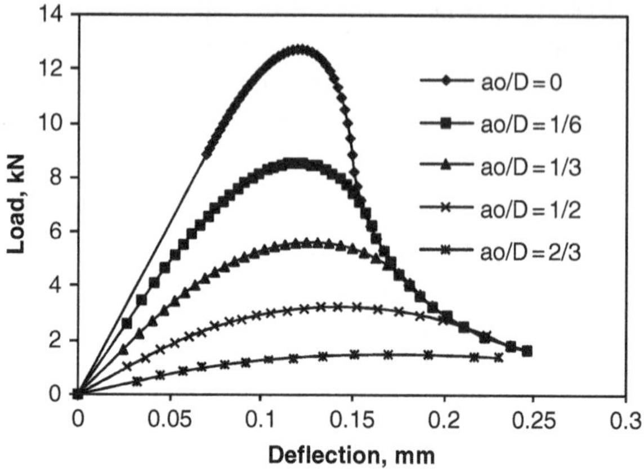

Fig. 3.18 Load vs. deflection curve for $D = 250\,\mathrm{mm}$

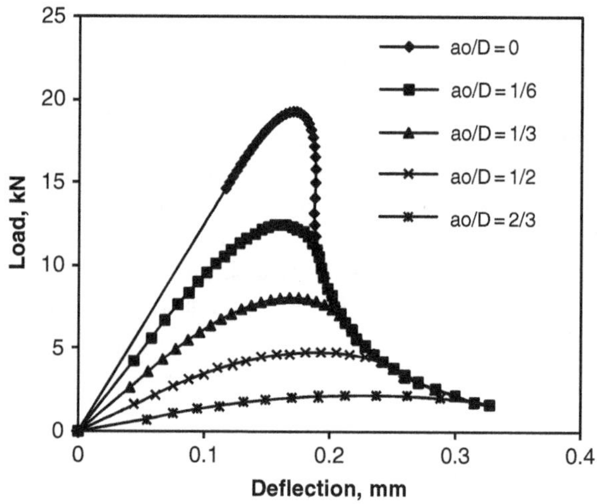

to 450 mm shown in Fig. 3.19, it is observed that the slope of softening branch is negative, meaning that the load will exhibit a discontinuity with a negative jump if the test is deflection controlled. This is called cusp catastrophe (Carpinteri 1990). As the depth of beam is further increased to 650 mm (Fig. 3.20), the formation of the cusp catastrophe spreads up to a_0/D ratio of 1/6 approximately. Therefore it is observed that either increase in size of the structure or decrease in a_0/D will show the increase in opposite patterns of brittleness.

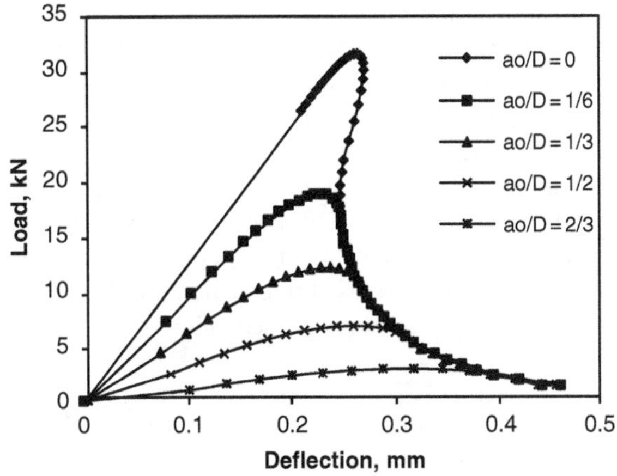

Fig. 3.19 Load vs. deflection curve for $D = 450\,\text{mm}$

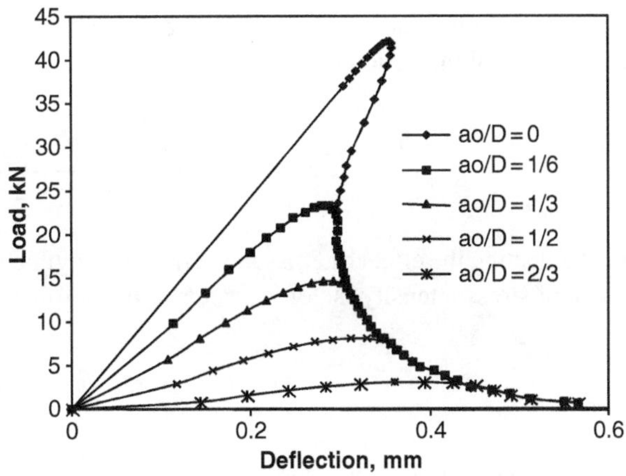

Fig. 3.20 Load vs. deflection curve for $D = 650\,\text{mm}$

3.4.7 Size-Scale Deviation From LEFM Concept

For geometrically similar structures, Hillerborg and co-workers (1976) initially put forward a nondimensional parameter commonly known as *brittleness number* β_B of concrete. In nonlinear fracture mechanics such as cohesive crack method, the parameter β_B is used to assess the applicability of LEFM to concrete. Mathematically, β_B is expressed as

$$\beta_B = \frac{D}{l_{ch}} \tag{3.50}$$

To show the deviation from LEFM for concrete structure, the numerical results with varying values of β_B are obtained. For the present study, geometrically similar three-point bending configuration beams with $B = 80\,\text{mm}$ and $S = 4D$ and the material properties $G_F = 150\,\text{N/m}$ and $f_t = 4.15\,\text{MPa}$ with linear softening function are considered. The maximum loads using cohesive crack model $P_{u,\text{CCM}}$ and well-known LEFM formula $P_{u,\text{LEFM}}$ are computed at different relative initial crack lengths such as $a_o/D = 1/6$, 1/3, and 1/2. Self-weight of the beam is also taken into consideration while calculating $P_{u,\text{LEFM}}$. The LEFM formula (Tada et al. 1985) for the standard TPBT is used to calculate the fracture toughness for known value of fracture energy G_{IC} as mentioned below:

$$K_I = \sigma_N \sqrt{D} k(\alpha) \tag{3.51}$$

$$k(\alpha) = \sqrt{\alpha} \frac{1.99 - \alpha(1-\alpha)(2.15 - 3.93\alpha + 2.7\alpha^2)}{(1+2\alpha)(1-\alpha)^{3/2}} \tag{3.52}$$

The maximum nominal stress in the beam due to external load $P_{u,\text{LEFM}}$ and self-weight of the structure is given by

$$P_{u,\text{LEFM}} = \frac{2\sigma_N B D^2}{3L} - \frac{w_g L}{2} \tag{3.53}$$

Neglecting the effect of Poisson's ratio as negligible effect in concrete, the critical value of stress intensity factor is evaluated using following relation:

$$K_{IC} = \sqrt{G_{IC} E} \tag{3.54}$$

The results of the ratio $P_{u,\text{CCM}}/P_{u,\text{LEFM}}$ with varying values of β_B are plotted in Fig. 3.21.

From the figure it is observed that the peak load predicted by the use of LEFM is normally overestimated as compared to the peak load (or true peak load) obtained using cohesive crack model. This deviation is higher at low value of *brittleness number* β_B. It infers that the LEFM is inapplicable to concrete of normal size of concrete elements. As described elsewhere (Petersson 1981), the LEFM is applicable to concrete in case the value of β_B is more than 10. For concrete, the value of l_{ch} normally varies between 200 and 300 mm, which implies that LEFM is applicable only when the beam depth is in-between 2 and 3 m. Recently, Bažant (2002) pointed out that the length of FPZ in normal concrete is in the order of 500 mm and depending upon the structure size D, different fracture mechanics concepts might be appropriate for analyzing the failure behavior. The classical form of LEFM may be approximately suitable for the ratio $D/FPZ \geq 100$, meaning that LEFM might be applied for normal concrete structures for $D \approx 50\,\text{m}$.

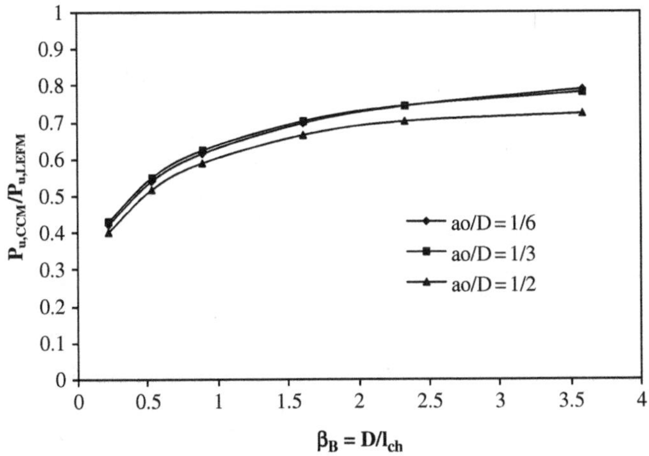

Fig. 3.21 Size-scale transition toward LEFM

3.4.8 Influence of Softening Function on Size-Effect Curve

Previous study (Planas and Elices 1990) showed that for the cohesive crack model, the generalized size-effect law might be expressed in the following form:

$$\frac{EG_{FC}}{K_{INu}^2} = 1 + C_1\left(\frac{l_{ch}}{D}\right) + C_2\left(\frac{l_{ch}}{D}\right)^2 + C_3\left(\frac{l_{ch}}{D}\right)^3 + \cdots \tag{3.55}$$

where C_i is the dimensionless coefficients depending on geometry and material properties, K_{INu} is the stress intensity factor computed for maximum actual loads, P_u and initial crack length. Equation (3.55) can be plotted on X–Y coordinates in which the left-hand terms and l_{ch}/D of right-hand terms represent Y and X coordinates, respectively. This equation is made to satisfy in such a manner that CCM and LEFM will yield the same result for infinitely large structures ($D \to \infty$). For geometrically similar beams with $B = 80$ mm, $S = 4D$, and the material properties $G_F = 150$ N/m and $f_t = 4.15$ MPa with linear softening, Petersson bilinear, exponential and nonlinear softening functions, the values of K_{INu} are determined using LEFM equations for the known values of peak load obtained using cohesive crack model and initial crack length. The self-weight of the specimen is also taken into account in the calculation. The size-effect curves are plotted in nondimensional forms of inverse of nominal stress intensity factor K_{INu} at peak load vs. inverse of characteristic dimension D at constant a_0/D ratio of 1/3 as shown in Fig. 3.22.

It is observed from the figure that the size-effect curves are in general influenced by the softening function of concrete. However, the Petersson bilinear and exponential softening relations yield almost the same result. It is also seen from the figure that as the size increases, all size-effect curves seems to be converging and for infinitely large structures ($D \to \infty$) the values of $EG_F/K_{INu}{}^2$ tend asymptotic value

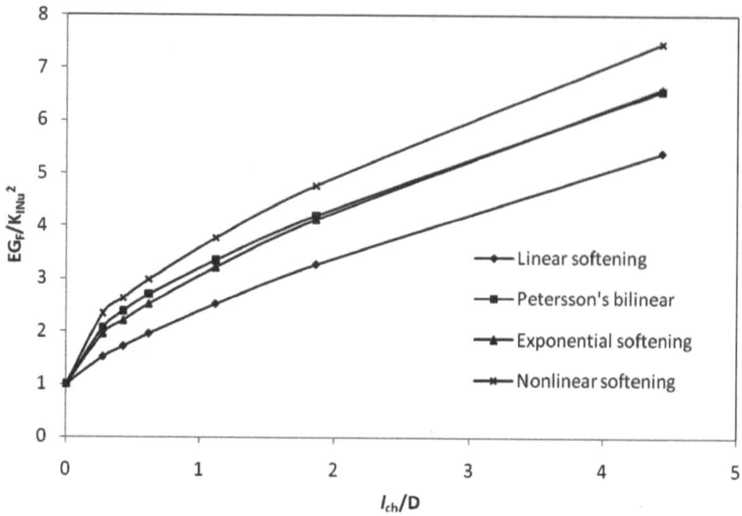

Fig. 3.22 Effect of softening function on the size-effect curves for a_o/D ratio = 1/3

to 1 exhibiting that fracture loads predicted by LEFM and CCM are the same for all size-effect curves.

3.5 Numerical Study Using Compact Tension (CT) Specimen

A comprehensive and systematic numerical study on the fracture parameters of concrete compact tension test specimens of different sizes using cohesive crack model is presented in this section. Though the fracture behavior of mode I loading remains similar to the TPBT specimen, in-depth analysis of the numerical results using various softening functions such as the linear (Hillerborg et al. 1976), bilinear (Petersson 1981, Wittmann et al. 1988, CEP-FIP Model Code-1990 1993), exponential (Karihaloo 1995), nonlinear (Reinhardt et al. 1986), and quasi-exponential (Planas and Elices 1990) is considered in the study. The material properties for a concrete mix for which $f_t = 3.21$ MPa, $E = 30$ GPa, and $G_F = 130$ N/m (Planas and Elices 1990) are taken in the present investigation. The value of ν is assumed to be 0.18. For CT test, specimen thickness $B = 100$ mm and size ranging $100 \le D \le 600$ mm are taken into consideration. The term $\{p_g\}$ in Eq. (3.13) due to self-weight of the structure is neglected in the analysis. The values $[K]$ and $\{C\}$ of Eq. (3.13) are computed using standard linear elastic finite element method for which the fracture nodes are shown in Fig. 3.23 and half of the specimen as discretized due to symmetry using four-node isoparametric quadrilateral elements is shown in Fig. 3.24. The specimen is divided into three bands of length $0.075D$, $0.225D$, and $0.3D$, respectively, along its longitudinal direction. The number of

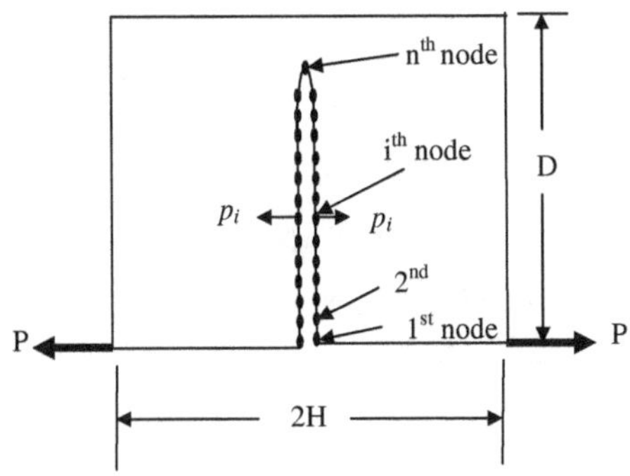

Fig. 3.23 Finite element pair of nodes along fracture line

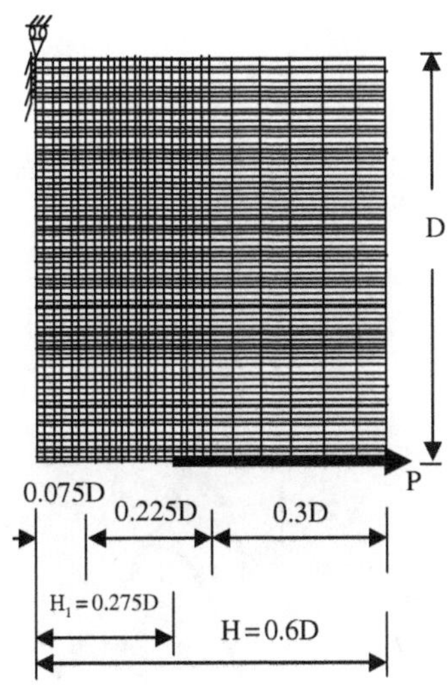

Fig. 3.24 Finite element mesh for half of the CT specimen

meshes in each band is taken to be 6, 18, and 6, respectively. The number of divisions along the depth of the specimen is fixed as 80. Three numbers of nodes from the top in horizontal direction and the top end in vertical direction are restrained against movement to allow the development of cohesive zone along potential fracture line. Then number of elements, number of degrees of freedom (d.o.f.), and number of fracture nodes along the potential fracture line become 2400, 5022, and 78, respectively.

3.5.1 Global P-COD Response Using Linear Softening Function

The numerical results of global response of P-COD curves with linear softening are presented in Figs. 3.25, 3.26, and 3.27.

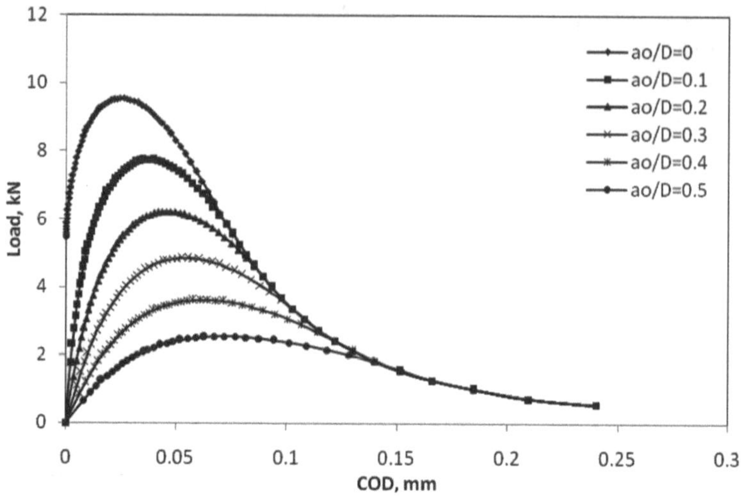

Fig. 3.25 Load–COD curve using linear softening for $D = 100$ mm

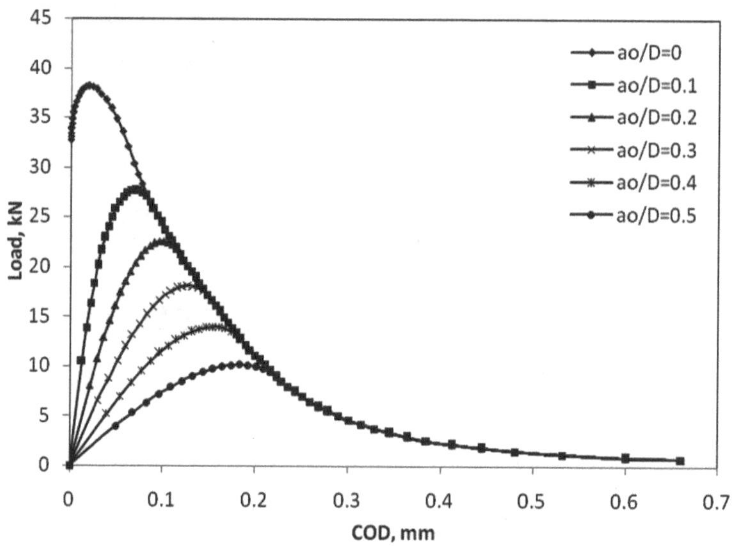

Fig. 3.26 Load–COD curve using linear softening for $D = 600$ mm

Figures 3.25 and 3.26 are plotted for $D = 100$ and 600 mm, respectively, for different initial notch length/depth (a_o/D) ratios ranging from 0 to 0.5, whereas Fig. 3.27 is plotted for constant a_o/D ratio $= 0.3$ for different size range ($100 \leq D \leq 600$ mm) specimens. It is observed that the stiffness and the load carrying of the specimen increase with decrease in the value of initial notch length/depth ratio for a given size or with increase in specimen size for a given value of a_o/D ratio. The slope of the softening branch increases by decreasing the value of a_o/D ratio for a

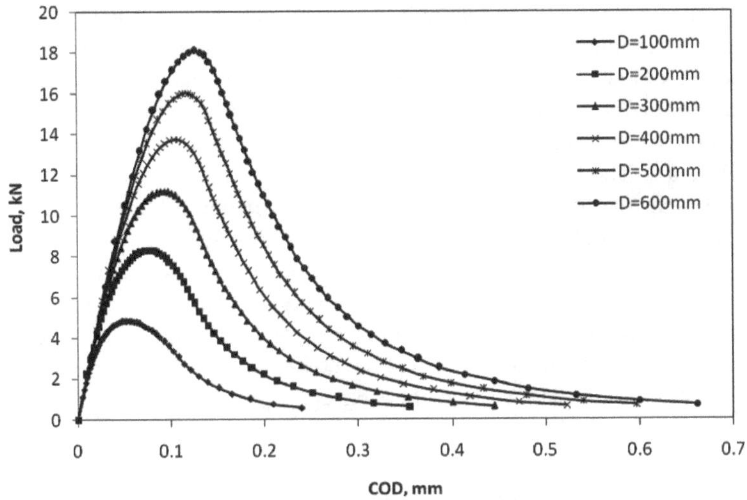

Fig. 3.27 Load–COD curve using linear softening for $100 \leq D \leq 600$ mm and $a_o/D = 0.3$

constant specimen size or by increasing specimen size at a constant value of a_o/D ratio. It is also noticed that the uncracked specimen becomes more unstable after the peak load is reached, whereas the specimen behaves in more ductile manner and is better controllable as the a_o/D ratio increases for a specified specimen size or the specimen size decreases at constant value of a_o/D ratio for cracked specimen. The terminal branch of the P-COD curves appears to be converging and is independent of the a_o/D ratio for a given size specimen which is caused by the assumption made in the analysis that the damage zone has zero thickness and concentrated on a line. Measuring P-COD curve, the snapback instability behavior shown in the case of P–δ (load–deflection) curve of TPBT specimen is avoided because in the P-COD curve, the crack opening displacement increases monotonically in softening zone during crack growth.

The cohesive crack behavior as discussed above and demonstrated in Figs. 3.25 and 3.26 may be practically different if the actual width of fracture process zone and the crack path are modeled accurately, since during crack propagation in concrete, a macro-crack has a tendency to follow a tortuous crack path due to various mechanisms of microcracking, crack bridging, aggregate interlocking, etc. The meandering of main crack depends on the concrete mix design and the properties of the aggregate particles such as size, texture, and angularity.

The cohesive crack model is a strictly uniaxial method considering zero width of the process zone and it does not allow the crack path to follow in an actual tortuous way. The assumptions made in the development of the model is different from the experimentally observed real behavior of fracture process of concrete and thus it allows the terminals of the softening branches to converge irrespective of the relative initial notch lengths. The real behavior of terminals of the softening branches may be evident from the experimental results (Nallathambi et al. 1984,

1985, Karihaloo 1995) of the three-point bending test specimens made of plain concrete with maximum aggregate size 20 mm having compressive strength 38 MPa and modulus of elasticity 33 GPa. Karihaloo (1995) presented the effect of relative notch length (a_o/D) on the typical load–deflection responses. For the beam dimensions span $S = 600$ mm, $D = 102$ mm, and $B = 80$ mm, the a_o/D ratios varied from 0.1 to 0.6. The load–deflection curves show that the terminals of the softening branches appear to be converging for relatively smaller difference in the values of a_o/D ratio. For higher difference in the a_o/D ratio (say, between 0.1 and 0.3 or 0.2 and 0.4 or 0.4 and 0.6), clear separations in the terminals of softening branches are observed. Further, the test results (Karihaloo 1995) confirm that load–deflection plots are also influenced by the maximum size of coarse aggregates, that is, the pre-peak behavior is more nonlinear and the post-peak behavior is less abrupt with larger tails when the maximum size of coarse aggregate is increased.

3.5.2 *Influence of Softening Functions on the Global P-COD Response*

The results of P-COD curves using various shapes of softening functions for compact tension specimens with a_0/D ratio $= 0.3$ are determined and presented in Figs. 3.28 and 3.29 for $D = 100$ and 600 mm, respectively.

The peak nominal stresses (P_u/BD) obtained using cohesive crack model with aforementioned six softening relations (except for linear softening) for 100- and 600-mm-size specimens are in the range of 0.407–0.438 and 0.245–0.268 MPa,

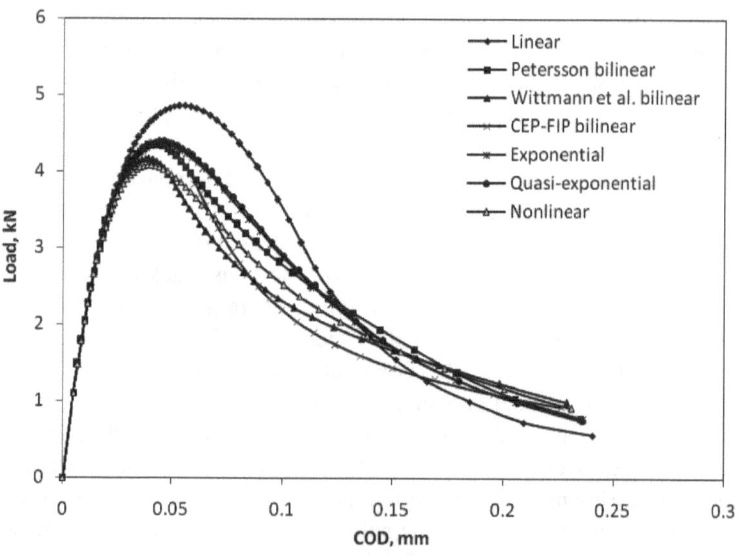

Fig. 3.28 Influence of softening curve on load–COD curve for $D = 100$ and $a_0/D = 0.3$

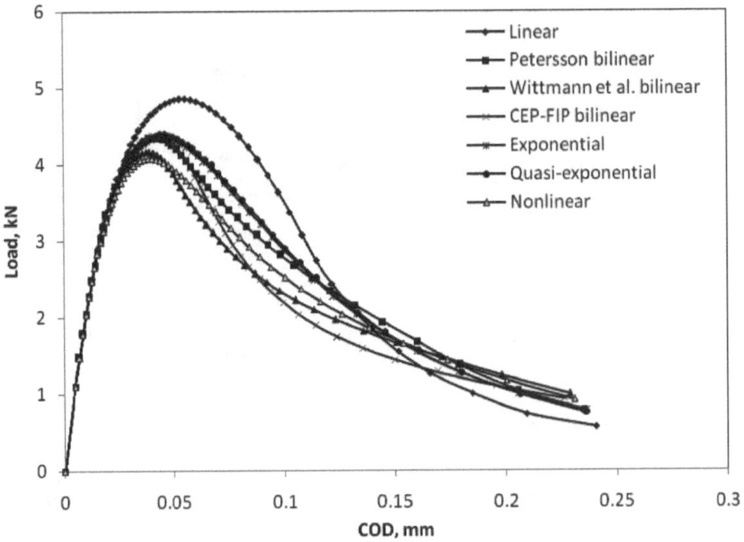

Fig. 3.29 Influence of softening curve on load–COD curve for $D = 600$ and $a_o/D = 0.3$

respectively, showing a less scatter in results and the difference between minimum and maximum peak nominal stresses for all sizes of specimens is less than 7.07 and 8.58%, respectively. The minimum and maximum nominal peak stresses are yielded by nonlinear softening and quasi-exponential softening functions, respectively, for both size-range specimens. The peak nominal stress using linear softening function for 100- and 600-mm-size specimens is 0.485 and 0.302 MPa, respectively. On an average the linear softening yields the peak load 11.75 and 15.89% larger than the bilinear curves, exponential, quasi-exponential, and nonlinear softening functions for the size range 100 and 600 mm, respectively. The post-peak response of P-COD curves with different softening curves (except linear softening) is slightly varied. Almost the same P-COD curves can be obtained using exponential and quasi-exponential softening functions.

3.5.3 Influence of Softening Functions on the Size-Scale Transition Toward LEFM

The numerical results with varying values of β_B are obtained to show the deviation from LEFM concept for geometrically similar normal size CT specimens having $B = 100$ mm, the material properties $G_F = 103$ N/m, and $f_t = 3.21$ MPa using all seven softening functions. The maximum load using cohesive crack model ($P_{u,CCM}$) or true peak load using well-known LEFM formula ($P_{u,LEFM}$) is computed at initial crack length/depth ratio such as $a_o/D = 0.3$. The LEFM formula (ASTM Standard 399 2006) for CT specimen is used to calculate the critical value of stress intensity

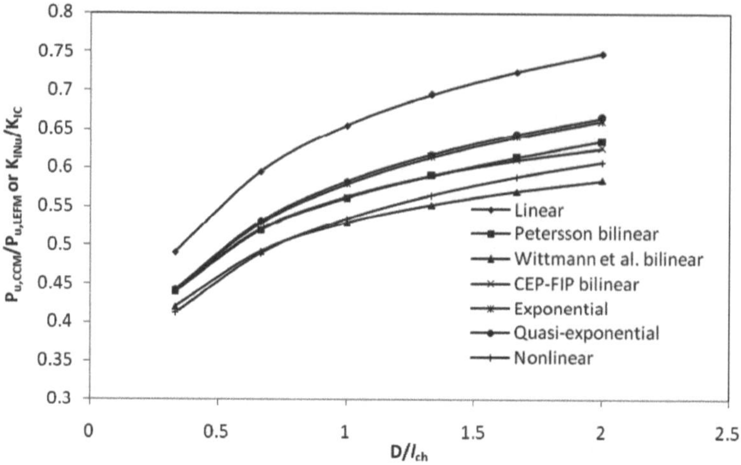

Fig. 3.30 Influence of softening curve on size-scale transition toward LEFM for $a_o/D = 0.3$

factor K_{IC} using Eq. (3.51) for known value of fracture energy G_{IC}. In Eq. (3.51), the $k(\alpha)$ is given by

$$k(\alpha) = \frac{(2+\alpha)\left[0.886 + 4.64\alpha - 13.32\alpha^2 + 14.72\alpha^3 - 5.6\alpha^4\right]}{(1-\alpha)^{3/2}} \quad (3.56)$$

where $\alpha = (a/D) = \alpha_o = (a_o/D)$, in Eq. (3.51) $\sigma_N = P/BD$ and Eq. (3.56) is valid for $0.2 \le \alpha \le 1$ within 0.5 % accuracy. The maximum nominal stress due to external load $P_{u,LEFM}$ of the structure is determined by $P_{u,LEFM} = \sigma_{Nu}BD(1/2)$. The value of K_{IC} is evaluated using Eq. (3.54). The result of the ratio $P_{u,CCM}/P_{u,LEFM}$ with varying values of β_B is plotted in Fig. 3.30.

Similar observation found in the case of TPBT specimen can be seen from Fig. 3.30. The figure further shows that the predicted ratio $P_{u,CCM}/P_{u,LEFM}$ using linear softening is higher than that using other six softening functions for the considered size range and it seems that the difference is more diverged as the size scale increases. On the other hand, the ratio of $P_{u,CCM}/P_{u,LEFM}$ for the bilinear curves, exponential, quasi-exponential, and nonlinear functions is relatively confined in narrow band for lower specimen size and seems to be slightly diverging for higher size scale. However, this ratio is almost the same for the exponential and quasi-exponential and also for Petersson's bilinear and CEB-FIP's bilinear softening functions. The minimum and maximum values of the ratios $P_{u,CCM}/P_{u,LEFM}$ for all the softening functions (except linear) are in the range of 0.412 (for nonlinear softening)–0.443 (for quasi-exponential softening) for $D = 100$ mm and in the range of 0.584 (for Wittmann et al. bilinear softening)–0.665 (for quasi-exponential softening) for $D = 600$ mm, respectively. Moreover, the ratio $P_{u,CCM}/P_{u,LEFM}$ for linear softening is 0.491 for $D = 100$ mm and 0.748 for $D = 600$ mm.

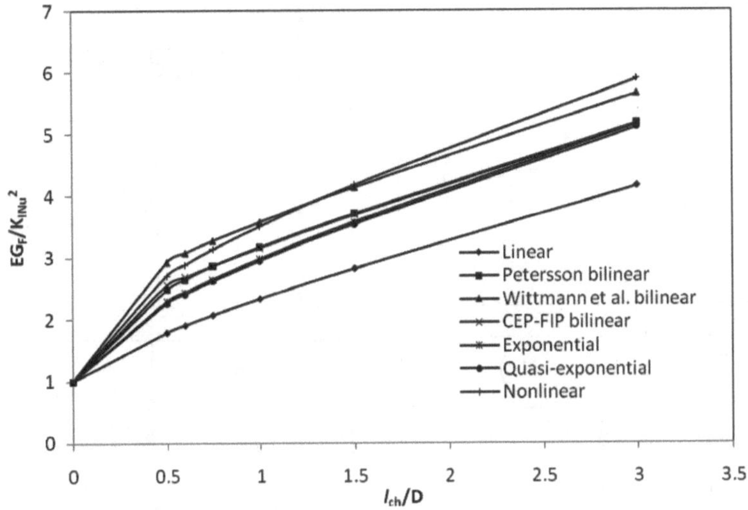

Fig. 3.31 Influence of softening curve on size-effect curve for $a_o/D = 0.3$

3.5.4 Influence of Softening Functions on the Size-Effect Curve

The size-effect curves are plotted using Eq. (3.55) in nondimensional forms of inverse of nominal stress intensity factor K_{INu} at peak load vs. inverse of characteristic dimension D at constant a_o/D ratio of 0.3 as shown in Fig. 3.31.

It is observed from the figure that the size-effect behavior is similar to the previous observation noticed in the case of TPBT.

3.5.5 Evolution of Fracture Zone

Development of fracture zone and stress distribution ahead of the initial crack tip of a CT specimen for $D = 600$ mm and a_o/D ratio $= 0.3$ with Wittmann et al.'s (1988) bilinear softening function are shown in Fig. 3.32 for four salient positions of load on the P-COD curve.

In the figure, the development of fracture zone is shown by filled-up solids. The length of fracture zone and the stress distribution ahead of the crack tip are represented (not to scale) for explaining the mechanism of process zone evolution. The fracture zone begins to form as soon as the structure is loaded. When the load position is approximately at 1, the fracture zone is small and the developed stress value in the process zone is always less than the tensile strength f_t of the material. This stress distribution does not show any stress singularity unlike in LEFM analysis. When the load position is at 2, i.e., at peak load, the length of fracture zone becomes about 75 mm and the stress value at the initial crack tip σ_{tip} achieves a value of $0.2423f_t$.

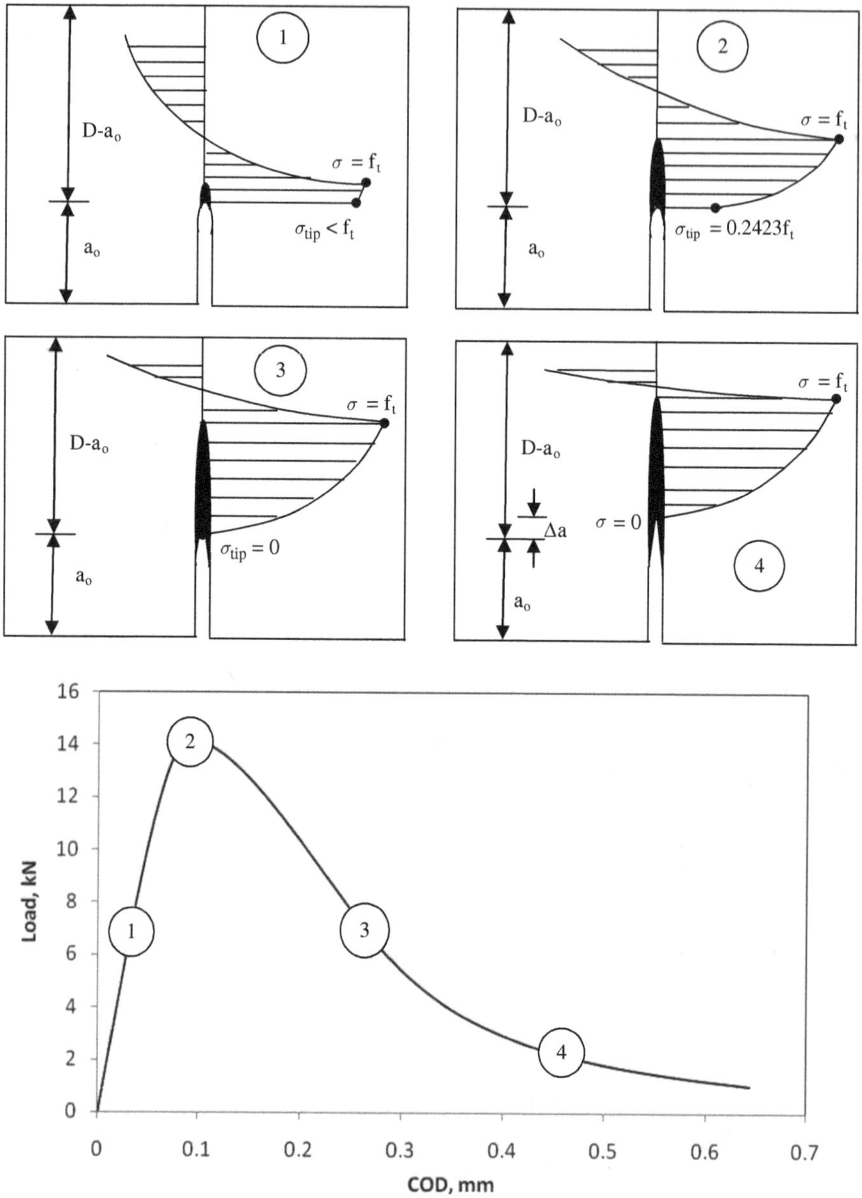

Fig. 3.32 Evolution of fracture zone and stress distribution ahead of the crack tip for different positions of load on the P-COD curve obtained from Wittmann et al.'s (1988) bilinear softening for $D = 600\,\text{mm}$ and $a_0/D = 0.3$

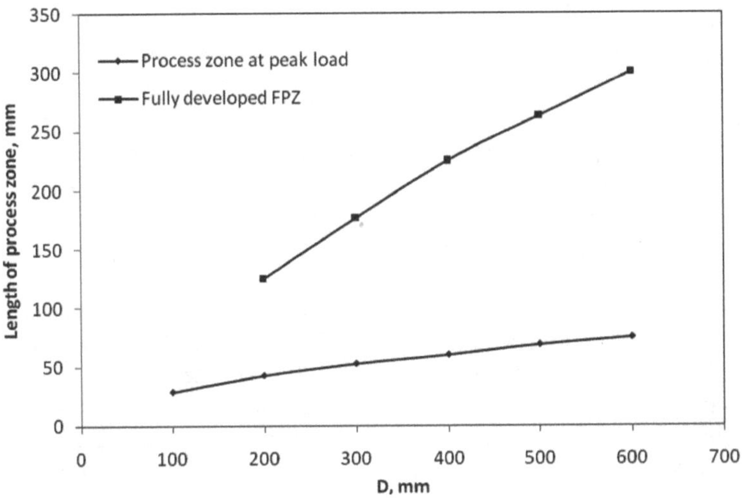

Fig. 3.33 Depth of fracture zone at peak load and fully developed FPZ for $a_o/D = 0.3$ obtained using Wittmann et al.'s (1988) bilinear softening function

The lengths of fracture zone corresponding to peak load and fully developed FPZ obtained using Wittmann et al.'s (1988) bilinear softening function for specimen size range 100–600 mm are presented in Fig. 3.33.

From the figure, it can be seen that fracture zone at peak load for $D = 100\,\text{mm}$ and a_o/D ratio $= 0.3$ becomes about 28.75 mm. In addition, the material is still able to transfer stress across the crack faces even up to peak load and on further loading of structure corresponding to load position 3, the stress-transferring capacity of concrete becomes zero. At this stage of loading, a full length of fracture zone, i.e., FPZ, can be observed. For $D = 600\,\text{mm}$, the length of FPZ is about 300 mm and is significant and variable with respect to specimen size as shown in Fig. 3.33. The fully developed FPZ for specimen size $D = 100\,\text{mm}$ is missing in the figure because it is not developed up to 78th node in the potential fracture line. On further loading of structure, say position 4, the stress-free crack is generated and the real crack has moved to a distance say Δa.

It is clear that the length of the fracture zone at peak load and FPZ varies with the specimen size significantly. As shown in Fig. 3.33, the fracture zones increase with increase in specimen size. However, the relative size of the fracture zone with respect to the specimen size decreases with increasing specimen size. This infers that the fracture behavior of normal concrete depends on the specimen size. For limiting case of infinitely large structure, the relative sizes of fracture zone and FPZ become very small and insignificant such that the LEFM can be applied. In general it can be stated that the local and global responses during crack initiation and its growth of cementitious materials are dependent on many factors. Hence, the influence of softening function and size effect on the important fracture parameters is presented in subsequent analyses.

3.5.6 Influence of Softening Functions on the Development of Fracture Process Zone

The length of fully developed fracture process zone is computed using all seven softening functions for the specimen size range $200 \leq D \leq 600$ mm because full length of FPZ is not developed even up to 78th node using some of the softening functions for 100-mm-size specimen. The length of FPZ/depth (FPZ/D) ratio shows a definite relationship with the brittleness of the structures as shown in Fig. 3.34; however, it also depends on the shape of the softening relations used.

Almost the same prediction of FPZ is seen using exponential or quasi-exponential softening functions of concrete. The length of FPZ obtained using the linear softening is the minimum as compared to that obtained using bilinear curves, exponential, quasi-exponential, and nonlinear functions for all size range. It is evident from the figure that the relative size of fracture process zone decreases as the size of the specimen is increased which is the main source of size effect in *quasibrittle* materials. The minimum and maximum values of the ratio FPZ/D for all the softening functions (except linear) are in the range of 0.575 (for Petersson's bilinear softening)–0.65 (for CEB-FIP's bilinear softening) for $D = 200$ mm and in the range of 0.4 (for Petersson's bilinear softening)–0.563 (for CEB-FIP's bilinear softening) for $D = 600$ mm, respectively. The difference between minimum and maximum fully developed FPZ is less than 11.54 and 28.95% for $D = 200$ and 600 mm, respectively. The ratio FPZ/D for linear softening is 0.438 for $D = 200$ mm and 0.238 for $D = 600$ mm. On an average the linear softening yields the FPZ 29.22 and 51.38% lower than the bilinear curves, exponential, quasi-exponential, and nonlinear softening functions for the size range 200 mm and 600 mm, respectively.

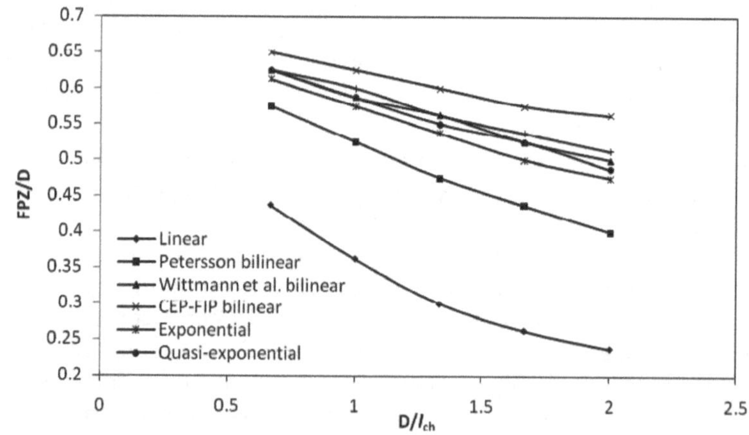

Fig. 3.34 Influence of softening curve on fully developed FPZ for $a_0/D = 0.3$

3.5.7 Influence of Softening Functions on the Load-Carrying Capacity Corresponding to the Fully Developed FPZ

The load corresponding to fully developed fracture process zone P_{FPZ} is calculated for the specimen size range $200 \le D \le 600$ mm using all the softening functions and results are plotted in the nondimensional ratio of $P_{\mathrm{FPZ}}/P_{\mathrm{u}}$ with respect to the brittleness of the structures as shown in Fig. 3.35.

It is clear from the figure that the ratio of $P_{\mathrm{FPZ}}/P_{\mathrm{u}}$ is an increasing function with the brittleness of *quasibrittle* structures. It implies that the value of $P_{\mathrm{FPZ}}/P_{\mathrm{u}}$ increases as the size of specimen increases. It is evident that the larger is the value of $P_{\mathrm{FPZ}}/P_{\mathrm{u}}$ on the P-COD curve, the closer is the load corresponding to fully developed FPZ to the peak load in softening zone. Cohesive crack model thus illustrates the variation from failure at relatively larger fracture process zone for small structures to the failure at a relatively smaller fracture process zone for large structures. The minimum and maximum values of the ratios $P_{\mathrm{FPZ}}/P_{\mathrm{u}}$ for all the softening functions (except linear) are in the range of 0.219 (for CEB-FIP's bilinear softening)–0.433 (for Petersson's bilinear softening) for $D = 200$ mm and in the range of 0.297 (for CEB-FIP's bilinear softening)–0.680 (for Petersson's bilinear softening) for $D = 600$ mm, respectively. The ratio $P_{\mathrm{FPZ}}/P_{\mathrm{u}}$ for linear softening is 0.773 for $D = 200$ mm and 0.918 for $D = 600$ mm, respectively. This means that the length of fully developed FPZ using linear softening is always smaller as compared to that predicted using bilinear curves, exponential, quasi-exponential, and nonlinear functions.

Larger disparity is observed in the predicted results by various softening functions of concrete as shown in Figs. 3.34 and 3.35. Since different softening functions

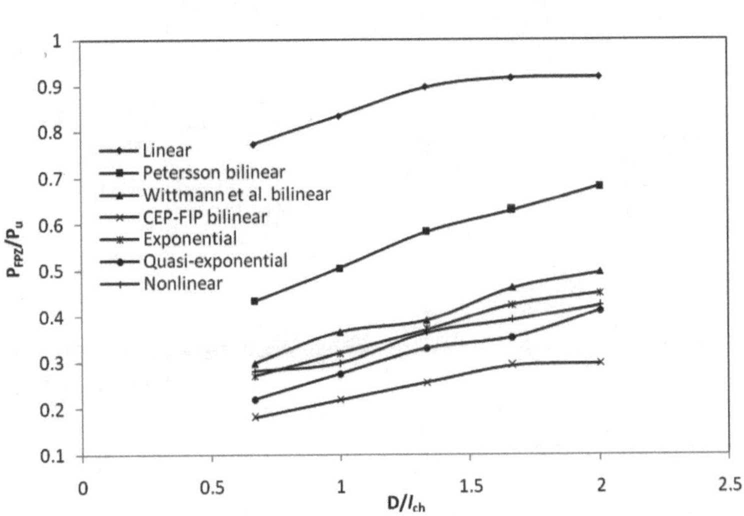

Fig. 3.35 Influence of softening curve on the ratio of load $P_{\mathrm{FPZ}}/P_{\mathrm{u}}$ for $a_{\mathrm{o}}/D = 0.3$

follow different shapes and different terminals, the shape of the softening branch of the cohesive law plays an important role in predicting the process zone and the corresponding load beyond the peak. So, it is quite natural that these results are relatively more affected by the different shapes of the softening function.

3.5.8 *Influence of Softening Functions on the Fictitious Crack Length Corresponding to Peak Load*

Fictitious crack length/depth (a_f/D) ratio at peak load is obtained using all seven softening functions and the variation of a_f/D ratio is plotted with brittleness of concrete as shown in Fig. 3.36.

It is observed from the figure that the a_f/D ratio at peak load maintains a definite relationship with brittleness of the structure and the influence of the shape of softening functions including linear one on the value of a_f/D ratio is relatively less. The relationship shows that for low brittleness the fictitious crack length at peak load is covered in almost whole ligament resulting in a ductile failure; on the other hand, the process zone decreases with increase in brittleness and it appears that $a_f/D \rightarrow a_o/D$ for $D/l_{ch} \rightarrow \infty$. The minimum and maximum values of a_f/D ratio for all the softening functions are in the range of 0.588 (for Wittmann et al.'s bilinear)–0.663 (for linear softening) for $D = 100$ mm and in the range of 0.425 (for Wittmann et al.'s bilinear and CEB-FIP's bilinear)–0.475 (for linear softening) for $D = 600$ mm, respectively.

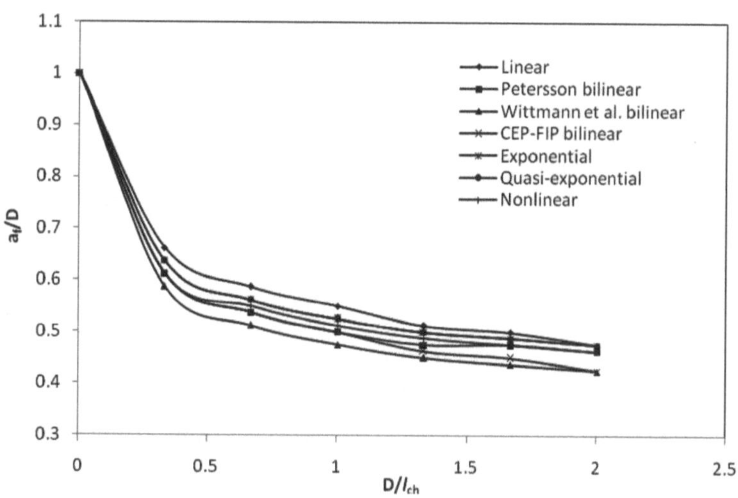

Fig. 3.36 Influence of softening curve on the fictitious crack length/depth (a_f/D) ratio for $a_o/D = 0.3$

3.5.9 Influence of Softening Functions on the $CTOD_c$

Critical value of crack-tip opening displacement ($CTOD_c$) corresponding to peak load is measured from the numerical results of cohesive crack model and plotted using each softening function as shown in Fig. 3.37.

In predicting the $CTOD_c$ using CEB-FIP Model Code-1990's (1993) bilinear softening, some oscillations are noticed beyond the specimen size of 300 mm. It is observed from the figure that in general the value of $CTOD_c$ is a function of brittleness and depends upon the shape of softening function of concrete except for the exponential and quasi-exponential softening functions which yield almost the same values of $CTOD_c$ for all size specimens. The value of $CTOD_c$ increases as the specimen size increases, meaning that the relative size of FPZ will be smaller for higher brittleness of the structures. The $CTOD_c$ obtained using linear softening is always greater than that obtained using bilinear curves, exponential, quasi-exponential, and nonlinear functions for all the size range. The minimum and maximum values of the ratios $l_{ch}/CTOD_c$ for all the softening functions (except linear) are in the range of 14668 (for quasi-exponential softening)–19220 (for Wittmann et al.' softening) for $D = 100$ mm and in the range of 7163 (for quasi-exponential softening)–10613 (for Wittmann et al.' softening) for $D = 600$ mm, respectively. The ratio $l_{ch}/CTOD_c$ for linear softening is 12428 for $D = 100$ mm and 6802 for $D = 600$ mm, respectively.

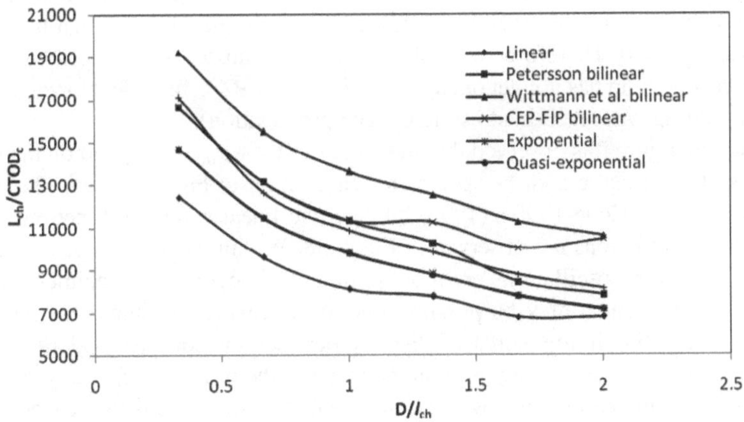

Fig. 3.37 Influence of softening curve on the $CTOD_c$ for $a_o/D = 0.3$

3.5.10 Influence of Softening Functions on the Local Stress Distribution at Notch Tip Corresponding to Peak Load

Influence of shape of softening curve and size effect on the notch-tip stress σ_{tip} distribution corresponding to peak load is shown in Fig. 3.38. The ordinate of figure is the value of $(f_t - \sigma_{tip})/f_t$ which may be termed as tip stress transfer ratio and the

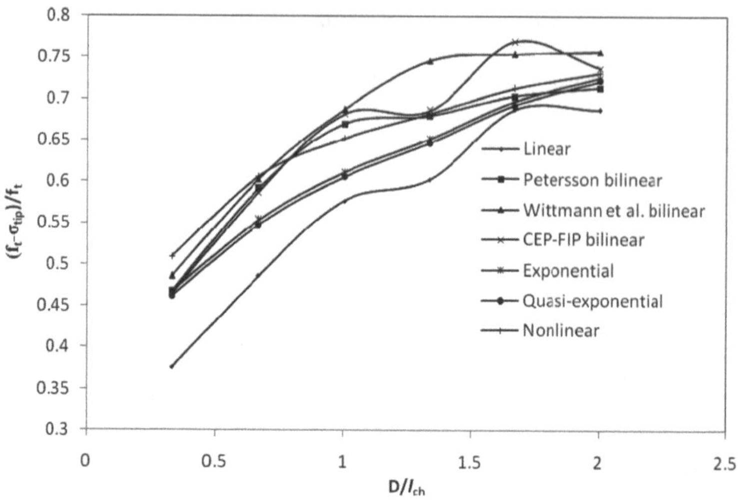

Fig. 3.38 Influence of softening curve on crack-tip stress at peak load for $a_o/D = 0.3$

abscissa is D/l_{ch}. It is evident that for $\sigma_{tip} = f_t$, i.e., the fracture node at notch tip is not yet opened, the tip stress is equal to the tensile strength of the material and for the other extreme the fracture process zone is fully developed when the notch-tip stress $\sigma_{tip} = 0$. Therefore value of the tip stress transfer ratio ranges between 0 (when the notch tip has not yet opened) and 1 (when FPZ is fully developed and any further loading will result in stress-free crack propagation).

From the figure it is observed that the values of $(f_t - \sigma_{tip})/f_t$ depend on the shape of the softening curve and the specimen size. Similar to Fig. 3.37, with the use of CEB-FIP Model Code-1990's (1993) bilinear and linear softening functions, some degrees of oscillations are observed in Fig. 3.38. Within the size range considered in the study, these oscillations are more pronounced beyond the specimen sizes of about 300 mm which may be possibly due to increasing specimen size and consequently resulting in the increased distance between the cohesive nodal forces. In general, there is an increasing trend in the value of the ordinates $(f_t - \sigma_{tip})/f_t$ as the specimen size increases. It shows that the size of fracture zone is influenced by the specimen size and shape of the softening relation used in the cohesive crack model. It is also observed that the stress-transferring capacity at the crack tip corresponding to the peak load decreases as the specimen size increases and as a result of which a real crack begins to propagate closer to the peak load for larger size specimens. It is also clear that the slow crack growth does not commence before the softening damage takes place. This behavior predicted using exponential and quasi-exponential softening functions is almost the same. The minimum and maximum values of $(f_t - \sigma_{tip})/f_t$ for all the softening functions (except linear) are in the range of 0.462 (for quasi-exponential softening)–0.510 (for nonlinear softening) for $D = 100$ mm and in the range of 0.732 (for quasi-exponential softening)–0.758 (Wittmann et al.'s bilinear softening) for $D = 600$ mm, respectively. The ratio $(f_t - \sigma_{tip})/f_t$ obtained using linear softening is 0.376 for $D = 100$ mm and 0.687 for $D = 600$ mm, respectively. It is revealed that the mean tip stress transfer ratio considering all softening functions (except linear) is 0.474 and 0.732 for $D = 100$ and 600 mm, respectively, which is always greater than that predicted using linear softening relation.

3.5.11 Comparison with Experimental Result

Compact tension specimens of different sizes made of various mix proportions and water/cement (W/C) ratios were tested by Wittmann et al. (1988) under three different loading rates. The tests were conducted on the CT specimens having measured standard dimensions of $a_o = 0.5D$, $D_1 = 125D$, $H = 0.6D$, and $H_1 = 0.5H$, in which the dimensions D and H_1 are not indicated in the figure (Wittmann et al. 1988) and they are carefully measured and found to be $2a_o$ and $0.5H$, respectively. Three sizes of specimens $D = 300$, 600, and 1200 mm with constant thickness $B = 120$ mm were tested and the mean curves of load–crack opening displacement in the plane of loading of loading points (P-COD) for specimens $D = 600$ and 1200 mm and load–crack mouth opening displacement (P-CMOD) for specimens of $D = 300$ mm were measured. The medium-size CT specimens, $D = 600$ mm, made of concrete with W/C ratio 0.43 and maximum aggregate size of 16 mm was considered as the reference specimen. The reference specimens were tested under loading rate of 0.2 mm/min for which the 28-day compressive strength of concrete was determined as 42.9 MPa. From fracture tests, the mean value of peak load and the specific fracture energy were obtained as 12.6 kN and 162 N/m, respectively. In order to calculate the essential parameters of a bilinear softening function using fictitious crack model, a method was also suggested by the authors (Wittmann et al. 1988). Besides the other important studies, it was concluded that load–displacement curves can be determined by means of finite element analysis with the assumption of bilinear softening functions and are in good agreement with those experimentally observed curves. The specific fracture energy can thus be calculated from the strain-softening relation that agrees well with the same determined directly from the P-COD curve. As such, the authors (Wittmann et al. 1988) suggested that the parameters of the bilinear softening function such as $G_F = 157$ N/m and $f_t = 4.4$ MPa can predict the fracture behavior of reference specimen in a better manner.

Here, a realistic comparison between the experimental and numerical results is carried out for the application of various softening functions in the cohesive crack model. For the numerical modeling, a concrete mix having maximum size of aggregate $d_a = 16$ mm and its properties $\nu = 0.18$ and $G_F = 157$ N/m is considered. The value of E is obtained as 34.9 GPa using CEB-FIP Model Code-1990 (1993) for a given value of compressive strength 42.9 MPa. Except for $H_1 = 0.5H$, a similar finite element discretization of the CT specimen as shown in Fig. 3.24 is done for determining the influence coefficients. All the seven softening functions are applied to determine the P-COD curves for the comparison with the mean P-COD curve obtained from the experiment for the reference specimen. The P-COD curves obtained from the numerical results and the test results are shown in Fig. 3.39.

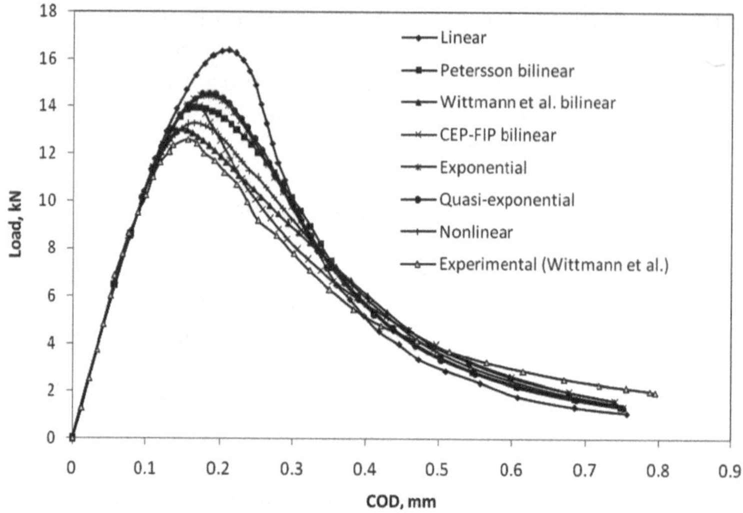

Fig. 3.39 Comparison of P-COD curves obtained from experimental results (Wittmann et al. 1988) and the numerical simulation using various softening relations for CT specimen having $D = 600\,\text{mm}$ and $a_o/D = 0.5$

It is observed that the ratios of peak load obtained from the numerical results to experimentally observed value are 1.301, 1.099, 1.036, 1.117, 1.148, 1.155, and 1.055 for linear, Petersson's (1981) bilinear, Wittmann et al.'s (1988) bilinear, CEB-FIP Model Code-1990's bilinear, exponential (Karihaloo 1995), quasi-exponential (Planas and Elices 1990), and nonlinear (Reinhardt et al. 1986) softening functions, respectively. It can be seen that except linear softening, the predicted values of the peak loads are overestimated by approximately 16%. The predicted peak loads are very close (within the accuracy level of 6%) to the experimental result for the Wittmann et al.'s (1988) bilinear and nonlinear (Reinhardt et al. 1986) softening functions. Furthermore, Petersson's (1981) bilinear, CEB-FIP Model Code-1990's bilinear, exponential (Karihaloo 1995), and quasi-exponential (Planas and Elices 1990) softening functions yield the peak loads within an accuracy range between 10 and 16%. The predicted value of the peak load using linear softening is overestimated by nearly 30%. It is further observed that the predicted post-peak response are scattered in a relatively narrow band and compares well with the experimentally observed response. However, the terminal portion of experimentally observed softening branch tends to deviate from the numerically predicted values even with all the softening functions used in the numerical modeling.

3.6 Size-Effect Study from Size-Effect Model

A similar investigation was carried out by Planas and Elices (1990) in which almost the same unstable fracture load was predicted between SEM and CCM for practical size range (100–400 mm) of pre-cracked concrete beam for TPBT geometry. It was observed from the size-effect study that fracture loads predicted by SEM can diverge about 31% for asymptotically large-size ($D \to \infty$) beam. The quasi-exponential softening function was used in the above study since the fracture criteria or the mathematical relationships may vary according to a particular stress–displacement softening function. With this intention, the nonlinear softening function with $c_1 = 3, c_2 = 6.93$, and w_c according to Eq. (3.38) is used in the present study.

3.6.1 Size-Effect Law for Size-Effect Model

The Bažant's SEM may be expressed in the following form (Planas and Elices 1990):

$$\frac{EG_{FB}}{K_{INu}^2} = 1 + \frac{\lambda_o}{D} \tag{3.57}$$

Assuming the G_F being denoted by G_{FC}, Eq. (3.57) may be transformed in the following form by making use of G_{FC}:

$$\frac{EG_{FC}}{K_{INu}^2} = \frac{G_{FC}}{G_{FB}}\left[1 + \frac{\lambda_o}{D}\frac{l_{ch}}{D}\right] \tag{3.58}$$

Equation (3.58) is now in the form of straight line variation on X–Y plot, having the common X and Y coordinates as in the case of size-effect law for CCM.

3.6.2 Size Effects and Fracture Load from CCM and SEM

Same concrete mix as used by Planas and Elices (1990) is taken in the present investigation for which $f_t = 3.21$ MPa, $E = 30$ GPa, and $G_{FC} = 103$ N/m. The value of ν is assumed to be 0.18 with nonlinear softening function as described in the foregoing section. For TPBT specimen of notched concrete beam with $B = 100$ mm, $100 \le D \le 400$ mm, and $S/D = 4$, the finite element analysis is carried out for determining the fracture peak load using cohesive crack model. Four-node isoparametric element was considered for finite element calculation. The half of the beam was discretized with 40 numbers of equal elements along the depth of the beam. The number of fracture nodes along the FPZ is 38.

The fracture parameters for SEM, G_{FB}, and c_f for each geometry ($a_o/D = 0.2$, 0.3, 0.4, and 0.5) are computed according to the RILEM method (1990) (see Sect. 2.7.4) and flowchart as shown in Fig. A.2. The values of K_{INu} were obtained using Eqs. (3.51) and (3.52) for the a_o and the P_u gained from CCM. Variations of nondimensional parameters G_{FB}/G_{FC} and c_f/l_{ch} obtained for each geometry $0.2 \le a_o/D \le 0.5$ are plotted in Fig. 3.40. It is observed from the figure that values of c_f/l_{ch} are less affected by the values of a_o/D ratios than G_{FB}/G_{FC}.

The details of specimen dimensions, material properties and calculation of size-effect parameters are presented in Table 3.2. The mean value of G_{FB} obtained from

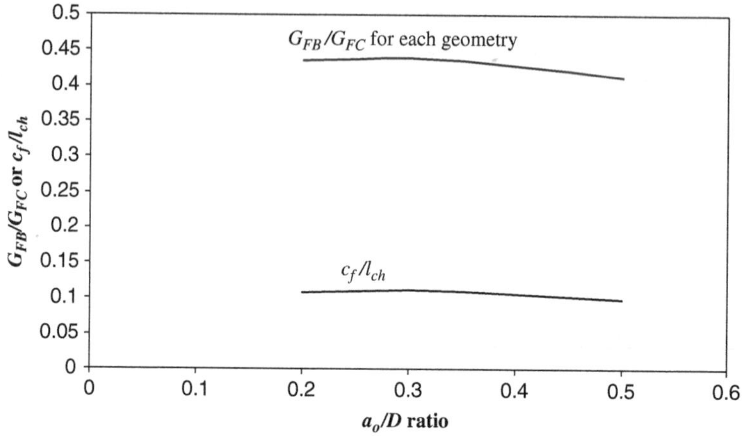

Fig. 3.40 Effect of geometry on the fracture parameters obtained from SEM

Table 3.2 Specimen details, material properties, and size-effect parameters

D (mm)	a_o/D	P_u(N)	K_{INu}(MPa mm$^{1/2}$)	G_{FB}[a]$/G_{FC}$	λ_o/l_{ch}	Difference in fracture load (%)
400	0.2	12321	30.76	0.43	0.535	−0.12
300	0.2	10278	29.06			−1.25
200	0.2	7808.5	26.59			−2.13
100	0.2	4711.4	22.37			−0.99
400	0.3	9393.4	30.92	0.43	0.521	0.02
300	0.3	7891.3	29.24			−1.09
200	0.3	6027.5	26.76			−2.06
100	0.3	3655.6	22.54			−1.04
400	0.4	6913.1	30.45	0.43	0.558	−0.53
300	0.4	5849.3	28.75			−1.59
200	0.4	4504.2	26.32			−2.17
100	0.4	2741	22.11			−0.84
400	0.5	4788.6	29.59	0.43	0.636	−1.37
300	0.5	4103.8	27.89			−2.19
200	0.5	3189.1	25.46			−2.42
100	0.5	1948.3	21.26			−0.55
∞	All			0.43		−34.43

[a]G_{FB} is the mean values of fracture energy for different geometries obtained from SEM

the fracture energy of each geometry is considered as a material property for comparison between CCM and SEM with regard to Eqs. (3.57) and (3.58). Thus mean value of G_{FB} is obtained as 44.22 N/m yielding the value of G_{FB}/G_{FC} of 0.43 to be used in Eq. (3.58).

The results of size effect obtained from CCM in nondimensional form (Eq. (3.55)) for some particular values of a_o/D ratios 0.3–0.5 are shown in Fig. 3.41.

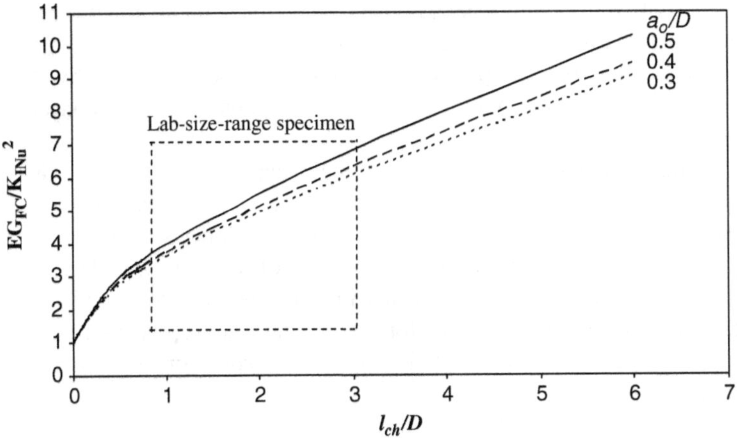

Fig. 3.41 Size-effect curves for CCM for TPBT specimens

It is found from the figure that the size-effect curves are dependent on the geometrical parameter a_o/D ratio, exhibiting all curves to be asymptotic to 1 for infinitely large-size structures ($D \rightarrow \infty$), and the fracture loads predicted by LEFM and FCM are the same.

The parameter λ_o/λ_{ch} in Eq. (3.58) for size-effect law of SEM is computed by least square fitting for a particular geometry ($0.2 \leq a_o/D \leq 0.5$) and is reported in Table 3.2. The nondimensional size-effect curves obtained from CCM and SEM for the geometries $a_o/D = 0.3$ and 0.5 are plotted in Fig. 3.42. It is noticed that the least square fitting over the size range taken into the investigation yields results of SEM

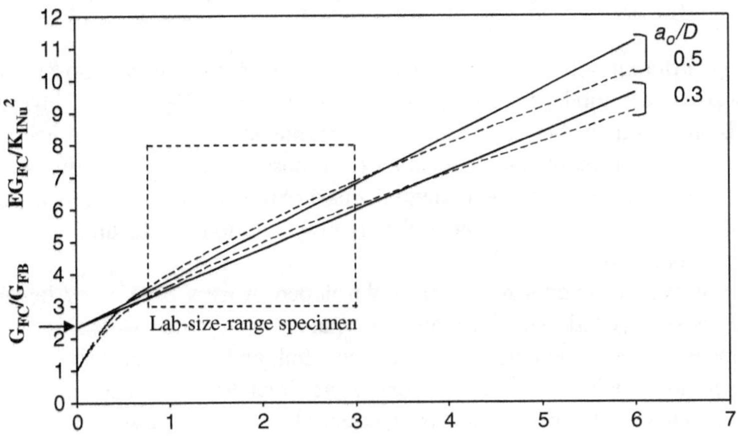

Fig. 3.42 Size-effect curves obtained from CCM and SEM

very close to CCM and within the practical size range the maximum difference in the predicted fracture load obtained from these two models is less than 2.5% (Table 3.2). However, it is also clear that predictions of the two models differed by about 34% for asymptotic size ($D \rightarrow \infty$), i.e., Bažant's SEM seems to be more conservative by about 34%.

The results obtained in the investigation are in agreement with the study reported by Planas and Elices (1990) and difference wherever appeared is due to the non-linear softening function used in the present study. Under such circumstances, if input parameters of CCM are known, the results of peak or unstable fracture load can also be calculated from SEM well within the acceptable error limits for the practical size range of beam depth for three-point bending geometry specimens. Thus, requirements of FEM calculation may be avoided for the above-mentioned size-range structures.

3.7 Closing Remarks

Within the limitation of the specimen sizes ($63 \leq D \leq 250$ mm) of TPBT taken into investigation, it is observed from the numerical result of CCM using different softening functions (except linear one) that the differences between the highest and lowest peak loads are less than 8%. On an average, the peak loads predicted using linear softening is approximately 10% larger than those obtained using other softening functions (bilinear curves, exponential, nonlinear). The different values of stress ratio at kink point do not affect the value of peak loads predicted by CCM; however, the softening branch becomes deeper and deeper as the values of stress ratio at kink points decrease.

The following remarks can be given from the numerical study on the cohesive crack fracture parameters using CT specimen of size range $100 \leq D \leq 600$ mm.

- Using different softening functions (except linear one), it was observed from cohesive crack model that the differences between the highest and lowest peak loads are less than about 9%. On an average, the peak load predicted using linear softening is approximately 16% larger than those obtained using other softening functions (bilinear curves, exponential, quasi-exponential, and nonlinear curves).
- Exponential and quasi-exponential softening relations yield almost the same fracture behavior.
- For a given specimen size, the fully developed process zone using linear softening is significantly smaller than that obtained using other softening functions (bilinear curves, exponential, quasi-exponential, and nonlinear curves). The linear softening yields the fracture process zone about 30 and 50% lower than do the bilinear curves, exponential, quasi-exponential, and nonlinear softening functions for the size range 200 and 600 mm, respectively.
- The linear softening predicts relatively higher value of crack-tip opening displacement and lower value of the tip stress transfer ratio at peak load than those

obtained using the bilinear curves, exponential, quasi-exponential, and nonlinear softening functions.

- The peak load gained from the test result is predicted by the cohesive crack model within about 6% of accuracy using Wittmann et al.'s (1988) bilinear and nonlinear (Reinhardt et al. 1986) softening functions, whereas those predictions vary between the accuracy range of 10 and 16% using Petersson's (1981) bilinear, CEB-FIP Model Code's (1990) bilinear, exponential (Karihaloo 1995), and quasi-exponential (Planas and Elices 1990) softening functions. The linear softening curve overestimates the experimentally observed peak load by about 30%.

References

ASTM International Standard E399-06 (2006) Standard Test Method for Linear-Elastic Method Plane-Strain Fracture Toughness K_{IC} of Metallic Materials. West Conshohocken, PA, Copyright ASTM International: 1–32.

Bažant ZP (2002) Concrete fracture models: Testing and practice. Eng Fract Mech 69: 165–205.

Bažant ZP, Becq-Giraudon E (2002) Statistical prediction of fracture parameters of concrete and implications for choice of testing standard. Cement Concr Res 32: 529–556.

Carpinteri A (1989a) Decrease of apparent tensile and bending strength with specimen size: Two different explanations based on fracture mechanics. Int J Solid Struct 25 (4): 407–429.

Carpinteri A (1989b) Cusp catastrophe interpretation of fracture instability. J Mech Phys Solid 37 (5): 567–582.

Carpinteri A (1989c) Post-peak and post-bifurcation analysis of cohesive crack propagation. Eng Fract Mech 32: 265–278.

Carpinteri A (1990) A catastrophe theory approach to fracture mechanics. Int J Fract 44: 57–69.

Carpinteri A, Colombo G (1989) Numerical analysis of catastrophic softening behaviour (snap-back instability). Comput Struct 31 (4): 607–636.

Carpinteri A, Cornetti P, Puzzi S (2006) Scaling laws and multiscale approach in the mechanics of heterogeneous and disordered materials. Appl Mech Rev ASME 59: 283–305.

CEB-Comite Euro-International du Beton-EB-FIP Model Code 1990 (1993) Bulletin D'Information No. 2123/214, Lausanne.

Gopalaratnam VS, Shah SP (1985) Softening response of plain concrete in direct tension. J Am Concr Inst 82: 310–323.

Guinea GV, Planas J, Elices M (1994) A general bilinear fit for the softening curve of concrete. Mater Struct 27 (2): 99–105.

Hillerborg A, Modeer M, Petersson PE (1976) Analysis of crack formation and crack growth in concrete by means of fracture mechanics and finite elements. Cement Concr Res 6: 773–782.

Karihaloo BL (1995) Fracture Mechanics and Structural Concrete, Concrete Design and Construction Series. Harlow, Essex, Longman Scientific & Technical.

Kumar S (2010) Behaviour of fracture parameters for crack propagation in concrete. Ph.D. Thesis submitted to Department of Civil Engineering, Indian Institute of Kharagpur, India.

Kumar S, Barai SV (2008) Cohesive crack model for the study of nonlinear fracture behaviour of concrete. J Inst Eng (India), CV 89 (Nov.): 7–15.

Kumar S, Barai SV (2009) Effect of softening function on the cohesive crack fracture parameters of concrete CT specimen. Sadhana-Acad Proc Eng Sci 36 (6): 987–1015.

Li YN, Liang RY (1994) Peak load determination in linear fictitious crack model. J Eng Mech ASCE 120 (2): 232–249.

MATLAB Version 7, The MathWorks Inc., Copyright 1984–2004.

Nallathambi P, Karihaloo BL, Heaton BS (1984) Effect of specimen and crack size, water/cement ratio and coarse aggregate texture upon fracture toughness of concrete. Magn Concr Res 36 (129): 227–236.

Nallathambi P, Karihaloo BL, Heaton BS (1985) Various size effects in fracture of concrete. Cement Concr Res 15: 117–126.

Petersson PE (1981) Crack growth and development of fracture zone in plain concrete and similar materials. Report No. TVBM-100, Lund Institute of Technology.

Planas J, Elices M (1990) Fracture criteria for concrete: Mathematical validations and experimental validation. Eng Fract Mech 35: 87–94.

Planas J, Elices M (1991) Nonlinear fracture of cohesive material. Int J Fract 51: 139–157.

Reinhardt HW, Cornelissen HAW, Hordijk DA (1986) Tensile tests and failure analysis of concrete. J Struct Eng ASCE 112 (11): 2462–2477.

RILEM Draft Recommendations (TC89-FMT) (1990) Size-effect method for determining fracture energy and process zone size of concrete. Mater Struct 23 (138): 461–465.

Roesler J, Paulino GH, Park K, Gaedicke C (2007) Concrete fracture prediction using bilinear softening. Cement Concr Compos 29: 300–312.

Tada H, Paris PC, Irwin G (1985) The Stress Analysis of Cracks Handbook. Hellertown, PA, Del Research Corporation.

Wittmann FH, Rokugo K, Bruhwiller E, Mihashi H, Simopnin P (1988) Fracture energy and strain softening of concrete as determined by compact tension specimens. Mater Struct 21 (1): 21–32.

Xu S, Reinhardt HW (1999) Determination of double-*K* criterion for crack propagation in quasi-brittle materials, Part II: Analytical evaluating and practical measuring methods for three-point bending notched beams. Int J Fract 98: 151–177.

Xu S, Zhang X (2008) Determination of fracture parameters for crack propagation in concrete using an energy approach. Eng Fract Mech 75: 4292–4308.

Chapter 4
Crack Propagation Study Using Double-*K* and Double-*G* Fracture Parameters

4.1 Introduction

This chapter is based on the works of Kumar and Barai (2008a–c, 2009a–f, 2010a, b) and Kumar (2010). In this chapter the application of universal form of weight function is introduced for determining the double-*K* fracture parameters using various test geometries. The definition of brittleness of concrete is mentioned using the double-*K* fracture parameters. An empirical formula for determining directly the effective crack length for CT specimen is highlighted. Furthermore, the size-effect prediction from the double-*K* fracture model is investigated. Finally, comprehensive numerical studies on the double-*K* and double-*G* fracture criteria are represented.

4.2 Weight Function Method

For a linear elastic cracked body subjected to any symmetrical mode I loading, it was proved (Bueckner 1970, Rice 1972) that if the SIF and the corresponding crack face displacement are known as functions of crack length, then the SIF for any other symmetrical loading acting on the same body can be directly determined. Assume that the K_r and K_s are the SIFs of a cracked body corresponding to two cases of mode I symmetrical loadings r and s respectively, in which r corresponds to the known reference value and s corresponds to any arbitrary loading case. According to weight function method value of K_s may directly be determined for the known value of K_r and the respective crack face displacement u_r as given below:

$$K_s = \int_0^a \sigma_s(x).m(x,a)\mathrm{d}x_s \tag{4.1}$$

The term $m(x, a)$ in Eq. (4.1) is known as weight function and expressed as

$$m(x,a) = \frac{E'}{2K_r}\frac{\partial u_r}{\partial a} \tag{4.2}$$

S. Kumar, S.V. Barai, *Concrete Fracture Models and Applications*,
DOI 10.1007/978-3-642-16764-5_4, © Springer-Verlag Berlin Heidelberg 2011

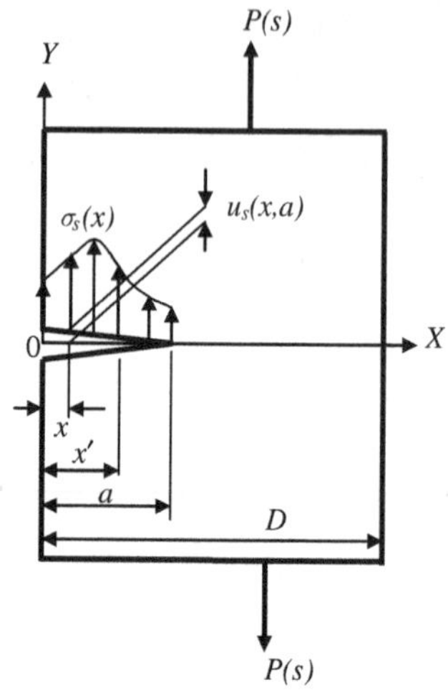

Fig. 4.1 The stress field induced along the prospective crack plane

where a is the crack length, $\sigma_s(x)$ is the stress distribution along the prospective crack line in the uncracked body under the external loading $P(s)$ for the loading case s as shown in Fig. 4.1, dx_s is the infinitesimal length along the crack surface, $E' = E$ for plane stress, and $E' = E/(1 - \upsilon^2)$ for plane strain. The $\sigma_s(x)$ can be determined either experimentally or numerically or analytically.

The meaning of weight function $m(x,a)$ may be interpreted as the contribution to the stress intensity factor due to a pair of point forces acting normal to the crack faces and located at a distance x within the crack length a. For a given cracked body, the weight function depends only on the geometry and is independent of the applied loads. Another advantage of the weight function is that the COD due to an arbitrary loading system can be calculated using the known weight function due to different loading system. Thus, for the known weight function $m(x,a)$, the COD $u_s(x,a)$ for the loading case s (Fig. 4.1) can be obtained as

$$u_s(x,a) = \frac{1}{E'} \int_x^a K_s(x') m(x,x') dx' \tag{4.3}$$

Using Eq. (2.39), it is possible to approximate the weight functions for mode I cracks with one four-term universal expression (Glinka and Shen 1991). In the four-term universal weight function, the number of unknown parameters are three (M_1, M_2, and M_3) which can be determined directly from three known reference SIFs without using the crack face displacement function. If the three reference SIFs, $K_{r_k}(k = 1,2,3)$ and the corresponding crack-line stress fields $\sigma_{r_k}(k = 1,2,3)$ are known, the simultaneous algebraic equations may be written in the following form:

$$K_{rk} = \int_0^a \sigma_{rk}(x) \frac{2}{\sqrt{2\pi(a-x)}} \begin{bmatrix} 1 + M_1(1-x/a)^{1/2} + M_2(1-x/a) \\ + M_3(1-x/a)^{3/2} \end{bmatrix} \mathrm{d}x \quad (4.4)$$

As an approximation, an additional condition for central through crack and double-edge cracks under symmetrical loading was used (Shen and Glinka 1991) as given below:

$$\frac{\partial u(x,a)}{\partial x} |_{x=0} = 0 \quad (4.5)$$

From Eqs. (4.2) and (4.5), it can be shown that

$$\frac{\partial m(x,a)}{\partial x} |_{x=0} = 0 \quad (4.6)$$

For the single-edge cracks, the following condition is found to be more appropriate (Fett et al. 1987, Shen and Glinka 1991):

$$\frac{\partial^2 u(x,a)}{\partial x^2} |_{x=0} = 0 \quad (4.7)$$

The following relation may be obtained using Eqs. (4.2) and (4.7):

$$\frac{\partial^2 m(x,a)}{\partial x^2} |_{x=0} = 0 \quad (4.8)$$

After satisfying Eqs. (4.6) and (4.8), one may get an additional condition which reduces the number of known reference SIFs.

4.3 Determination of Universal Weight Function for Edge Cracks in Finite Width Plate

In the present work, first of all an attempt is made to derive the universal weight function using the reference SIFs for the edge crack in a finite width of plate. The two reference SIFs for two different loadings as shown in Figs. 4.2 and 4.3 along with Eq. (4.8) are used to obtain the desired weight function. For the first reference case (Fig. 4.2), an edge crack in a finite width of plate subjected to uniform stress σ_o is taken. The standard equation of SIF for this case is written as

$$K_I = F_1 \sigma_o \sqrt{\pi a} \quad (4.9)$$

in which

$$F_1 = 1.122 - 0.231a/D + 10.55(a/D)^2 - 21.71(a/D)^3 + 30.382(a/D)^4 \quad (4.10)$$

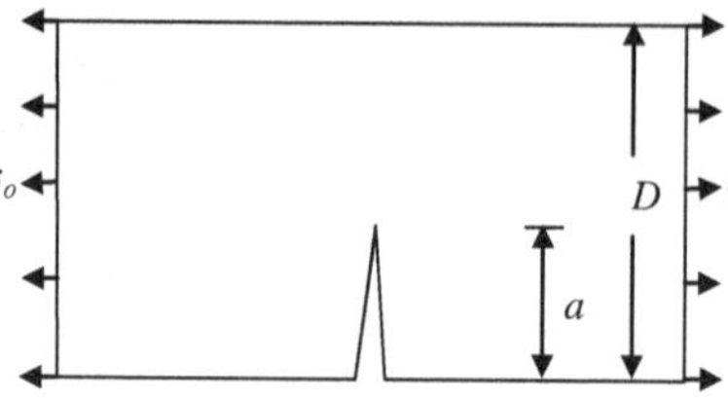

Fig. 4.2 Single-edge cracked specimen of finite width subjected to uniform loading

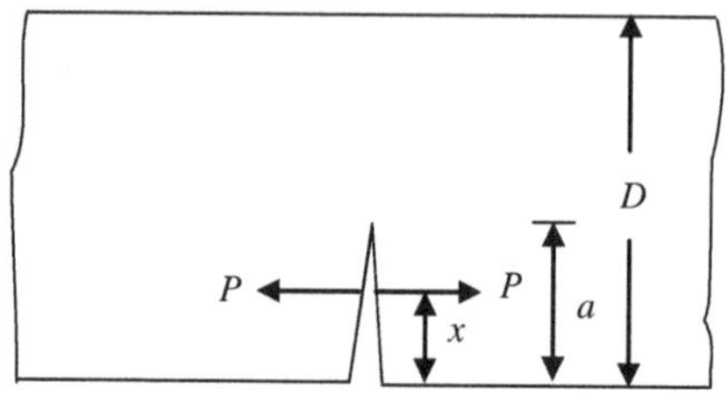

Fig. 4.3 Single-edge cracked specimen of finite width subjected to a pair of normal forces

Equation (4.10) has an accuracy of 0.5% for $a/D \leq 0.6$. The SIF for the case of an edge crack subjected to a pair of normal point forces P acting at a distance x as shown in Fig. 4.3 is expressed (Tada et al. 1985) as

$$K_{\mathrm{I}} = \frac{2P}{\sqrt{\pi a}} F(x/a, a/D) \tag{4.11}$$

where $F(x/a, a/D)$ is Green's function and given as

$$\begin{aligned} F(x/a, a/D) = {} & \frac{3.52(1 - x/a)}{(1 - a/D)^{3/2}} - \frac{4.35 - 5.28x/a}{(1 - a/D)^{1/2}} \\ & + \left\{ \frac{1.30 - 0.30(x/a)^{3/2}}{\sqrt{[1 - (x/a)^2]}} + 0.83 - 1.76x/a \right\} \\ & \times [1 - (1 - x/a)a/D] \end{aligned} \tag{4.12}$$

A pair of unit normal loads acting at the edge of the plate is considered to be second reference case of loading. Putting $P = 1$ (unit) and $x = 0$ in Eqs. (4.11) and (4.12), the SIF per unit thickness is then expressed as

$$K_{\mathrm{I}} = F_2 \frac{2}{\sqrt{\pi a}} \tag{4.13}$$

$$F_2 = \frac{3.52}{(1 - a/D)^{3/2}} - \frac{4.35}{(1 - a/D)^{1/2}} + 2.13(1 - a/D) \tag{4.14}$$

From Eqs. (4.4), (4.9), and (4.10), it is expressed as

$$\frac{\pi F_1}{\sqrt{2}} = 2 + M_1 + 2M_2/3 + M_3/2 \tag{4.15}$$

Using Eqs. (4.4), (4.13), and (4.14), following equation can be obtained as

$$\sqrt{2}F_2 = 1 + M_1 + M_2 + M_3 \tag{4.16}$$

Equations (4.4) and (4.8) yield

$$M_2 = 3 \tag{4.17}$$

After solving Eqs. (4.15), (4.16), and (4.17), the following expressions for parameters of weight function are obtained:

$$\begin{aligned} M_2 &= 3 \\ M_3 &= 2\sqrt{2}F_2 - 2\pi F_1 \\ M_1 &= \sqrt{2}F_2 - 1 - M_2 - M_3 \end{aligned} \tag{4.18}$$

Thus, the three parameters of weight function for the edge cracks in a finite width of plate are expressed by expression (4.18) and then the four-term universal weight function can be characterized by the following expression.

$$m(x,a) = \frac{2}{\sqrt{2\pi(a-x)}} \begin{bmatrix} 1 + M_1(1-x/a)^{1/2} + M_2(1-x/a) \\ +M_3(1-x/a)^{3/2} \end{bmatrix} \tag{4.19}$$

The nondimensional forms of Tada Green's function and the derived equivalent weight function having coefficients M_1, M_2, and M_3 determined using Eq. (4.19) are compared. It is found that the maximum absolute error up to $a/D = 0.2$ is less than 10% and this error increases for deeper cracks.

From the above comparison it is noticed that the standard procedures to find out the parameters of the universal weight function for the case of edge cracks in finite width of plate is not suitable. One of the obvious reasons may be attributed to the geometrical factor expressed in Eq. (4.10) which is suitable for $a/D \leq 0.6$. Therefore, an alternative procedure is to be sought to get the correct values of the parameters of the universal weight function for the assumed cracked geometry. As a result, it is proposed to fit the Tada Green's function using least square technique for determining the different parameters of the universal weight function.

4.3.1 Four-Term Universal Weight Function

The form of four-term universal weight function is expressed in Eq. (4.19). In this method, the values of the Tada Green's function at different values of *a/D* (ranging $0 \leq a/D < 1$) and *x/D* (ranging $0 \leq x/D < 1$) are computed. Then the values of three parameters M_1, M_2, and M_3 are determined using least square technique. From the least square fitting it was found that at every point of *a/D* ratio (for values

Table 4.1 Coefficients of four-term weight function parameters M_1, M_2, and M_3

i	a_i	b_i	c_i	d_i	e_i	f_i
1	0.0572011	–0.8741603	4.0465668	–7.89441845	7.8549703	–3.18832479
2	0.4935455	4.43649375				
3	0.340417	–3.9534104	16.1903942	–16.0958507	14.6302472	–6.1306504

$0 \le x/D < 1$), the values of M_1, M_2, and M_3 could be represented as a function of *a/D* ratio in the following form:

$$M_i = \frac{1}{(1-a/D)^{3/2}}\begin{bmatrix} a_i + b_i a/D + c_i(a/D)^2 + d_i(a/D)^3 \\ +e_i(a/D)^4 + f_i(a/D)^5 \end{bmatrix}, \text{for } i = 1 \text{ and } 3 \tag{4.20a}$$

and

$$M_i = \left[a_i + b_i a/D\right], \text{for } i = 2 \tag{4.20b}$$

It is noticed from Eq. (4.20b) that M_2 can be represented simply by straight line equation. The values of coefficients a_i, b_i, c_i, ..., f_i are given in Table 4.1. For an edge crack in the finite width plate, the accuracy of the weight function is verified with respect to standard Green's function and their comparison is plotted in Figs. 4.4, 4.5, and 4.6 for different *a/D* ratios 0.2–0.6, 0.8, and 0.95, respectively.

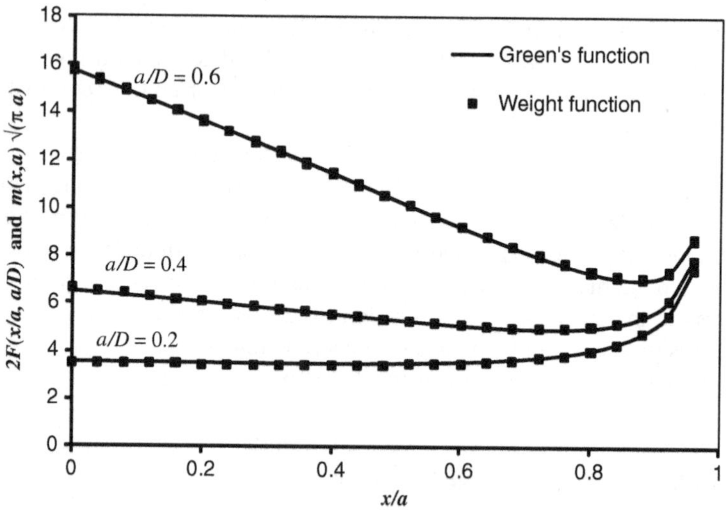

Fig. 4.4 Comparison of four-term weight function with Tada Green's function for *a/D* ratios 0.2, 0.4, and 0.6

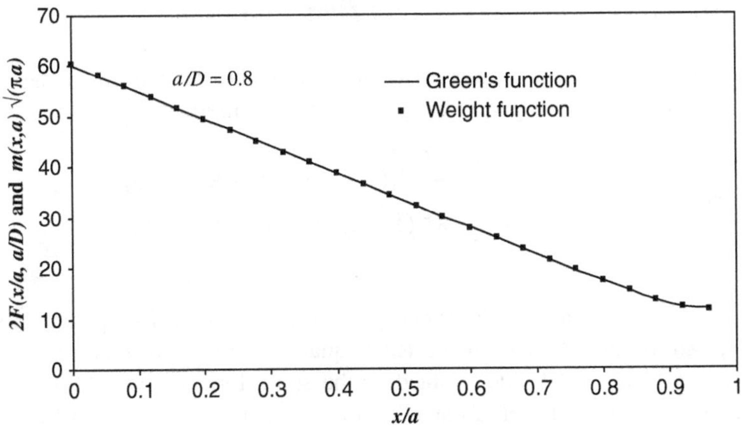

Fig. 4.5 Comparison of four-term weight function with Tada Green's function for *a/D* ratio 0.8

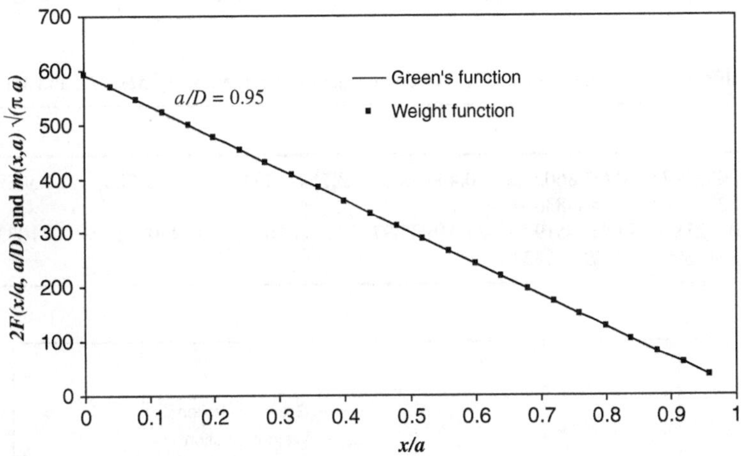

Fig. 4.6 Comparison of four-term weight function with Tada Green's function for *a/D* ratio 0.95

From the analysis of the results of the weight function in the form of Eq. (4.19) having four terms, it is observed that the largest absolute error with respect to Tada Green's function equation is less than 2% for $0 \leq a/D \leq 0.95$ in the range of $0 \leq x/a \leq 0.96$. For $x/a \geq 0.96$ that is very near to the crack tip, the maximum error will slightly increase and this will be less than 3% (exactly 2.912%) at $x/a = 0.98$ which occurs for $a/D = 0.80$.

4.3.2 Five-Term Universal Weight Function

If the desired accuracy is further to be improved, the same form of the weight function with five terms as given in Eq. (4.21) may be used:

$$m(x,a) = \frac{2}{\sqrt{2\pi(a-x)}}\begin{bmatrix} 1 + M_1(1-x/a)^{1/2} + M_2(1-x/a) \\ + M_3(1-x/a)^{3/2} + M_4(1-x/a)^2 \end{bmatrix} \tag{4.21}$$

In the similar manner as described previously, the values of parameters M_1, M_2, M_3, and M_4 are obtained using least square technique. In this case, the values of M_1 and M_3 can be determined in the same form of Eq. (4.20a) (i = 1 and 3), whereas the values of M_2 and M_4 are calculated in the form of Eq. (4.20b) (i = 2 and 4) representing simply equation of straight lines. The coefficients $a_i, b_i, c_i, \ldots, f_i$ of these parameters M_1, M_2, M_3, and M_4 are presented in Table 4.2. The weight function with five terms and Tada Green's function are compared graphically in Figs. 4.7, 4.8, and 4.9. It is noticed that with universal weight function, the

Table 4.2 Coefficients of five-term weight function parameters M_1, M_2, M_3, and M_4

i	a_i	b_i	c_i	d_i	e_i	f_i
1	–0.000824975	0.6878602	0.4942668	–3.25418434	3.4426983	–1.3689673
2	0.782308	–3.0488836				
3	–0.3049218	13.4186519	–23.31662697	35.51066606	–34.440981408	14.10339412
4	0.28347699	–7.378355423				

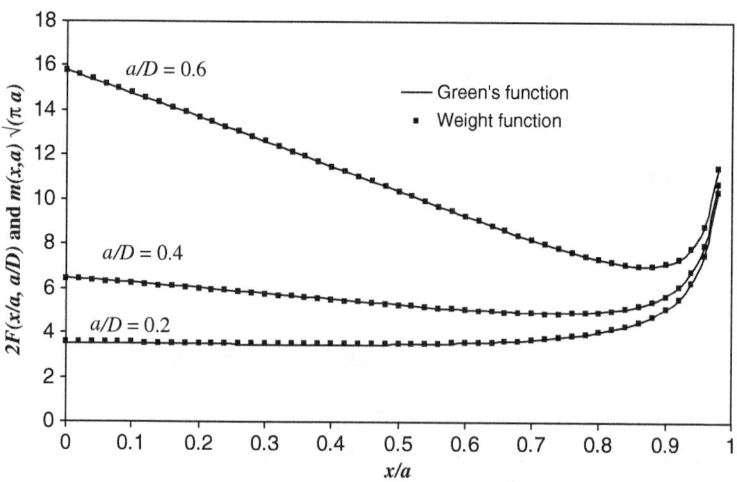

Fig. 4.7 Comparison of five-term weight function with Tada Green's function for *a/D* ratios 0.2, 0.4, and 0.6

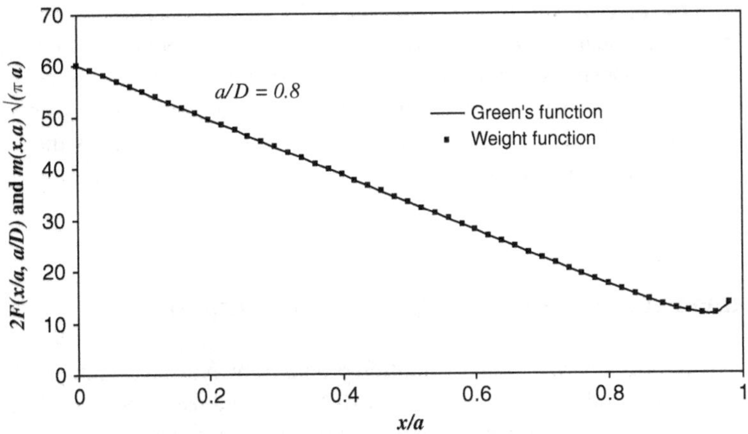

Fig. 4.8 Comparison of five-term weight function with Tada Green's function for *a/D* ratio 0.8

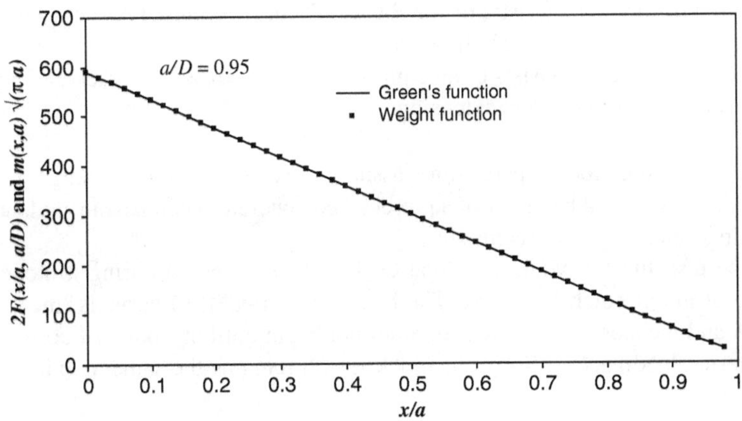

Fig. 4.9 Comparison of five-term weight function with Tada Green's function for *a/D* ratio 0.95

maximum absolute difference in the result is less than 1% for $0.2 < a/D \leq 0.95$ and it is less than 1.5% for $0 \leq a/D \leq 0.2$ in the range of crack length of $0 \leq x/a \leq 0.98$. The range of *a/D* ratio less than 0.2 is not of practical interest in engineering analysis.

Therefore, the universal form of weight function having four terms can be used with less than 3% difference with respect to standard Green's function, whereas this difference is still less than 1.5% with five-term weight function. Hence, in engineering application for an edge-cracked plate with finite width, the four-term universal weight function may be used as a replacement of Tada Green's function without much appreciable error. For higher degree of desired accuracy, the five-term weight function may be used as an alternative solution of SIF. The replacement of Tada Green's function by using universal weight function not only reduces the time of integration for determination of SIF but also will provide a closed-form solution for SIF up to practical cases of polynomial local stress distribution along the crack line. Since the cohesive stress distribution in the fracture process zone of *quasibrittle* materials like concrete is taken into account, the application of such weight functions in the determination of SIF in concrete structure will simplify the integral expression and consequently avoid the need of complicated numerical integration technique.

4.4 Linear Asymptotic Superposition Assumption

Linear asymptotic superposition assumption proposed by Xu and Reinhardt (1999b) enables one to introduce the concept of LEFM for analyzing the complete fracture process and calculating the double-K fracture parameters of concrete. The hypotheses of the assumption are given below:

1. The nonlinear characteristic of the P-CMOD curve is caused by fictitious crack extension in front of a stress-free crack.
2. An effective crack consists of an equivalent elastic stress-free crack and equivalent elastic fictitious crack extension.

The linear asymptotic superposition assumption can be explained using Fig. 4.10, which shows a typical behavior of a pre-cracked concrete specimen during loading–unloading–reloading in a fracture test.

In the test, the P-CMOD envelope OABCDE is observed during a monotonic loading at initial notch length a_0. The behavior of P-CMOD curve is linear up to point A and the slow crack extension does not begin until this point of corresponding external loading P_{ini}. Since the crack length remains the same as initial crack

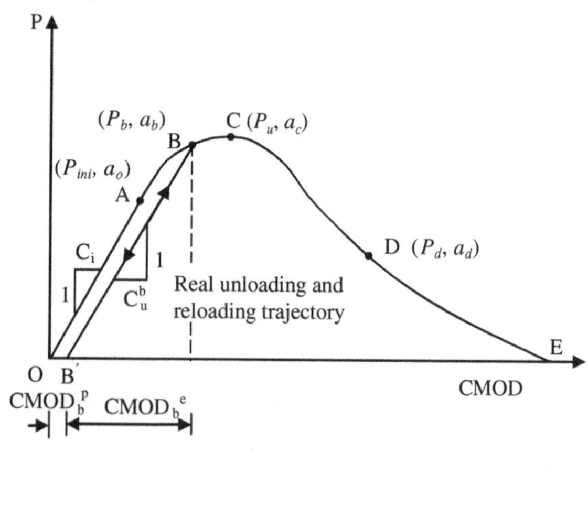

Fig. 4.10 Real unloading and reloading trajectory at an arbitrary point B

length a_o in the region OA, the LEFM can be directly applied for determining the stress intensity factor and the initiation toughness K_{IC}^{ini} of the material is obtained by using the initial cracking load P_{ini} and a_o in the LEFM formula. Beyond point A, the LEFM cannot be directly applied. Consider a point B on the curve, where the external load acting is P_b and the effective crack length is a_b. The difference of a_b and a_o is termed as fictitious crack extension. Now the unloading from point B is carried out until the load value becomes zero. Further, reloading is done considering an effective initial crack length a_b such that the P-CMOD curve will follow the path B′BCDE, which means that the LEFM principles can be applicable at point B for determining the stress intensity factor using the load value P_b and initial equivalent crack length value a_b. Since the nonlinearity on the segment AB is caused by both the inelastic part CMOD_p^b of the total CMOD_b and the difference of elastic compliance between the initial elastic compliance C_i and the reloading elastic compliance C_u^b, according to the first hypothesis, the effect of the CMOD_p^b should be considered for reasonable evaluation of the a_b. If only the elastic part CMOD_b^e or the compliance C_u^b is used, it may lead to an underestimate in the value of a_b because of the nonlinearity in the portion AB.

According to the second hypothesis, an elastic progressively extending crack consists of an equivalent elastic stress-free crack and an equivalent elastic fictitious crack extension. In this case, the progressively extending crack somewhat inelastic in nature should be equivalent to an elastic crack taking the influence of the inelastic part CMOD_p^b into account. If the equivalent elastic crack length at point B is a_b and the unloading trajectory meets the point O having zero inelastic deformation after unloading, then the unloading trajectory is termed as fictitious trajectory as shown in Fig. 4.11. In this way the inelastic deformation is taken care of by the secant compliance C_s. When reloading is carried out, the reloading trajectory will follow-up the same as the unloading trajectory. The compliance value changes from C_i to C_s when the equivalent elastic crack length changes from a_o to a_b. Therefore, two specimens A and B made of same materials and having the same geometrical dimensions except for different equivalent elastic crack lengths a_o and a_b, respectively, could be imagined in the fracture tests. Then the possible P-CMOD curves for these test specimens may be OABCDE for the specimen A and OBCDE for the specimen B. The P-CMOD curve with the path OBCDE for specimen B could be assumed to be measured which should have been pre-cracked to grow the equivalent elastic crack to a desirable length a_b. Beyond point B, with the increase of external load, the crack will further develop and the P-CMOD curves obtained from both specimens A and B are the same. The difference between the initial crack lengths a_o and a_b in the specimen A and B will only result in the difference between the two trajectories OAB and OB for the respective specimens until the load P_b is reached. It is pointed out that the fictitious crack extension $\Delta a_b = a_b - a_o$ is not stress free but some cohesive forces are acting on it.

For progressive crack growth, it is assumed that a series of specimens made of same material having same geometrical dimensions are pre-cracked to grow the different equivalent elastic cracks such as a_o, a_1, a_2, a_3, ... and the load is

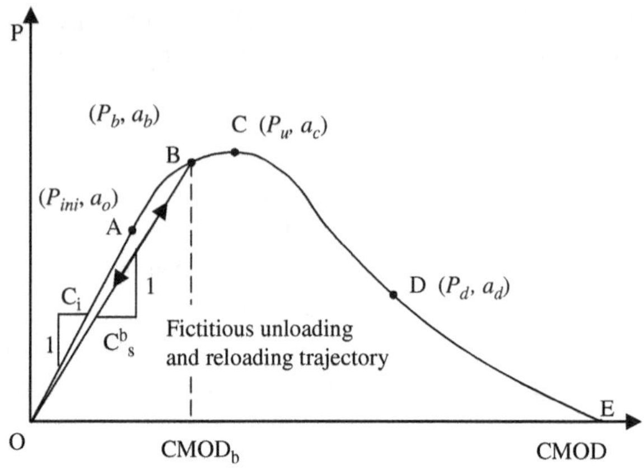

Fig. 4.11 Fictitious unloading and reloading trajectories at an arbitrary point B

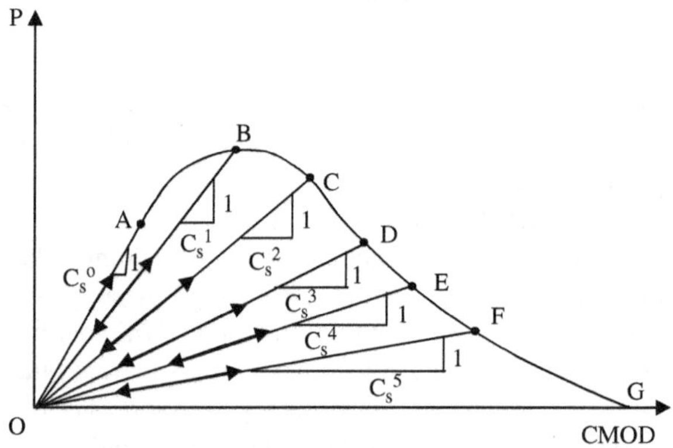

Fig. 4.12 Asymptotic superposition of initially linear P-CMOD curves

applied on each specimen until nonlinear deformation appears on the measured P-CMOD curve. This mechanism is demonstrated in Fig. 4.12. As a result, a series of P-CMOD curves OA, OB, OC, OD, ... will be linear. These lines are drawn in the same coordinate or are asymptotically superimposed together with an origin O and the end points A, B, C, D, E, and so on. When the number of lines is good enough to yield a smooth curve, it is expected that this envelope will coincide with the measured P-CMOD curve of a specimen with an initial crack length a_0. This assumption is termed as linear asymptotic superimposition assumption.

All the secant lines OA, OB, OC, OD, ... are linear segments of a measured P-CMOD curve and after the end points of these lines the slow crack extension appears. Hence, the linear asymptotic superimposition assumption makes use of the LEFM principles at the end point of these segments and a complete fracture process with nonlinear characteristics can be solved using series of linear elastic fracture mechanics cases.

4.5 Determination of Double-*K* Fracture Parameters

4.5.1 Effective Crack Extension for the Analytical Method and Weight Function Approach

First, the P-CMOD plot from the standard three-point bending and four-point bending tests or P-COD plot from the standard CT test is obtained. Then using linear asymptotic superposition assumption, the equivalent elastic crack length a_c at maximum load is be solved by using LEFM formulae for each type of specimen geometry as mentioned below.

4.5.1.1 For TPBT with S/D = 4 Using LEFM Formulae

$$\text{CMOD} = \frac{6PSa}{BD^2E}V(\beta) \tag{4.22}$$

$$V(\beta) = 0.76 - 2.28\beta + 3.87\beta^2 - 2.04\beta^3 + \frac{0.66}{(1-\beta)^2} \tag{4.23}$$

where $\beta = (a + H_o)/(D + H_o)$, $a = a_c$ equivalent elastic crack length at maximum load P_u, H_o is the thickness of the clip gauge holder (Tada et al. 1985). The measured initial compliance C_i from the P-CMOD curve is used to calculate the Young's modulus E as per the RILEM (1990) formula:

$$E = \frac{6Sa_o}{C_iBD^2}V(\beta_o) \tag{4.24}$$

In Eq. (4.24), a_o is the initial notch length and $\beta_o = (a_o + H_o)/(D + H_o)$.

4.5.1.2 For Standard CT Specimen Using LEFM Equations

Similar to Eq. (4.22), COD for CT specimen is expressed as (Murakami 1987)

$$\text{COD} = \frac{P}{BE}V(\alpha) \tag{4.25}$$

$$V(\alpha) = \begin{bmatrix} 2.163 + 12.219\alpha - 20.065\alpha^2 \\ -0.9925\alpha^3 + 20.609\alpha^4 - 9.9314\alpha^5 \end{bmatrix} \times \left(\frac{1+\alpha}{1-\alpha}\right)^2 \tag{4.26}$$

where $\alpha = a/D$, $a = a_c$ equivalent elastic crack length at maximum load P_u. The empirical expression (4.26) is valid within 0.5% accuracy for $0.2 \leq \alpha, 0.975$. The value E is calculated using the P-COD curve as

$$E = \frac{V_1(\alpha_o)}{C_iB} \tag{4.27}$$

4.5.1.3 For Four-Point Bending Test (FPBT) with $S/D = 4$ and $L = D$ Using LEFM Formulae

Like three-point bending test, the fracture behavior of the four-point bending test can be easily determined, so two different loading conditions with the same specimen geometry in all respect can be obtained for precise comparison of loading effect on fracture parameters (Tada et al. 1985). The loading and support conditions along with the dimensions of the four-point bending test considered in the study are shown in Fig. 4.13, where the symbols B, D, and S are the width, depth, and support span, respectively, with $S/D = 4$, moment arm $L = D$, and the load span $= 2D$.

For the pure bending case (also for FPBT), the following formulae (Tada et al. 1985) are used to calculate the equivalent crack extension:

$$\text{CMOD} = \frac{12PLa}{BD^2E}V(\beta) \tag{4.28}$$

$$V(\beta) = 0.8 - 1.7\beta + 2.4\beta^2 + \frac{0.66}{(1-\beta)^2} \tag{4.29}$$

Equation (4.29) has better than 1% accuracy for any value of β. Similarly, the measured initial compliance C_i from the P-CMOD curve can be used to calculate the E as per the following formula:

$$E = \frac{12La_o}{C_iBD^2}V(\beta_o) \tag{4.30}$$

It was concluded (Karihaloo and Nallathambi 1991) that almost the same value of E might be obtained from P-CMOD curve, load–deflection curve, and compressive cylinder test. Hence, in case initial compliance is not known, the value of E determined using compressive cylinder tests may be used to obtain the critical crack length of the specimen.

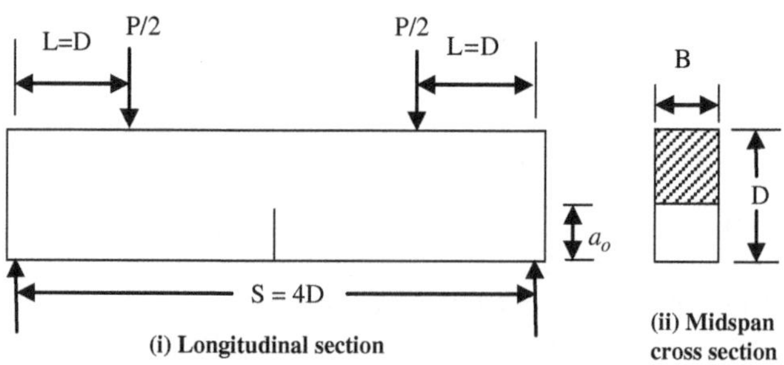

Fig. 4.13 Dimensions and loading schemes of four-point bending test specimen

4.5.2 *Effective Crack Extension for Simplified Method*

4.5.2.1 For TPBT with S/*D* = 4

Solution of highly nonlinear equations for determining the effective crack extension requires either trial-and-error approach or some specialized numerical methods, hence for calculation of CMOD for the standard TPBT specimens, the following empirical expression was proposed by Xu and Reinhardt (2000):

$$\mathrm{CMOD} = \frac{P}{BE}\left[3.70 + 32.60tg^2\left(\frac{\pi}{2}\beta\right)\right] \tag{4.31}$$

where $\beta = (a+H_o)/(D+H_o)$. The trial-and-error method is not required for solution of Eq. (4.31) and it can be solved directly. As compared to Eq. (4.23), Eq. (4.31) yields less than 2% error between $0.2 \le \beta \le 0.7$ and when β is 0.80, the maximum error is less than 3.5%. Using Eq. (4.31), the length of effective crack extension is expressed by simple equation as

$$a = \frac{2(D+H)}{\pi}\mathrm{arctg}\sqrt{\frac{BEC}{32.6} - 0.1135} - H_o \tag{4.32}$$

in which *C* is the compliance = CMOD/*P*.

4.5.2.2 For Standard CT Specimen

An empirical formula was also proposed by Xu and Reinhardt (1999c) for determining the effective crack length for standard CT specimen within 1% accuracy for $0.2 \le \alpha \le 0.62$ and within 3% accuracy for $0.2 \le \alpha \le 0.7$. An improvement over this empirical formula was proposed by Kumar and Barai (2009a). The improved empirical relation is valid for $0.2 \le \alpha \le 0.9$ within 0.7% accuracy as compared with Eq. (4.26) and is given as below:

$$\begin{aligned}\mathrm{COD} &= \frac{P}{BE}\left[-10.4595 + 12.4573\left(\frac{1-0.047\alpha}{1-\alpha}\right)^2\right], && \text{for } 0.2 \le \alpha < 0.6\\ &= \frac{P}{BE}\left[3.4127 + 3.7226\left(\frac{1+\alpha}{1-\alpha}\right)^2\right], && \text{for } 0.6 \le \alpha < 0.8\\ &= \frac{P}{BE}\left[-8.4875 + 3.8564\left(\frac{1+\alpha}{1-\alpha}\right)^2\right], && \text{for } 0.8 \le \alpha \le 0.9\end{aligned} \tag{4.33}$$

Equation (4.33) is used in the present study.

4.5.3 Calculation of Double-K Fracture Parameters

The cohesive stress distribution in the fictitious crack zone gives rise to cohesion toughness as a part of total toughness of the cracked body. Superposition method is used in order to calculate the SIF at the tip of the effective crack length K_{I}. According to this scheme, total stress intensity factor K_{I} is equal to the summation of SIF caused due to external load $K_{\mathrm{I}}^{\mathrm{P}}$ and the SIF contributed by cohesive stress $K_{\mathrm{I}}^{\mathrm{C}}$ (Jenq and Shah 1985b, Xu and Reinhardt 1999b) as shown in Fig. 4.14. The value of K_{I} is expressed in the following expression:

$$K_{\mathrm{I}} = K_{\mathrm{I}}^{\mathrm{P}} + K_{\mathrm{I}}^{\mathrm{C}} \tag{4.34}$$

After determining the critical effective crack extension at unstable condition of loading, the two parameters ($K_{\mathrm{IC}}^{\mathrm{ini}}$ and $K_{\mathrm{IC}}^{\mathrm{un}}$) of double-$K$ fracture model are determined using LEFM formulae as given in the following equations.

4.5.3.1 For TPBT with S/D = 4 Using LEFM Formulae

The SIF is calculated using Eqs. (3.51) and (3.52), in which α is a/D and σ_{N} is the nominal stress in the beam due to external load P and self-weight of the structure which is given by (Tada et al. 1985)

$$\sigma_{\mathrm{N}} = \frac{3S}{4BD^2}[2P + w_{\mathrm{g}}S] \tag{4.35}$$

In Eq. (4.35), w_{g} is the self-weight per unit length of the beam.

4.5.3.2 For Standard CT Specimen Using ASTM Standard 399-06

The SIF is calculated using Eqs. (3.51) and (3.56) (ASTM International Standard E399-06 2006).

4.5.3.3 For FPBT Geometry with $S/D = 4$ and $L = D$ Using LEFM Formulae

The SIF created at the fictitious crack tip due to self-weight of the beam is equivalent to the SIF due to a concentrated load of the magnitude half of the total load caused by the self-weight, acting at the midspan of the beam. Hence, in addition to the external load P acting on the four-point bending test, the influence due to a concentrated load equal to $w_{\mathrm{g}}.S/2$ and acting at the midspan (resembling the three-point bending consideration) is also effective if the self-weight of the beam is taken into consideration. The SIF for the FPBT, K_{I} due to external load and self-weight of the beam at any loading stage can be calculated using the following procedures.

The SIF due to external load P on the FPBT specimen $K_{\mathrm{I}}^{\mathrm{E}}$ (Murakami 1987) is expressed as

$$K_{\mathrm{I}}^{\mathrm{E}} = \sigma_{\mathrm{NE}}\sqrt{D}k_1(\alpha) \tag{4.36}$$

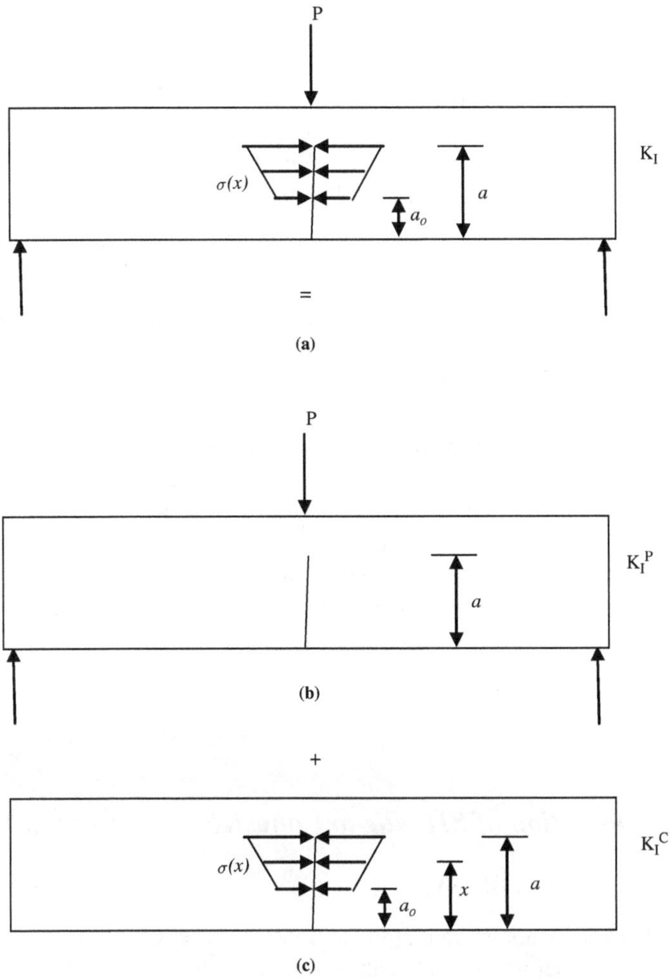

Fig. 4.14 Calculation of SIF using superposition method

$$\sigma_{\mathrm{NE}} = \frac{3PL}{bD^2} \tag{4.37}$$

$$k_1(\alpha) = \sqrt{\pi\alpha}\left\{\begin{array}{l} 1.122 - 1.121\alpha + 3.740\alpha^2 \\ +3.873\alpha^3 - 19.05\alpha^4 + 22.55\alpha^5 \end{array}\right\} \tag{4.38}$$

Equation (4.38) has better than 1% accuracy for $\alpha \leq 0.7$. The contribution of SIF due to self-weight of the specimen $K_{\mathrm{I}}^{\mathrm{S}}$ (considering TPBT for $S/D = 4$) is expressed as

$$K_{\mathrm{I}}^{\mathrm{S}} = \sigma_{\mathrm{NS}}\sqrt{D}k_2(\alpha) \tag{4.39}$$

$$\sigma_{NS} = \frac{3w_g S^2}{4bD^2} \tag{4.40}$$

where $k_2(\alpha)$ is obtained using expression (3.52) and σ_{NE} and σ_{NS} are the nominal stresses in the beam due to external load P and self-weight of the structure, respectively. The effective stress intensity factor K_I of the FPBT loading condition can be obtained as the summation of the values K_I^E and K_I^S. Hence

$$K_I = \sqrt{D}\,[\sigma_{NE}k_1(\alpha) + \sigma_{NS}k_2(\alpha)] \tag{4.41}$$

The respective LEFM equations can be used for calculation of unstable fracture toughness K_{IC}^{un} at the tip of effective crack length a_c in which $a = a_c$ and P is the maximum load P_u for TPBT, CT test, and FPBT specimen geometries. The initiation toughness K_{IC}^{ini} is calculated using those LEFM equations if the initial cracking load P_{ini} at initial crack tip is known. Generally, two methods are employed to determine the value of P_{ini}. The first is the direct measurement method in which the P_{ini} is directly obtained using experiments. In the second method, the starting point of nonlinearity in P-CMOD or P-COD curve is considered to be P_{ini}. Since it is difficult to detect the crack initiation load from experimental approach, an inverse analytical method is used to calculate the value of K_{IC}^{ini} and according to this, the following relation can be employed:

$$K_{IC}^{ini} = K_{IC}^{un} - K_{IC}^{C} \tag{4.42}$$

4.5.4 Determination of SIF due to Cohesive Stress (K_I^C) in FPZ

4.5.4.1 Cohesive Stress Distribution

It is a well-known phenomenon that in the fracture process zone, a nonlinear stress profile can develop before the peak load depending on type of cohesive laws (Zi and Belytschko 2003). However, the bilinear softening law is often used for all the practical purposes in case of normal concrete. It was reported in the study (Xu and Reinhardt 1998) that the shapes of the distribution of cohesive stress ahead of the crack tip were not sensitive to the overall value of stress intensity factor at the crack tip and linear or bilinear shape of cohesive stress distribution could be assumed depending upon the external loading on the structures. Further, the linear stress distribution was considered for loading stage up to the peak load and the corresponding critical effective crack extension a_c for the development of the double-K fracture model (Xu and Reinhardt 1999b). This assumption might lead to the simplification in numerical computations without affecting much on the final values of fracture parameters. Moreover, the preceding assumption seems to be reasonable because up to the peak load, the process zone length is relatively small and mainly the initial portion of the softening branch is used. The cohesive stress acting in the fictitious fracture zone on TPBT, CT test, and FPBT specimens is idealized as a series of pair normal forces subjected to single-edge cracked specimen of finite width as

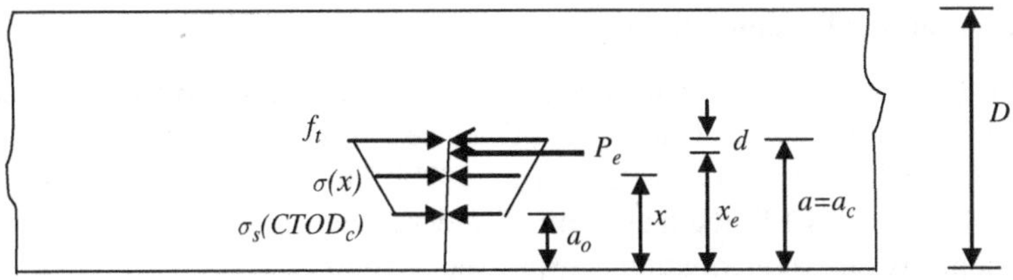

Fig. 4.15 Distribution of cohesive stress in the fictitious crack zone at critical load

shown in Fig. 4.3. The linearly varying distribution of cohesive stress is presented in Fig. 4.15.

The value of cohesion toughness K_{IC}^{C} is negative because of closing stress in fictitious crack zone. However, the absolute value of K_{IC}^{C} is taken as a contribution of the total fracture toughness. The same computation method (Jenq and Shah 1985b) of K_{I}^{C} has been also used in the analytical method (Xu and Reinhardt 1999b, c). For simplified method the value of K_{I}^{C} is computed using a resultant pair of concentrated force as shown in Fig. 4.15. During loading of specimen, the critical condition is achieved at maximum load value and the corresponding critical crack-tip opening displacement ($\mathrm{CTOD_c}$). In Fig. 4.15, $\sigma_s(\mathrm{CTOD_c})$ is the cohesive stress at the tip of initial notch at $\mathrm{CTOD_c}$ and $\sigma(x)$ is expressed as

$$\sigma(x) = \sigma_s(\mathrm{CTOD_c}) + \frac{x - a_o}{a - a_o}[f_t - \sigma_s(\mathrm{CTOD_c})] \tag{4.43}$$

Equation (4.43) is expressed in nondimensional form of crack length, $U = x/a$, as

$$\sigma(U) = \sigma_s(\mathrm{CTOD_c}) + \frac{U - a_o/a}{1 - a_o/a}[f_t - \sigma_s(\mathrm{CTOD_c})], \quad \text{for } a_o/a \le U \le 1 \text{ or } 0 \le \mathrm{CTOD} \le \mathrm{CTOD_c} \tag{4.44}$$

The value of $\sigma_s(\mathrm{CTOD_c})$ is calculated by using softening functions of concrete.

4.5.4.2 Determination of $\mathrm{CTOD_c}$

Since it is difficult to measure the values of $\mathrm{CTOD_c}$ directly, for practical purposes the value of crack opening displacement is monitored. At critical condition, the value of crack opening displacement becomes critical value of $\mathrm{COD_c}$ and for the known value of COD the crack opening displacement within the crack length COD(x) is computed using the following expression (Jenq and Shah 1985a):

$$\mathrm{COD}(x) = \mathrm{COD_c}\left\{(1 - x/a)^2 + (1.081 - 1.149a/D)[x/a - (x/a)^2]\right\}^{1/2} \tag{4.45}$$

In Eq. (4.45), the values of $x = a_o$ and $a = a_c$ are used for calculation of $\mathrm{CTOD_c}$.

4.5.4.3 Analytical Method for K_{IC}^{C}

The stress intensity factor for an edge crack in a finite width of plate subjected to a pair of normal point forces P acting at a distance x (Fig. 4.3) can be obtained using Eqs. (4.11) and (4.12) (Xu and Reinhardt 1999b). Putting $U = x/a$ and $\alpha = a/D$ in Eqs. (4.11) and (4.12), the final form of integral equation for cohesive toughness due to cohesive stress distribution is expressed as follows:

$$K_{IC}^{C} = \int_{a_0/a}^{1} 2\sqrt{\frac{a}{\pi}}\sigma(U)F(U,\alpha)\mathrm{d}\alpha \tag{4.46}$$

$$F(U,\alpha) = \frac{3.52(1-U)}{(1-\alpha)^{3/2}} - \frac{4.35-5.28U}{(1-\alpha)^{1/2}} + \left\{\frac{1.30-0.30U^{3/2}}{\sqrt{[1-U^2]}} + 0.83 - 1.76U\right\}[1-(1-U)\alpha] \tag{4.47}$$

At critical condition the value of a is taken to be a_c. The integration of Eq. (4.47) is done by using Gauss–Chebyshev quadrature method because of the existence of singularity at the integral boundary.

4.5.4.4 Simplified Approach for K_{IC}^{C}

According to the simplified method, the evaluation of K_{IC}^{C} is done using a calibration function $Z(x_e/a, a_o/a)$ with Eq. (4.11) (Xu and Reinhardt 2000). The distributed cohesive force $\sigma(x)$ is converted into a single equivalent force P_e per unit thickness as shown in Fig. 4.15. The absolute value of the K_{IC}^{C} is then expressed as

$$K_{IC}^{C} = Z\left(\frac{x_e}{a}, \frac{a_o}{a}\right)\frac{2P_e}{\sqrt{\pi a}}F(x_e/a, a/D) \tag{4.48}$$

Putting $U_e = x_e/a$, $\alpha_o = a_o/D$, and $\alpha = a/D$ in Eq. (4.48), the nondimensional value of $K_{IC}^{C}/f_t\sqrt{D}$ is expressed as

$$\frac{K_{IC}^{C}}{f_t\sqrt{D}} = Z\left(U_e, \frac{\alpha_o}{\alpha}\right)\frac{2P_e}{f_t\sqrt{\pi a D}}F(U_e, \alpha) \tag{4.49}$$

For trapezoidal cohesive stress distribution, using a nondimensional factor γ as

$$\gamma = \frac{\sigma_s(\mathrm{CTOD_c})}{f_t} \tag{4.50}$$

Eq. (4.44) can be represented in nondimensional form as

$$\frac{\sigma(U)}{f_t} = \gamma + \frac{(U - \alpha_o/\alpha)}{(1 - \alpha_o/\alpha)}[1 - \gamma] \tag{4.51}$$

From Fig. 4.15, the value of resultant force P_e can be written as

$$P_e = \frac{f_t}{2}(1 + \gamma)(a - a_o), \text{ for } a \le a_c \tag{4.52}$$

In nondimensional form, Eq. (4.49) is expressed as

$$\frac{2P_e}{f_t\sqrt{\pi a D}} = (1 + \gamma)(1 - \alpha_o/\alpha)\sqrt{\alpha/\pi} \tag{4.53}$$

It is shown in Fig. 4.15 that d is the centroid of the resultant cohesive force P_e measured from the extending crack tip. Hence d is calculated as

$$d = \frac{(1 + 2\gamma)}{3(1 + \gamma)}(a - a_o) \tag{4.54}$$

and

$$x_e = a - d = a - \frac{(1 + 2\gamma)}{3(1 + \gamma)}(a - a_o) \tag{4.55}$$

Also,

$$U_e = \frac{x_e}{a} = \frac{1}{3(1 + \gamma)}\left\{2 + \gamma + (1 + 2\gamma)\frac{\alpha_o}{\alpha}\right\} \tag{4.56}$$

Finally, the calibration function was empirically determined in the following form:

$$Z(U_e, \alpha_o/\alpha) = \frac{6(1.025 - 0.1\gamma)}{1 + 1.83(\alpha_o - 0.2)}\left(\frac{\alpha_o}{\alpha}\right)^p\sqrt{\frac{\alpha}{\pi}}U_e^{-0.2}, \text{ for } 0.2 \le \alpha_o \le 0.8 \tag{4.57}$$

where $p = 1.5(\alpha_o - 0.2) + 0.8$ for $0.2 \le \alpha_o \le 0.6$; $p = 3(\alpha_o - 0.6) + 1.4$ for $0.6 \le \alpha_o \le 0.7$; and $p = 6(\alpha_o - 0.7) + 1.7$, for $0.7 \le \alpha_o \le 0.8$.

The value of $F(U_e, \alpha)$ in Eq. (4.49) is computed using Eq. (4.47), and then the value of K_{IC}^C is obtained using Eqs.(4.49) and (4.57). A detailed comparison between the computed values of nondimensional parameter $K_{IC}^C/f_t\sqrt{D}$ using *analytical method* and *simplified approach* for $0.6 \le \gamma \le 0.7$ and $0.2 \le \alpha_o \le 0.8$ and α up to 0.975 can be found elsewhere (Xu and Reinhardt 2000). It was shown from comparison that the maximum error between the calculated results using simplified approach is less than 3.6%.

4.5.4.5 The Weight Function Approach for K_{IC}^{C}

In the weight function method, the K_{IC}^{C} is determined using Eq. (4.1) due to trapezoidal cohesive stress distribution (Fig. 4.15) at the unstable fracture condition. According to this method, first of all the parameters of the weight function are determined by the derived Eq. (4.19) or (4.21) depending upon the number of terms in the weight function expression. The value of $\sigma(x)$ in Eq. (4.1) is replaced by Eq. (4.43), hence the closed-form expression of K_{IC}^{C} can be obtained. The value of K_{IC}^{C} using four-term weight function is expressed in the following form:

$$K_{IC}^{C} = \int_{a_o}^{a} \left\{ \sigma_s(\mathrm{CTOD_c}) + \frac{x - a_o}{a - a_o}[f_t - \sigma_s(\mathrm{CTOD_c})] \right\} \times \frac{2}{\sqrt{2\pi(a-x)}} \begin{bmatrix} 1 + M_1(1 - x/a)^{1/2} \\ + M_2(1 - x/a) + M_3(1 - x/a)^{3/2} \end{bmatrix} \mathrm{d}x \tag{4.58}$$

After integration of Eq. (4.58) the closed-form solution of K_{IC}^{C} is determined as

$$K_{IC}^{C} = \frac{2}{\sqrt{2\pi a}} \left\{ A_1 a \left[2s^{1/2} + M_1 s + \frac{2}{3} M_2 s^{3/2} + \frac{M_3}{2} s^2 \right] + A_2 a^2 \left[\frac{4}{3} s^{3/2} + \frac{M_1}{2} s^2 + \frac{4}{15} M_2 s^{5/2} + \frac{M_3}{6} \left\{ 1 - (a_o/a)^3 - 3s a_o/a \right\} \right] \right\} \tag{4.59}$$

where $A_1 = \sigma_s(\mathrm{CTOD_c})$, $A_2 = \dfrac{f_t - \sigma_s(\mathrm{CTOD_c})}{a - a_o}$ and $s = (1 - a_o/a)$. At the critical effective crack extension, a is equal to a_c in Eq. (4.59). Similarly, the closed-form expression for K_{IC}^{C} using five-term weight function is derived and expressed as

$$K_{IC}^{C} = \frac{2}{\sqrt{2\pi a}} \left\{ A_1 a \left[2s^{1/2} + M_1 s + \frac{2}{3} M_2 s^{3/2} + \frac{M_3}{2} s^2 + \frac{2}{5} M_4 s^{5/2} \right] + A_2 a^2 \begin{bmatrix} \frac{4}{3} s^{3/2} + \frac{M_1}{2} s^2 + \frac{4}{15} M_2 s^{5/2} + \frac{4}{35} M_4 s^{7/2} \\ + \frac{M_3}{6} \left\{ 1 - (a_o/a)^3 - 3s a_o/a \right\} \end{bmatrix} \right\} \tag{4.60}$$

where the values of A_1, A_2, and s are the same as indicated above. After determining the value of K_{IC}^{C} using any of the above-mentioned three methods, the double-K fracture parameters K_{IC}^{ini} and K_{IC}^{un} can be evaluated as described in the previous sections. A flowchart for computing double-K fracture parameters can be seen in Appendix (Fig. A.3).

4.6 Determination of Double-G Fracture Parameters

The double-G fracture model introduced by Xu and Zhang (2008) is based on the energy release rate approach. Apart from the external load, crack length, and specimen geometry, the fracture energy release rate approach also depends on the deformation characteristic (Young's modulus) of the material. Hence ductility property is associated with energy release rate approach unlike that of stress intensity factor-based models such as two-parameter fracture model, size-effect model, effective crack model, double-K fracture model, and the K_{R}-curve approach based on cohesive stress distribution. As defined in Chap. 2 (see Sect. 2.7.8), the double-G fracture criterion is similar to the double-K fracture criterion except that the former is based on the energy release rate concept and the latter is based on the stress intensity factor approach. Hence with the similar assumptions as of double-K fracture model, the double-G fracture model consists of two characteristic fracture parameters: the initiation fracture energy release $G_{\mathrm{IC}}^{\mathrm{ini}}$ and the unstable fracture energy release $G_{\mathrm{IC}}^{\mathrm{un}}$. The cohesive stress provides additional resistance to the stable crack propagation in terms of cohesive breaking energy $G_{\mathrm{I}}^{\mathrm{C}}$ until the critical condition is achieved. At the onset of unstable crack propagation or critical condition, the total energy release $G_{\mathrm{IC}}^{\mathrm{un}}$ consists of initiation fracture energy release $G_{\mathrm{IC}}^{\mathrm{ini}}$ and critical value of the cohesive breaking energy $G_{\mathrm{IC}}^{\mathrm{C}}$. The analytical method (Xu and Zhang 2008) used for determining double-G fracture parameters is briefly summarized as follows.

4.6.1 Unstable Fracture Energy Release

According to this approach, for a cracked elastic body subjected to external force P, the energy release rate G is expressed as

$$G = \frac{P^2}{2B}\frac{\mathrm{d}C}{\mathrm{d}a} \tag{4.61}$$

where $\mathrm{d}C/\mathrm{d}a$ is the change in compliance with respect to crack extension.

4.6.1.1 For TPBT Specimen Geometry

Based on linear asymptotic superposition assumption for three-point bending geometry, the energy release rate during crack propagation is determined using the compliance relation of Tada et al. (1985). Considering $\alpha = a/D$, the compliance C for this case can be expressed as

$$C = \frac{\delta}{P} = \frac{3S^2}{2BD^2E}V(\alpha) \tag{4.62}$$

$$V(\alpha) = \left(\frac{\alpha}{1-\alpha}\right)^2\left[5.58 - 19.57\alpha + 36.82\alpha^2 - 34.94\alpha^3 + 12.77\alpha^4\right] \tag{4.63}$$

The empirical expression (4.63) is valid within 1% accuracy for any value of α. The value of dC/da in Eq. (4.61) is computed as below:

$$\frac{\mathrm{d}C}{\mathrm{d}a} = \frac{1}{D}\frac{\mathrm{d}C}{\mathrm{d}\alpha} = \frac{3S^2}{2BD^3E}V'(\alpha) \tag{4.64}$$

$$\begin{aligned} V'(\alpha) = {} & \frac{2\alpha}{(1-\alpha)^3}\left[5.58 - 19.57\alpha + 36.82\alpha^2 - 34.94\alpha^3 + 12.77\alpha^4\right] \\ & + \left(\frac{\alpha}{1-\alpha}\right)^2\left[-19.57 + 73.64\alpha - 104.82\alpha^2 + 51.08\alpha^3\right] \end{aligned} \tag{4.65}$$

Considering the influence of the self-weight of the beam, the unstable fracture energy release $G_{\mathrm{IC}}^{\mathrm{un}}$ can be written using Eqs. (4.61) ,(4.64) and (4.65) for $P = P_{\mathrm{u}}$, $a = a_{\mathrm{c}}$, and $\alpha = \alpha_{\mathrm{c}}$:

$$G_{\mathrm{IC}}^{\mathrm{un}} = \frac{3\left(P_{\mathrm{u}} + w_{\mathrm{g}}S/2\right)^2 S^2}{4B^2D^3E}V'(\alpha_{\mathrm{c}}) \tag{4.66}$$

4.6.1.2 For CT Specimen Geometry

Similar to that of TPBT geometry, the value of dC/da for CT specimen is obtained using Eqs. (4.25) and (4.26):

$$\frac{\mathrm{d}C}{\mathrm{d}a} = \frac{1}{D}\frac{\mathrm{d}C}{\mathrm{d}\alpha} = \frac{1}{EBD}V'(\alpha) \tag{4.67}$$

$$\begin{aligned} V'(\alpha) = {} & \frac{4(1+\alpha)}{(1-\alpha)^3}\begin{bmatrix} 2.163 + 12.219\alpha - 20.065\alpha^2 - 0.9925\alpha^3 \\ + 20.609\alpha^4 - 9.9314\alpha^5 \end{bmatrix} \\ & + \left(\frac{1+\alpha}{1-\alpha}\right)^2\begin{bmatrix} 12.219 - 40.13\alpha - 2.9775\alpha^2 \\ +82.436\alpha^3 - 49.657\alpha^4 \end{bmatrix} \end{aligned} \tag{4.68}$$

Then the unstable fracture energy release $G_{\mathrm{IC}}^{\mathrm{un}}$ can be written using Eqs. (4.61), (4.67), and (4.68) for $P = P_{\mathrm{u}}$, $a = a_{\mathrm{c}}$, and $\alpha = \alpha_{\mathrm{c}}$:

$$G_{\mathrm{IC}}^{\mathrm{un}} = \frac{P_{\mathrm{u}}^2}{2B^2DE}V'(\alpha_{\mathrm{c}}) \tag{4.69}$$

4.6.2 Determination of Critical Cohesive Breaking Energy

The concept of energy dissipation in the FPZ is demonstrated in Figs. 4.16 and 4.17.

The local cohesive breaking energy $g_{\mathrm{f}}(x)$ can be explained using Fig. 4.16, in which the crack opening displacement at certain location x in the FPZ increases from zero to w_x and the corresponding cohesive stress drops down from f_{t} to $\sigma(w_x)$. Two conditions, i.e., CTOD$_{\mathrm{c}} \le w_{\mathrm{s}}$ (Fig. 4.16a) and $w_{\mathrm{s}} \le$ CTOD$_{\mathrm{c}} \le w_{\mathrm{c}}$ (Fig. 4.16b), may

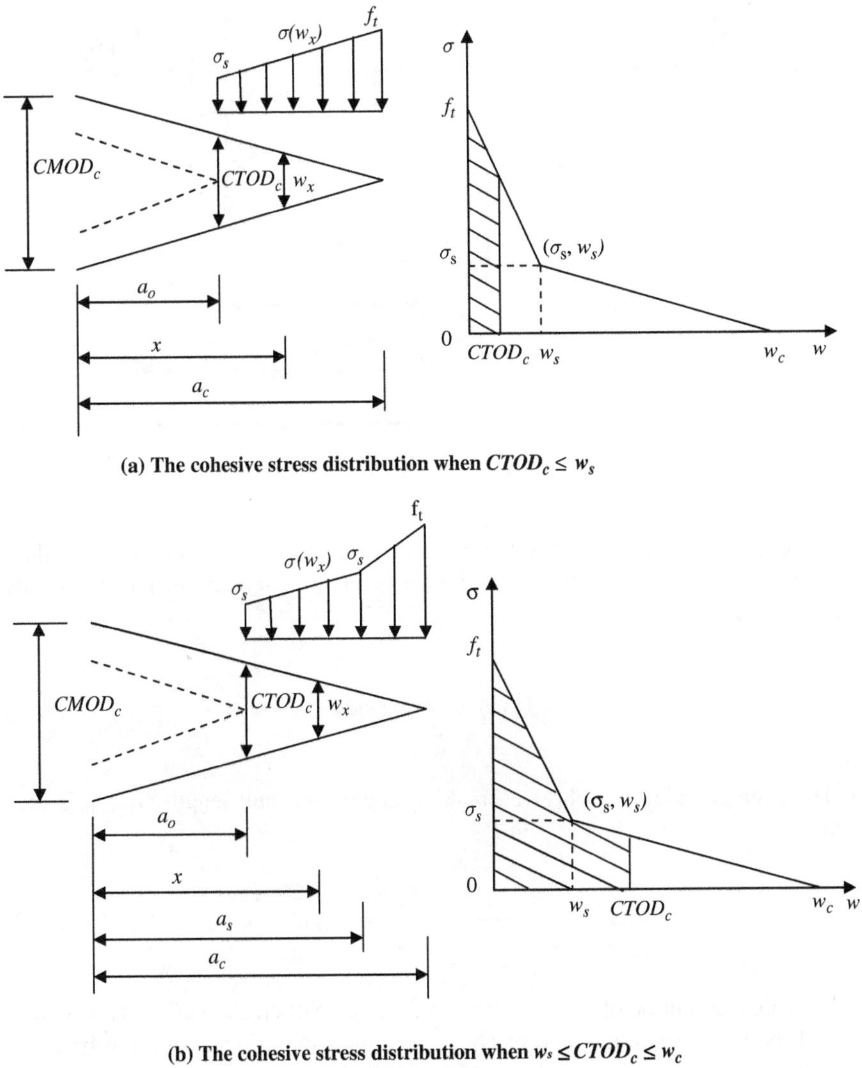

Fig. 4.16 The cohesive stress distribution along the FPZ

arise at the notch tip while using modified bilinear softening function. According to fracture energy definition by Hillerborg, the local cohesive breaking energy $g_f(x)$ at any location x along the FPZ and corresponding COD w_x is defined as

$$g_f(x) = \int_0^{w_x} \sigma(w)\mathrm{d}w \tag{4.70}$$

The distribution of local cohesive breaking energy can be obtained according to Hillerborg's concept at different locations along the FPZ as shown in Fig. 4.17. The

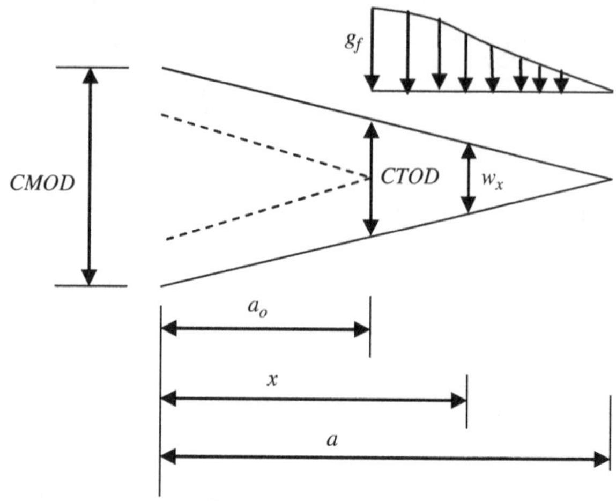

Fig. 4.17 The local breaking energy due to cohesive stress distribution along the FPZ

total energy dissipation at any value of crack extension a is the sum of the values of local cohesive energy in the whole process zone $(a - a_{\mathrm{o}})$ and is mathematically expressed as

$$\Pi(x) = \int_{a_{\mathrm{o}}}^{a} g_{\mathrm{f}}(x)\mathrm{d}x \tag{4.71}$$

The average value of cohesive breaking energy per unit length $G_{\mathrm{I-coh}}$ is then defined as

$$G_{\mathrm{I\text{-}coh}} = \frac{1}{a - a_{\mathrm{o}}} \int_{a_{\mathrm{o}}}^{a} \int_{0}^{w} \sigma(w_x)\mathrm{d}w\,\mathrm{d}x \tag{4.72}$$

At critical condition of unstable fracture, the value of crack extension a becomes a_c and the corresponding value of $\mathrm{G_{I-coh}}$ becomes the critical cohesion breaking energy $G_{\mathrm{IC}}^{\mathrm{C}}$:

$$G_{\mathrm{IC}}^{\mathrm{C}} = \frac{1}{a - a_{\mathrm{o}}} \int_{a_{\mathrm{o}}}^{a_{\mathrm{c}}} \int_{0}^{w} \sigma(w_x)\mathrm{d}w\,\mathrm{d}x \tag{4.73}$$

The limits of inner integral of Eq. (4.73) are taken to be 0–$\mathrm{CTOD_c}$ for the case of $\mathrm{CTOD_c} \leq w_{\mathrm{s}}$ (Fig. 4.16a). For the second case, i.e., $w_{\mathrm{s}} \leq \mathrm{CTOD_c} \leq w_{\mathrm{c}}$ (Fig. 4.16b), Eq. (4.73) can be expanded in the following form:

$$G_{\mathrm{IC}}^{\mathrm{C}} = \frac{1}{a - a_{\mathrm{o}}} \left[\int_{a_s}^{a_{\mathrm{c}}} \int_{0}^{w_s} \sigma(w_x)\mathrm{d}w\,\mathrm{d}x + \int_{a_{\mathrm{o}}}^{a_{\mathrm{s}}} \int_{w_s}^{\mathrm{CTOD_c}} \sigma(w_x)\mathrm{d}w\,\mathrm{d}x \right] \tag{4.74}$$

The w_x in Eqs. (4.73) and (4.74) can be determined using Eq. (4.45) for the known value of $CMOD_c$ and a_c at onset of unstable fracture. Once the value of w_x is known, the value of $\sigma(w_x)$ is evaluated using a particular softening relation of concrete.

4.6.3 Determination of Initial Fracture Energy Release

The value of G_{IC}^{ini} is determined using Eqs. (4.66) and (4.69) for three-point bending and compact tension specimens, respectively, for $P = P_{ini}$ and $\alpha = \alpha_o$, in which the P_{ini} is the crack initiation load measured experimentally. If the value of P_{ini} is not known in advance, an inverse analysis similar to double-K fracture parameters is employed for evaluating the value of G_{IC}^{ini} as per the following equation:

$$G_{IC}^{ini} = G_{IC}^{un} - G_{IC}^{C} \tag{4.75}$$

4.6.4 Determination of Effective Double-K Fracture Parameters

Analytical foundation of double-K fracture criterion and double-G fracture criterion is based on the linear asymptotic superposition assumption which means that linear elastic fracture mechanics relationship is still valid at all point of P-CMOD envelope. Therefore, it is convenient to obtain the effective double-K fracture parameters, i.e., effective initiation fracture toughness $\overline{K_{IC}^{ini}}$ and effective unstable fracture toughness $\overline{K_{IC}^{un}}$ in terms of equivalent stress intensity factors using double-G fracture model:

$$\begin{aligned} \overline{K_{IC}^{ini}} &= \sqrt{EG_{IC}^{ini}} \\ \overline{K_{IC}^{un}} &= \sqrt{EG_{IC}^{un}} \end{aligned} \tag{4.76}$$

Equation (4.76) is useful for performing the comparative study and relationship between the two fracture criteria: double-K fracture model and double-G fracture model. The mathematical procedures for computing double-G fracture parameters can be seen in a flowchart as presented in Appendix (Fig. A.4).

4.7 Application of Weight Function Approach for Double-K Fracture Parameters

4.7.1 Comparison with TPBT Geometry

The experimental results of standard TPBT (Refai and Swartz 1987) which contain complete P-CMOD curves are taken into consideration. Because of the initial compliance, maximum load and corresponding CMOD from the tests are needed in the calculation of double-K fracture parameters; Xu and Reinhardt (1999b) used the above test results (Refai and Swartz 1987). In the present work, the same test results are therefore considered for the validation of weight function method to determine the values of K_{IC}^{ini} and K_{IC}^{un}. These parameters are also calculated using the existing analytical method (Xu and Reinhardt 1999b) for the comparison purpose. For computation of K_{IC}^{C} in analytical method, the numerical integration of the expression of K_{IC}^{C} is done using the Gauss–Chebyshev quadrature technique. The two series of test results (Refai and Swartz 1987), namely series B and series C, are used to determine the values of double-K fracture model. Concrete mix proportion (by mass) of both the beam series was 1:2.468:3.585 [cement (C):fine aggregate (FA):coarse aggregate (CA)] with water–cement ratio of 0.50 (by mass). The nominal size of coarse aggregate was 19 mm. The cylinder compressive strength and the modulus of elasticity for test series B were found to be 53.1 MPa and 38.4 GPa, respectively, whereas these values for test series C were 54.4 MPa and 39.3 GPa, respectively.

The dimensions of test series beams B and C were $76 \times 203 \times 762$ mm ($B \times D \times S$) and $76 \times 350 \times 1143$ mm ($B \times D \times S$), respectively. In the experiment (Refai and Swartz 1987), the initial pre-cracked crack lengths were measured directly using a dye-penetrant technique for 10 specimens in the test series B and 14 specimens in the test series C. Based on test results, the compliance calibration curves at the maximum load were derived using regression technique. These expressions are useful to calculate the lengths of initial pre-cracked cracks for other specimens. In the work (Xu and Reinhardt 1999b), the initial compliance C_i for each specimen from both the test series was measured carefully. From the measured values of C_i, the values of initial pre-cracked crack lengths were calculated using the linear asymptotic superposition assumption. Further, it was shown that the computed values of a_o/D using initial compliance (Eq. (4.24)) and the linear asymptotic superposition assumption were in close agreement with those obtained in Refai and Swartz (1987). Hence, in the present calculation the measured value of C_i and computed values of a_o/D are kept as the standard initial inputs. Then the value of effective crack length for each specimen is evaluated with Eq. (4.22) using trial-and-error method and the linear asymptotic superposition assumption at peak load and corresponding CMOD.

A complete program for the computations was developed using MATLAB to characterize the universal weight functions and evaluate the double-K fracture parameters for standard three-point bending test using existing analytical method and weight function method. The nonlinear softening relation, Eq. (3.36), needs the value of material constants c_1, c_2, and w_c as well as the tensile strength of concrete f_t should be known. In the present calculation of double-K fracture parameters, the same (Xu and Reinhardt 1999b) values of $c_1 = 3$, $c_2 = 7$, and $w_c = 160\,\mu m$ are considered. Tensile strength of concrete is computed using the relation $f_t = 0.4983\sqrt{f_c'}$ MPa (Karihaloo and Nallathambi 1991). The computed values of a_c, CTOD$_c$, K_{IC}^{C}, and the double-K fracture parameters K_{IC}^{ini} and K_{IC}^{un} for test series B and C are presented in Tables 4.3 and 4.4, respectively. The values of cohesive toughness for both the test series are determined employing the existing

Table 4.3 Comparison of double-*K* fracture parameters for test series B [dimensions of beam 76 × 203 × 762 mm ($B \times D \times S$), $f_c = 53.1$ MPa, $E = 38.4$ GPa, $H_O = 3.2$ mm]

Specimen No.	$C_i \times 10^{-6}$ (mm/N)	a_o/D	P_u (N)	$CMOD_c$ (μm)	$CTOD_c$ (μm)	a_c/D	K_{IC}^{un} (MPa m$^{1/2}$)	Using numerical integration method $[K_{IC}^{C}]_{NI}$ (MPa m$^{1/2}$)	$[K_{IC}^{ini}]_{NI}$ (MPa m$^{1/2}$)	Using four-term weight function method $[K_{IC}^{C}]_{WF4}$ (MPa m$^{1/2}$)	$[K_{IC}^{ini}]_{WF4}$ (MPa m$^{1/2}$)	Using five-term weight function method $[K_{IC}^{C}]_{WF5}$ (MPa m$^{1/2}$)	$[K_{IC}^{ini}]_{WF5}$ (MPa m$^{1/2}$)	Comparison with results of Xu and Reinhardt (1999b) K_{IC}^{un}/K_{IC}^{un*}	$[K_{IC}^{ini}]_{NI}/K_{IC}^{ini*}$
B1	6.289	0.383	5523	45.9	12.678	0.436	1.361	0.577	0.784	0.572	0.789	0.578	0.783	1.000	0.986
B3	8.575	0.442	4365	51.1	12.976	0.499	1.311	0.612	0.699	0.605	0.706	0.613	0.698	1.002	0.986
B4	4.489	0.319	5612	43.4	17.892	0.422	1.331	0.839	0.492	0.837	0.494	0.839	0.492	1.001	0.970
B5	16.498	0.558	3207	89.5	21.080	0.641	1.633	0.765	0.868	0.754	0.879	0.766	0.867	0.999	0.983
B7	29.189	0.648	2249	89.4	12.772	0.691	1.462	0.546	0.916	0.534	0.928	0.547	0.915	1.003	0.993
B8	29.942	0.636	2227	80	11.537	0.677	1.356	0.535	0.821	0.523	0.833	0.536	0.82	1.001	0.992
B9	44.903	0.706	1537	92.8	10.450	0.741	1.347	0.513	0.834	0.500	0.847	0.514	0.833	1.001	0.989
B10	30.534	0.654	2004	77.4	9.581	0.686	1.285	0.477	0.808	0.466	0.819	0.478	0.807	0.999	0.988
B11	82.623	0.775	980	116.6	10.351	0.809	1.424	0.529	0.895	0.515	0.909	0.530	0.894	1.001	0.992
B12	82.623	0.775	891	120.6	12.185	0.820	1.433	0.636	0.797	0.621	0.812	0.638	0.795	1.002	0.881
B13	126.63	0.815	579	130.2	10.931	0.858	1.432	0.667	0.765	0.653	0.779	0.670	0.762	1.012	1.011
B15	6.061	0.376	5033	43.1	13.216	0.441	1.263	0.654	0.609	0.649	0.614	0.654	0.609	0.999	0.920
B16	4.265	0.309	5790	44	18.894	0.419	1.357	0.865	0.492	0.863	0.494	0.865	0.492	0.999	0.965
B17	5.612	0.362	5166	53.3	20.972	0.476	1.438	0.889	0.549	0.886	0.552	0.890	0.548	1.001	0.975
B18	10.46	0.478	4053	65.5	17.541	0.554	1.467	0.716	0.751	0.708	0.759	0.717	0.75	0.999	0.980
B19	11.493	0.495	3919	58	11.822	0.539	1.348	0.537	0.811	0.529	0.819	0.538	0.81	0.999	0.987
B20	9.428	0.459	4187	61.4	17.303	0.538	1.430	0.728	0.702	0.721	0.709	0.729	0.701	1.000	0.980
B21	25.146	0.626	2450	86.8	13.973	0.675	1.473	0.587	0.886	0.574	0.899	0.588	0.885	1.002	0.991
B22	26.044	0.631	2450	107.4	20.262	0.703	1.687	0.728	0.959	0.714	0.973	0.729	0.958	1.001	0.988
B24	32.869	0.665	1982	86.6	11.235	0.703	1.381	0.519	0.862	0.507	0.874	0.521	0.86	1.004	0.997
B25	23.799	0.617	2784	94	15.710	0.668	1.608	0.588	1.02	0.576	1.032	0.589	1.019	1.000	0.989
B26	82.14	0.774	1559	329.9	43.150	0.853	3.216	0.871	2.345	0.856	2.36	0.873	2.343	0.999	0.992
B27	69.998	0.756	1203	130.5	13.866	0.799	1.585	0.593	0.992	0.579	1.006	0.595	0.99	0.992	0.989
B28	54.76	0.730	1403	141.5	19.510	0.794	1.749	0.740	1.009	0.725	1.024	0.742	1.007	1.002	0.990
B29	63.887	0.748	1069	171.1	25.300	0.834	1.890	0.968	0.922	0.952	0.938	0.969	0.921	1.003	0.989

Table 4.3 (continued)

Specimen No.	$C_i \times 10^{-6}$ (mm/N)	a_0/D	P_u (N)	$CMOD_c$ (μm)	$CTOD_c$ (μm)	a_c/D	K_{IC}^{un} (MPa m$^{1/2}$)	Using numerical integration method: $[K_{IC}^{C}]_{NI}$ (MPa m$^{1/2}$)	$[K_{IC}^{ini}]_{NI}$ (MPa m$^{1/2}$)	Using four-term weight function method: $[K_{IC}^{C}]_{WF4}$ (MPa m$^{1/2}$)	$[K_{IC}^{ini}]_{WF4}$ (MPa m$^{1/2}$)	Using five-term weight function method: $[K_{IC}^{C}]_{WF5}$ (MPa m$^{1/2}$)	$[K_{IC}^{ini}]_{WF5}$ (MPa m$^{1/2}$)	Comparison with results of Xu and Reinhardt (1999b): K_{IC}^{un}/K_{IC}^{un*}	$[K_{IC}^{ini}]_{NI}/K_{IC}^{ini*}$
B30	58.411	0.738	1292	145.5	19.899	0.805	1.756	0.771	0.985	0.756	1.000	0.773	0.983	1.005	0.995
B31	8.979	0.45	4855	82.8	27.498	0.564	1.805	0.878	0.927	0.871	0.934	0.879	0.926	0.999	0.981
B32	35.025	0.673	2138	101.5	13.168	0.713	1.559	0.525	1.034	0.513	1.046	0.527	1.032	1.000	0.990
B33	5.836	0.369	4832	64	27.481	0.521	1.551	1.067	0.484	1.063	0.488	1.067	0.484	1.001	0.964
B34	18.249	0.575	2494	90.3	22.459	0.678	1.516	0.911	0.605	0.899	0.617	0.912	0.604	1.001	0.976
B35	63.887	0.748	891	107.8	13.907	0.811	1.328	0.790	0.538	0.774	0.554	0.792	0.536	1.004	0.985
B36	8.081	0.431	4409	62.8	20.784	0.534	1.481	0.843	0.638	0.837	0.644	0.844	0.637	1.001	0.979
B37	9.203	0.455	4676	73.8	22.707	0.551	1.664	0.799	0.865	0.792	0.872	0.800	0.864	1.000	0.983
B38	26.044	0.631	2539	100	17.158	0.689	1.628	0.639	0.989	0.626	1.002	0.641	0.987	1.001	0.989
B39	19.758	0.588	2539	87.8	19.471	0.672	1.498	0.796	0.702	0.783	0.715	0.797	0.701	1.001	0.980
B40	11.134	0.49	3830	80.8	24.418	0.599	1.633	0.884	0.749	0.875	0.758	0.885	0.748	1.002	0.983
B41	47.459	0.713	1514	158.5	25.984	0.798	1.926	0.884	1.042	0.869	1.057	0.886	1.04	1.004	0.992
B42	43.808	0.703	1515	184.6	34.536	0.811	2.139	1.073	1.066	1.057	1.082	1.074	1.065	1.004	0.993
B43	242.77	0.863	579	182	7.195	0.879	1.815	0.359	1.456	0.350	1.465	0.360	1.455	1.012	1.010
B44	20.479	0.594	3073	95.5	18.464	0.657	1.680	0.654	1.026	0.642	1.038	0.655	1.025	1.002	0.991
B45	15.713	0.55	3563	80	16.421	0.608	1.580	0.618	0.962	0.608	0.972	0.619	0.961	1.001	0.988
Mean					17.969		1.581	0.711	0.870	0.701	0.880	0.713	0.869	1.001	0.983
Std. dev.					7.116		0.326	0.166	0.306	0.167	0.308	0.166	0.305	0.003	0.021
Cov					0.396		0.206	0.234	0.351	0.238	0.349	0.233	0.352	0.003	0.022

Refai and Swartz (1987)

Table 4.4 Comparison of double-*K* fracture parameters for test series C [dimensions of beam 76 × 305 × 1143 mm ($B \times D \times S$), $f_c = 54.4$ MPa, $E = 39.3$ GPa, $H = 3.2$ mm]

								Using numerical integration method		Using four-term weight function method		Using five-term weight function method		Comparison with results of Xu and Reinhardt (1999b)	
Specimen No.	$C_i \times 10^{-6}$ (mm/N)	a_0/D	P_u (N)	$CMOD_c$ (μm)	$CTOD_c$ (μm)	a_c/D	K_{IC}^{un} (MPa m$^{1/2}$)	$[K_{IC}^{C}]_{NI}$ (MPa m$^{1/2}$)	$[K_{IC}^{ini}]_{NI}$ (MPa m$^{1/2}$)	$[K_{IC}^{C}]_{WF4}$ (MPa m$^{1/2}$)	$[K_{IC}^{ini}]_{WF4}$ (MPa m$^{1/2}$)	$[K_{IC}^{C}]_{WF5}$ (MPa m$^{1/2}$)	$[K_{IC}^{ini}]_{WF5}$ (MPa m$^{1/2}$)	K_{IC}^{un}/K_{IC}^{un*}	$[K_{IC}^{ini}]_{NI}/K_{IC}^{ini*}$
C1	6.737	0.403	6547	100.4	40.273	0.552	1.939	1.253	0.686	1.247	0.692	1.254	0.685	1.002	0.972
C2	8.081	0.437	6057	106.8	39.136	0.575	1.955	1.203	0.752	1.194	0.761	1.203	0.752	1.002	0.973
C3	8.216	0.44	5879	96.0	33.428	0.562	1.815	1.115	0.700	1.106	0.709	1.116	0.699	1.001	0.971
C4	10.101	0.478	5612	98.8	29.150	0.575	1.815	0.968	0.847	0.958	0.857	0.969	0.846	1.001	0.979
C5	14.815	0.546	4543	100.8	22.040	0.612	1.714	0.795	0.919	0.782	0.932	0.796	0.918	1.002	0.986
C6	12.391	0.515	4543	108.5	31.150	0.623	1.793	1.067	0.726	1.054	0.739	1.068	0.725	1.001	0.976
C7	20.202	0.597	3385	107	21.117	0.664	1.627	0.830	0.797	0.815	0.812	0.832	0.795	1.001	0.984
C8	25.14	0.631	3207	120.4	20.394	0.688	1.730	0.757	0.973	0.741	0.989	0.758	0.972	1.002	0.991
C9	26.308	0.638	2450	124	25.011	0.725	1.652	1.025	0.627	1.007	0.645	1.026	0.626	1.004	0.981
C10	28.286	0.648	2494	130.9	24.977	0.729	1.714	0.973	0.741	0.956	0.758	0.975	0.739	1.002	0.981
C11	60.236	0.745	1514	185	25.198	0.814	1.976	0.964	1.012	0.945	1.031	0.966	1.01	1.004	0.990
C12	62.974	0.75	1269	142.3	17.461	0.807	1.614	0.871	0.743	0.852	0.762	0.873	0.741	1.005	0.991
C13	77.577	0.772	846	168.9	22.326	0.852	1.768	1.217	0.551	1.198	0.57	1.220	0.548	1.007	0.989
C14	131.42	0.82	868	198.2	15.809	0.861	1.974	0.752	1.222	0.735	1.239	0.755	1.219	1.002	0.992
C15	8.797	0.453	4899	99.4	35.983	0.598	1.742	1.283	0.459	1.273	0.469	1.284	0.458	1.002	0.954
C16	11.862	0.507	5077	120.3	35.922	0.622	1.979	1.090	0.889	1.078	0.901	1.091	0.888	1.001	0.979
C17	15.713	0.556	4276	98.4	20.442	0.617	1.656	0.772	0.884	0.759	0.897	0.773	0.883	1.002	0.987
C19	15.056	0.549	4498	101.6	22.074	0.615	1.718	0.796	0.922	0.783	0.935	0.797	0.921	1.002	0.987
C20	11.672	0.504	4676	96.4	26.952	0.600	1.681	0.988	0.693	0.976	0.705	0.989	0.692	1.001	0.973
C21	9.428	0.465	5879	82.8	22.136	0.537	1.660	0.816	0.844	0.807	0.853	0.817	0.843	0.999	0.979
C22	6.017	0.381	7660	68.4	21.925	0.455	1.653	0.819	0.834	0.814	0.839	0.820	0.833	1.000	0.982
C23	5.963	0.379	6013	64	23.891	0.487	1.446	1.044	0.402	1.039	0.407	1.045	0.401	1.000	0.953
C24	7.634	0.426	6124	64.4	17.143	0.485	1.461	0.743	0.718	0.736	0.725	0.744	0.717	0.999	0.988
C25	127.77	0.818	668	135	10.105	0.853	1.510	0.705	0.805	0.688	0.822	0.707	0.803	1.005	0.994
C26	13.468	0.529	4276	76	14.797	0.576	1.413	0.674	0.739	0.663	0.75	0.675	0.738	1.000	0.983

Table 4.4 (continued)

								Using numerical integration method		Using four-term weight function method		Using five-term weight function method		Comparison with results of Xu and Reinhardt (1999b)	
Specimen No.	$C_i \times 10^{-6}$ (mm/N)	a_o/D	P_u (N)	$CMOD_c$ (μm)	$CTOD_c$ (μm)	a_c/D	K_{IC}^{un} (MPa m$^{1/2}$)	$[K_{IC}^{C}]_{NI}$ (MPa m$^{1/2}$)	$[K_{IC}^{ini}]_{NI}$ (MPa m$^{1/2}$)	$[K_{IC}^{C}]_{WF4}$ (MPa m$^{1/2}$)	$[K_{IC}^{ini}]_{WF4}$ (MPa m$^{1/2}$)	$[K_{IC}^{C}]_{WF5}$ (MPa m$^{1/2}$)	$[K_{IC}^{ini}]_{WF5}$ (MPa m$^{1/2}$)	K_{IC}^{un}/K_{IC}^{un*}	$[K_{IC}^{ini}]_{NI}/K_{IC}^{ini*}$
C27	22.761	0.616	2895	87.2	13.268	0.657	1.371	0.652	0.719	0.638	0.733	0.654	0.717	1.003	0.989
C28	26.487	0.639	2628	97.6	14.960	0.686	1.438	0.707	0.731	0.691	0.747	0.708	0.73	1.004	0.989
C29	22.222	0.612	2984	86	12.817	0.651	1.369	0.629	0.740	0.615	0.754	0.630	0.739	1.002	0.988
C30	22.082	0.611	3118	92.5	14.635	0.655	1.450	0.667	0.783	0.653	0.797	0.669	0.781	1.001	0.986
C31	79.403	0.774	1203	119.1	8.4780	0.795	1.416	0.492	0.924	0.479	0.937	0.494	0.922	1.002	0.994
C32	5.163	0.352	7571	72.2	29.090	0.468	1.697	1.053	0.644	1.050	0.647	1.054	0.643	1.001	0.973
C33	15.238	0.551	3697	99.3	24.557	0.641	1.594	0.980	0.614	0.966	0.628	0.981	0.613	1.003	0.976
C34	45.634	0.712	1514	95.5	10.909	0.750	1.260	0.664	0.596	0.648	0.612	0.666	0.594	1.002	0.985
Mean					22.653		1.655	0.890	0.765	0.877	0.777	0.891	0.763	1.002	0.982
Std. dev.					8.354		0.195	0.203	0.160	0.204	0.161	0.203	0.160	0.002	0.010
cov					0.369		0.118	0.228	0.2091	0.233	0.207	0.228	0.209	0.002	0.010

Refai and Swartz (1987)

analytical method using Eq. (4.46), four-term weight function method using Eq. (4.59) and five-term weight function using Eq. (4.60). Using Eq. (4.42), the value of K_{IC}^{ini} is then computed in each case. In Tables 4.3 and 4.4, the cohesive toughness determined using numerical integration method, four-term weight function method, and five-term weight function method are denoted as $[K_{IC}^{C}]_{NI}$, $[K_{IC}^{C}]_{WF4}$, and $[K_{IC}^{C}]_{WF5}$, respectively, and the corresponding computed values of initiation toughness are denoted as $[K_{IC}^{ini}]_{NI}$, $[K_{IC}^{ini}]_{WF4}$, and $[K_{IC}^{ini}]_{WF5}$. The computed values of K_{IC}^{un} and $[K_{IC}^{ini}]_{NI}$ are compared with the corresponding values of fracture parameters presented by Xu and Reinhardt (1999b), which are denoted as K_{IC}^{un*} and K_{IC}^{ini*}, respectively, in Tables 4.3 and 4.4 for both the test series B and C. Comparison is made in terms of mean values of the ratios K_{IC}^{un}/K_{IC}^{un*} and $[K_{IC}^{ini}]_{NI}/K_{IC}^{ini*}$ for each test data in the above test series. From overall observation it is found that the present values of K_{IC}^{un} and K_{IC}^{ini} agree very closely with the respective values of K_{IC}^{un*} and K_{IC}^{ini*} evaluated by Xu and Reinhardt (1999b) well within approximately 2% of accuracy.

From Tables 4.3 and 4.4 it is observed that the average value of K_{IC}^{un} for the test series B and C is 1.581 and 1.655 MPa m$^{1/2}$, respectively. The standard deviation and the coefficient of variation of the K_{IC}^{un} for the test series B are 0.326 MPa m$^{1/2}$ and 0.206, respectively, whereas those for test series C are 0.195 MPa m$^{1/2}$ and 0.118, respectively. Hence, no appreciable difference in the mean values of K_{IC}^{un} between the test series B and C is observed. The mean values of $[K_{IC}^{C}]_{NI}$, $[K_{IC}^{C}]_{WF4}$, and $[K_{IC}^{C}]$ for test series B are 0.711, 0.701, and 0.713 MPa m$^{1/2}$, respectively, while those for test series C are 0.890, 0.877, and 0.891 MPa m$^{1/2}$, respectively. As compared to $[K_{IC}^{C}]_{NI}$, the predicted values of $[K_{IC}^{C}]_{WF4}$ for test series B and C are 1.41 and 1.46% less, respectively. However, almost same values of $[K_{IC}^{C}]_{NI}$ and $[K_{IC}^{C}]_{WF5}$ are obtained for both the test series. Similarly, it is also observed that with respect to $[K_{IC}^{ini}]_{NI}$ the computed values of $[K_{IC}^{ini}]_{WF4}$ for test series B and C are 1.15 and 1.57%, respectively, on higher side. However, no appreciable difference is observed between the computed values of $[K_{IC}^{ini}]_{NI}$ and $[K_{IC}^{ini}]_{WF5}$ for both the test series.

4.7.2 Comparison with CT and WST Geometries

Some experimental results on CT and WST specimens available in literature are considered to evaluate the double-K fracture parameters for the validation and comparison of *weight function method* with the analytical method. Test results on compact tension test reported by Brühwiler and Wittmann (1990) and on wedge splitting test presented by Zhao et al. (1991) and Zhu (1997) were used to determine the double-K fracture parameters in the work (Xu and Reinhardt 1999c). The above experimental results along with tests results reported by Kim et al. (1997) and Xu et al. (2006) are considered for the present comparative study. The value of E obtained from the cylinder tests provided by Brühwiler and Wittmann (1990), Zhao et al. (1991), and Kim et al. (1997) is taken in the respective calculation. On the other hand, measured values of E from P-COD curves as reported in Xu and Reinhardt (1999c) for the data series of Zhu (1997) are used in the present evaluation. The values of initial compliance reported by Xu et al. (2006) are used to calculate the values of E for this data series. Except for the test series reported in Kim et al. (1997), the relation $f_t = 0.4983\sqrt{f_c'}$ (Karihaloo and Nallathambi 1991) for determining the value of f_t and the material constants $c_1 = 3$, $c_2 = 7$, and $w_c = 160\,\mu\text{m}$ for normal concrete (Reinhardt et al. 1986) are taken in the calculation of K_{IC}^{C}. For test series reported in Kim et al. (1997), the value of COD at peak load is read carefully from the P-COD curves. The values of tensile strength and material constants c_1, c_2, and w_c for each specimen of different types of concrete are determined using the following expressions given by Xu (1999) and CEB-FIP Model code-1990 (1993):

$$\begin{aligned}
c_1 &= 9(d_{\max}/8)^{0.75} \\
c_2 &= (0.92 - d_{\max}/400)\lambda \\
w_c &= \alpha_F G_F/f_{tm} \\
\lambda &= 10 - [f_{ck}/(2f_{cko})]^{0.7} \\
\alpha_F &= \lambda - d_{\max}^{0.9}/8 \\
G_F &= [0.0204 + 0.0053 d_{\max}^{0.95}/8](f_{cm}/f_{cmo})^{0.7}
\end{aligned} \tag{4.77}$$

where f_{tm} is the mean tensile strength (MPa); f_{cm} is the mean compressive strength (MPa); $f_{cmo} = 10$ MPa; G_F is the fracture energy (N/mm); d_a is the maximum size of aggregates (mm); f_{ck} is the characteristic strength of concrete (MPa); $f_{cko} = 10$ MPa. According to CEB-FIP Model Code-1990 (1993), the relation $f_{cm} = f_{ck} + 8$ is taken and the mean tensile strength is calculated using the following expression:

$$f_{tm} = \alpha_{fct,m}(f_{ck}/f_{cko})^{2/3} \tag{4.78}$$

where $\alpha_{fct,m} = 1.40$ MPa. The material properties for the data series reported in Kim et al. (1997) obtained in the above manner is reported in Table 4.5.

Table 4.5 Material properties for WST specimens for $d_a = 10$ mm

Specimen No.	Cylinder compressive strength, f_c (MPa)	E (MPa)	f_{tm} using Eq. (4.78) (MPa)	c_1 using Eq. (4.77)	c_2 using Eq. (4.77)	w_c using Eq. (4.77) (μm)
L60	31.27	29940	2.46	1.18	7.95	188
L40	47.41	33720	3.50	1.18	7.51	166
L20	82.8	40570	5.34	1.18	6.70	140
G60	32.57	28790	2.55	1.18	7.92	185
G40	45.85	33400	3.40	1.18	7.55	167
G20	85.68	39570	5.49	1.18	6.64	138
R60	34.87	26510	2.71	1.18	7.85	181
R40	55.28	31580	3.94	1.18	7.32	159
R30	66.91	34350	4.57	1.18	7.04	150
R20	88.80	38140	5.64	1.18	6.57	137

Kim et al. (1997)

All the notations of the fracture parameters are taken as the same of the previous comparison with TPBT. The computed values of K_{IC}^{un} and $[K_{IC}^{ini}]_{NI}$ are compared with the corresponding values of fracture parameters presented in Xu and Reinhardt (1999c), which are denoted as K_{IC}^{un*} and K_{IC}^{ini*}, respectively, in Tables 4.6, 4.7, and 4.8 for the test series of Brühwiler and Wittmann (1990), Zhao et al. (1991), and Zhu (1997). From overall observation it is found that the present values of unstable fracture toughness K_{IC}^{un} agree with those of K_{IC}^{un*} evaluated by Xu and Reinhardt (1999c) well within approximately 7% of accuracy. However, excepting few cases, viz. specimen No. KZP-7, KZP-11, KZP-19, KZP-16, KZP-1, KZP-4, KZP-9, KPZ-17, and WP-6 of test series reported in Zhu (1997), the values of initiation toughness $[K_{IC}^{ini}]_{NI}$ obtained in the present calculation are in good agreement with those of K_{IC}^{ini*} within 15% of accuracy. It is observed that the value of cohesive toughness $[K_{IC}^{C}]_{NI}$ obtained in the present case is relatively on higher side than is that presented by Xu and Reinhardt (1999c). Experience shows that both the fracture characteristic values of double-K fracture parameters are sensitive to the values of numerical approximations in different material properties and geometrical factors such as the values of a_c/D ratio and the Green's function. Particularly a minor discrepancy in the value of a_c/D leads to relatively higher error in the values of the fracture parameters. Especially, the value of initiation toughness is dependent on both the computed values, i.e., unstable fracture toughness and cohesive toughness, which are indirectly related to many material and geometrical parameters. Therefore, the value of initiation toughness is relatively more influenced by those material and geometrical properties than is the value of unstable fracture toughness. The values of a_c/D ratio in the present case are evaluated applying linear elastic fracture mechanics equations used for the standard compact tension test, whereas those values are gained in Xu and Reinhardt (1999c) using standard equations or empirical equations depending upon whether the test specimen is standard or nonstandard. Apart from the other factors, the higher discrepancy between the calculated values of $[K_{IC}^{ini}]_{NI}$ and K_{IC}^{ini*} for the test series reported in Zhu (1997) is mainly attributed to different equations used for the respective evaluation of effective crack length/depth ratio, a_c/D.

From the evaluated results vide Tables 4.6, 4.7, 4.8, 4.9, and 4.10, it is observed that the values of $[K_{IC}^{C}]_{WF4}$ are somewhat less than those of $[K_{IC}^{C}]_{NI}$ and the values of $[K_{IC}^{C}]_{WF5}$ is almost the same as those of $[K_{IC}^{C}]_{NI}$ in all the test specimens. As compared to $[K_{IC}^{C}]_{NI}$, the maximum absolute difference in predicted values of $[K_{IC}^{C}]_{WF4}$ for test series of Brühwiler and Wittmann (1990) is 1.460% for specimen No. 2, Zhao et al. (1991) is 1.136% for specimen No. V4-WS, Zhu (1997) is 1.924% for specimen No. KZP-21, Kim et al. (1997) is 1.721% for specimen No. G20, and Xu et al. (2006) is 1.494% for specimen No. WS70-1. The corresponding values of $[K_{IC}^{ini}]_{WF4}$ are slightly on the higher side.

Xu and Reinhardt (1999c) reported that the value of K_{IC}^{ini} is independent of the specimen size but the value of K_{IC}^{un} appears to be slightly size dependent using the middle size compact tension test results of Brühwiler and Wittmann (1990). From the test results on wedge splitting test specimens of Zhao et al. (1991), it was shown that both the parameters K_{IC}^{ini} and K_{IC}^{un} are independent of specimen thickness, whereas they are dependent on the volume of the specimen. Xu and Reinhardt (1999c) also reported that there is almost no significant influence of the initial notch

Table 4.6 Comparison of double-K fracture parameters K_{IC}^{ini} and K_{IC}^{un} of CT specimens ($f_c = 42.92$ MPa, $E = 31.0$ GPa)

								Using numerical integration method		Using four-term weight function method		Using five-term weight function method		Comparison with results of Xu and Reinhardt (1999c)	
Specimen No.	Specimen dimension, $D \times 2H \times B$(m)	a_0/D	P_u (N)	COD_c (μm)	$CTOD_c$ (μm)	a_c/D	K_{IC}^{un} (MPa m$^{1/2}$)	$[K_{IC}^{C}]_{NI}$) (MPa m$^{1/2}$)	$[K_{IC}^{ini}]_{NI}$ (MPa m$^{1/2}$)	$[K_{IC}^{C}]_{WF4}$ (MPa m$^{1/2}$)	$[K_{IC}^{ini}]_{WF4}$ (MPa m$^{1/2}$)	$[K_{IC}^{C}]_{WF5}$ (MPa m$^{1/2}$)	$[K_{IC}^{ini}]_{WF5}$ (MPa m$^{1/2}$)	K_{IC}^{un}/K_{IC}^{un*}	$[K_{IC}^{ini}]_{NI}/K_{IC}^{ini*}$
1	1.2×1.44×0.12	0.5	22150	356.1	97.871	0.590	2.215	1.432	0.783	1.413	0.803	1.434	0.781	1.0	0.904
2	1.2×1.44×0.12	0.5	21250	309.1	77.487	0.572	1.989	1.267	0.722	1.249	0.741	1.269	0.720	1.0	0.907
3	1.2×1.44×0.12	0.5	21000	378.2	112.484	0.610	2.266	1.622	0.645	1.601	0.666	1.624	0.642	1.122	0.853
4	1.2×1.44×0.12	0.5	20550	372.7	111.317	0.611	2.228	1.637	0.591	1.616	0.612	1.639	0.589	1.0	0.868
5	1.2×1.44×0.12	0.5	20320	310.9	81.914	0.582	1.967	1.362	0.605	1.343	0.624	1.364	0.603	1.0	0.886
6	1.2×1.44×0.12	0.5	19000	291.5	76.996	0.582	1.842	1.376	0.466	1.357	0.486	1.378	0.464	1.0	0.855
7	0.6×0.72×0.12	0.5	13000	216.7	61.221	0.597	1.884	1.114	0.769	1.100	0.784	1.116	0.768	1.0	0.913
8	0.6×0.72×0.12	0.5	12900	227.3	66.688	0.606	1.940	1.184	0.755	1.170	0.770	1.186	0.754	1.0	0.909
9	0.6×0.72×0.12	0.5	12700	201.5	54.792	0.588	1.781	1.054	0.727	1.040	0.741	1.055	0.725	1.0	0.913
10	0.6×0.72×0.12	0.5	12250	225.8	68.088	0.614	1.899	1.246	0.652	1.232	0.667	1.248	0.651	1.0	0.909
11	0.6×0.72×0.12	0.5	12150	219.7	65.502	0.610	1.859	1.224	0.636	1.209	0.650	1.225	0.634	1.0	0.892
12	0.6×0.72×0.12	0.5	12150	197.0	54.508	0.592	1.728	1.086	0.642	1.072	0.656	1.088	0.641	1.0	0.900
13	0.3×0.36×0.12	0.5	7780	141.7	42.430	0.612	1.692	0.914	0.778	0.904	0.788	0.915	0.777	1.0	0.934
14	0.3×0.36×0.12	0.5	7600	122.7	33.835	0.591	1.524	0.807	0.717	0.797	0.727	0.808	0.716	1.0	0.935
15	0.3×0.36×0.12	0.5	7260	150.0	47.969	0.632	1.720	1.037	0.683	1.026	0.694	1.038	0.6820	1.0	0.920
16	0.3×0.36×0.12	0.5	7080	125.8	37.104	0.607	1.514	0.903	0.612	0.893	0.622	0.904	0.611	1.0	0.919
17	0.3×0.36×0.12	0.5	7080	116.7	32.698	0.595	1.440	0.834	0.606	0.824	0.616	0.835	0.605	1.0	0.924
18	0.3×0.36×0.12	0.5	6950	137.5	43.255	0.626	1.602	1.005	0.596	0.995	0.607	1.006	0.595	1.0	0.910
													Mean	1.007	0.903

Brühwiler and Wittmann (1990)

Table 4.7 Comparison of double-*K* fracture parameters K_{IC}^{ini} and K_{IC}^{un} of WST specimens ($f_c = 30.47$ MPa, $E = 31.6$ GPa)

Specimen No.	Specimen dimension, $D \times 2H \times B$(m)	a_o/D	P_u (N)	COD_c (μm)	$CTOD_c$ (μm)	a_c/D	K_{IC}^{un} (MPa m$^{1/2}$)	Using numerical integration method $[K_{IC}^{C}]_{NI}$ (MPa m$^{1/2}$)	$[K_{IC}^{ini}]_{NI}$ (MPa m$^{1/2}$)	Using four-term weight function method $[K_{IC}^{C}]_{WF4}$ (MPa m$^{1/2}$)	$[K_{IC}^{ini}]_{WF4}$ (MPa m$^{1/2}$)	Using five-term weight function method $[K_{IC}^{C}]_{WF5}$ (MPa m$^{1/2}$)	$[K_{IC}^{ini}]_{WF5}$ (MPa m$^{1/2}$)	Comparison with results of Xu and Reinhardt (1999c) K_{IC}^{un}/K_{IC}^{un*}	$[K_{IC}^{ini}]_{NI}/K_{IC}^{ini*}$
T1-WS	0.2×0.2×0.15	0.5	4685	131.9	49.564	0.707	1.560	1.152	0.408	1.143	0.417	1.152	0.4079	1.073	0.895
T2-WS	0.2×0.2×0.2	0.5	6003	121.9	45.423	0.702	1.459	1.1287	0.331	1.120	0.339	1.129	0.3300	1.073	0.855
T3-WS	0.2×0.2×0.3	0.5	10519	121.9	43.726	0.681	1.528	0.9897	0.5391	0.981	0.547	0.990	0.5384	1.068	0.911
T4-WS	0.2×0.2×0.45	0.5	11889	129.6	50.121	0.725	1.467	1.3037	0.163	1.295	0.172	1.304	0.1628	1.076	0.755
													Mean	1.073	0.854
V1-WS	0.15×0.15×0.15	0.5	3488	90.9	33.581	0.697	1.271	0.9947	0.277	0.987	0.284	0.994	0.2761	1.072	0.839
V2-WS	0.2×0.2×0.2	0.5	6003	121.9	45.423	0.702	1.459	1.1287	0.331	1.120	0.339	1.129	0.3300	1.073	0.855
V3-WS	0.3×0.3×0.3	0.5	16008	162.4	55.975	0.661	1.731	1.037	0.693	1.029	0.702	1.039	0.6919	1.064	0.909
V4-WS	0.45×0.45×0.45	0.5	36137	185.7	56.987	0.619	1.762	0.960	0.802	0.949	0.813	0.961	0.8012	1.054	0.878
													Mean	1.066	0.870

Zhao et al. (1991)

Table 4.8 Comparison of double-K fracture parameters K_{IC}^{ini} and K_{IC}^{un} of WST specimens ($f_c = 41.80$ MPa)

Specimen No.	Specimen dimension, $D \times 2H \times B$(m)	E (MPa)	a_c/D	P_u (N)	COD_c (μm)	$CTOD_c$ (μm)	a_c/D	K_{IC}^{un}(MPa m$^{1/2}$)	Using numerical integration method: $[K_{IC}^{C}]_{NI}$ (MPam$^{1/2}$)	Using numerical integration method: $[K_{IC}^{ini}]_{NI}$ (MPa m$^{1/2}$)	Using four-term weight function method: $[K_{IC}^{C}]_{WF4}$ (MPa m$^{1/2}$)	Using four-term weight function method: $[K_{IC}^{ini}]_{WF4}$ (MPa m$^{1/2}$)	Using five-term weight function method: $[K_{IC}^{C}]_{WF5}$ (MPa m$^{1/2}$)	Using five-term weight function method: $[K_{IC}^{ini}]_{WF5}$ (MPa m$^{1/2}$)	Comparison with results of Xu and Reinhardt (1999c): K_{IC}^{un}/K_{IC}^{un*}	Comparison with results of Xu and Reinhardt (1999c): $[K_{IC}^{ini}]_{NI}/K_{IC}^{ini*}$
KZP-7	0.2×0.2×0.1	35378	0.3	6500	70.0	37.451	0.506	1.430	1.136	0.294	1.135	0.295	1.136	0.294	1.030	0.624
KZP-11	0.2×0.2×0.1	35268	0.3	6300	62.0	32.271	0.487	1.307	1.063	0.245	1.063	0.246	1.063	0.245	1.025	0.527
KZP-19	0.2×0.2×0.1	35665	0.3	6250	75.0	41.307	0.529	1.478	1.244	0.234	1.242	0.236	1.244	0.234	1.034	0.571
														Mean	1.030	0.589
KZP-16	0.2×0.2×0.1	30258	0.35	6500	110.0	54.433	0.563	1.725	1.153	0.572	1.150	0.575	1.154	0.571	1.042	0.805
KZP-1	0.2×0.2×0.1	35749	0.4	5750	73.7	29.206	0.542	1.421	0.893	0.527	0.889	0.531	0.894	0.527	1.038	0.788
KZP-4	0.2×0.2×0.1	32801	0.4	5200	91.5	39.646	0.584	1.493	1.081	0.412	1.076	0.417	1.081	0.412	1.047	0.756
KZP-9	0.2×0.2×0.1	32070	0.4	5700	83.0	33.184	0.545	1.425	0.894	0.531	0.890	0.536	0.894	0.531	1.038	0.794
KZP-17	0.2×0.2×0.1	31932	0.4	5400	113.0	51.016	0.609	1.710	1.195	0.515	1.189	0.520	1.196	0.514	1.053	0.814
WP-6	0.2×0.2×0.1	32523	0.4	5150	90.0	38.806	0.581	1.464	1.068	0.395	1.063	0.400	1.069	0.395	1.046	0.745
														Mean	1.044	0.784
KZP-6	0.2×0.2×0.1	35135	0.5	4500	90.0	27.515	0.617	1.474	0.819	0.655	0.811	0.663	0.820	0.654	1.054	0.876
KZP-10	0.2×0.2×0.1	27333	0.5	4540	109.5	32.244	0.607	1.425	0.740	0.685	0.732	0.693	0.741	0.684	1.052	0.879
KZP-20	0.2×0.2×0.1	36065	0.5	4680	105.0	34.345	0.640	1.688	0.923	0.764	0.915	0.773	0.924	0.764	1.060	0.911
														Mean	1.055	0.889
KZP-12	0.2×0.2×0.1	34263	0.6	3090	93.8	19.602	0.677	1.326	0.666	0.660	0.654	0.671	0.667	0.659	1.067	0.909
KZP-9	0.2×0.2×0.1	34342	0.6	2800	118.0	29.716	0.721	1.515	0.913	0.602	0.901	0.614	0.914	0.601	1.076	0.956
KZP-21	0.2×0.2×0.1	35069	0.6	3330	85.0	15.348	0.655	1.287	0.546	0.742	0.535	0.752	0.547	0.741	1.063	0.898
														Mean	1.069	0.921

Zhu (1997)

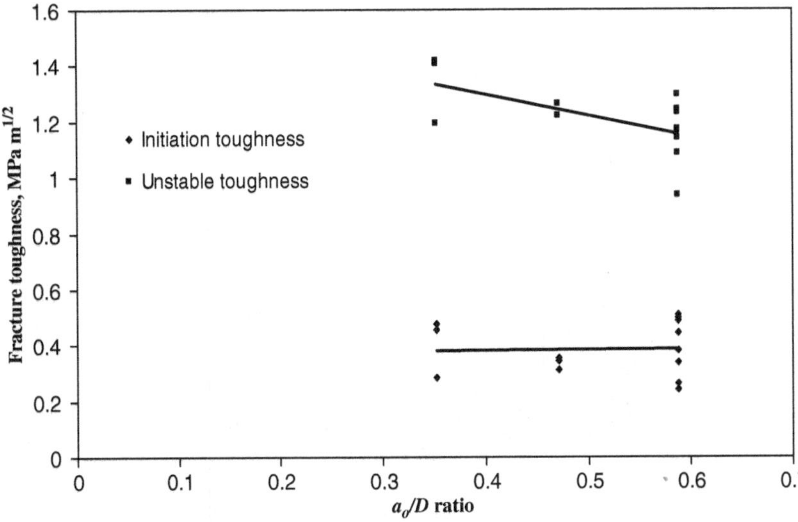

Fig. 4.18 Variation of double-*K* fracture parameters with a_0/D ratio for test series (Xu et al. 2006)

length/depth (a_0/D) ratio on the calculated values of fracture parameters K_{IC}^{ini} and K_{IC}^{un} using test results of Zhu (1997).

Since the double-*K* fracture parameters obtained using five-term weight function are in very close agreement with those determined using analytical method, the results obtained using the former method are considered for further analysis. The effect of initial notch length to depth (a_0/D) ratio on the double-*K* fracture parameters K_{IC}^{ini} and K_{IC}^{un} for the test series of Xu et al. (2006) vide Table 4.10 is shown in Fig. 4.18. It can be noticed from the figure that there is no significant influence of a_0/D ratio on the values of K_{IC}^{ini} and K_{IC}^{un}. This effect is similar to the observation made by Xu and Reinhardt (1999c) using the test results of data series of Zhu (1997).

Calculated values of double-*K* fracture parameters from test series of Kim et al. (1997) (vide Table 4.9) using five-term weight function are plotted with respect to their values of tensile strength as shown in Fig. 4.19. It is seen from the figure that the values of K_{IC}^{un} will increase gradually for corresponding increase in tensile strength of concrete while ignoring the factors such as types of coarse aggregates used in the concrete mixes. However, no significant difference in the values of K_{IC}^{ini} up to normal strength of concrete is observed.

For high-strength concrete (specimen Nos. L20, G20, and R20), the values of K_{IC}^{ini} will tend to approach the values of K_{IC}^{un}. It is a known fact that most aggregates are stronger than the cement-based matrix in normal strength concrete and consequently the fracture process normally takes place around the aggregates. On the other hand, a crack will run through the aggregates in high-strength concrete and therefore the mechanical interactions between aggregate particles and cement-based

Table 4.9 Comparison of double-*K* fracture parameters K_{IC}^{ini} and K_{IC}^{un} of WST specimens

Specimen No.	Specimen dimension, $D \times 2H \times B$ (m)	a_o/D	P_u (kN)	COD_c (μm)	$CTOD_c$ (μm)	a_c/D	K_{IC}^{un} (MPa m$^{1/2}$)	Using numerical integration method		Using four-term weight function method		Using five-term weight function method	
								$[K_{IC}^{C}]_{NI}$ (MPa m$^{1/2}$)	$[K_{IC}^{ini}]_{NI}$ (MPa m$^{1/2}$)	$[K_{IC}^{C}]_{WF4}$ (MPa m$^{1/2}$)	$[K_{IC}^{ini}]_{WF4}$ (MPa m$^{1/2}$)	$[K_{IC}^{C}]_{WF5}$ (MPa m$^{1/2}$)	$[K_{IC}^{ini}]_{WF5}$ (MPa m$^{1/2}$)
L60	0.2×0.24×0.12	0.5	3.85	65.222	18.319	0.594	0.957	0.560	0.397	0.554	0.403	0.560	0.397
L40	0.2×0.24×0.12	0.5	4.14	65.672	19.005	0.602	1.063	0.834	0.229	0.826	0.238	0.835	0.228
L20	0.2×0.24×0.12	0.5	7.84	61.118	4.367	0.505	1.433	0.262	1.171	0.259	1.175	0.263	1.170
G60	0.2×0.24×0.12	0.5	3.64	62.060	16.847	0.588	0.883	0.557	0.326	0.551	0.332	0.558	0.325
G40	0.2×0.24×0.12	0.5	3.87	64.561	19.178	0.609	1.022	0.853	0.169	0.844	0.177	0.853	0.168
G20	0.2×0.24×0.12	0.5	7.40	64.725	9.598	0.523	1.431	0.558	0.873	0.548	0.883	0.559	0.872
R60	0.2×0.24×0.12	0.5	3.39	68.188	19.693	0.602	0.869	0.644	0.224	0.637	0.231	0.645	0.224
R40	0.2×0.24×0.12	0.5	3.91	65.184	18.661	0.600	0.994	0.920	0.073	0.910	0.083	0.921	0.073
R30	0.2×0.24×0.12	0.5	5.23	61.496	13.365	0.552	1.116	0.722	0.394	0.711	0.405	0.723	0.393
R20	0.2×0.24×0.12	0.5	8.24	67.243	2.944	0.502	1.492	0.171	1.321	0.162	1.323	0.171	1.321

Kim et al. (1997)

Table 4.10 Comparison of double-*K* fracture parameters K_{IC}^{ini} and K_{IC}^{un} of WST specimens ($f_c = 0.8 f_{cu} = 38.368$ MPa)

Specimen No.	Specimen dimension, $D \times 2H \times B$ (m)	C_i(m/N)$\times 10^{-6}$	a_0/D	P_u (N)	COD_c (μm)	$CTOD_c$ (μm)	a_c/D	K_{IC}^{un}(MPa m$^{1/2}$)	Using numerical integration method		Using four-term weight function method		Using five-term weight function method	
									$[K_{IC}^{C}]_{NI}$ (MPa m$^{1/2}$)	$[K_{IC}^{ini}]_{NI}$ (MPa m$^{1/2}$)	$[K_{IC}^{C}]_{WF4}$ (MPa m$^{1/2}$)	$[K_{IC}^{ini}]_{WF4}$ (MPa m$^{1/2}$)	$[K_{IC}^{C}]_{WF5}$ (MPa m$^{1/2}$)	$[K_{IC}^{ini}]_{WF5}$ (MPa m$^{1/2}$)
WS110-1	0.17×0.2×0.2	2.551	0.353	10711	66.3	31.202	0.536	1.409	0.9553	0.4540	0.9530	0.4562	0.955	0.454
WS110-2	0.17×0.2×0.2	2.580	0.353	9563	55.2	25.247	0.521	1.195	0.912	0.283	0.911	0.285	0.913	0.283
WS110-3	0.17×0.2×0.2	2.281	0.353	11017	58.8	27.332	0.529	1.416	0.942	0.474	0.940	0.476	0.942	0.474
WS90-1	0.17×0.2×0.2	4.868	0.471	6656	74.4	26.664	0.625	1.219	0.909	0.310	0.903	0.316	0.910	0.309
WS90-2	0.17×0.2×0.2	5.228	0.471	6771	79.5	28.248	0.621	1.221	0.880	0.342	0.873	0.348	0.880	0.341
WS90-3	0.17×0.2×0.2	4.433	0.471	6962	69.9	24.920	0.623	1.263	0.908	0.356	0.901	0.362	0.908	0.355
WS70-1	0.17×0.2×0.2	8.165	0.588	4093	57.6	12.941	0.674	0.939	0.678	0.261	0.667	0.271	0.678	0.261
WS70-2	0.17×0.2×0.2	8.187	0.588	4361	74.1	18.537	0.700	1.139	0.801	0.338	0.791	0.349	0.802	0.338
WS70-3	0.17×0.2×0.2	8.293	0.588	4935	80.4	19.563	0.693	1.240	0.745	0.495	0.735	0.505	0.746	0.494
WS70-4	0.17×0.2×0.2	8.191	0.588	3978	80.4	21.611	0.722	1.174	0.935	0.239	0.924	0.250	0.936	0.238
WS70-5	0.17×0.2×0.2	7.980	0.588	4476	79.8	20.635	0.709	1.231	0.851	0.380	0.841	0.391	0.852	0.380
WS70-6	0.17×0.2×0.2	8.332	0.588	4935	86.1	21.630	0.701	1.297	0.787	0.510	0.777	0.521	0.787	0.510
WS70-7	0.17×0.2×0.2	8.228	0.588	4973	72.6	16.634	0.679	1.164	0.677	0.487	0.667	0.497	0.677	0.487
WS70-8	0.17×0.2×0.2	8.313	0.588	4820	67.5	14.917	0.671	1.088	0.643	0.445	0.633	0.455	0.644	0.445

Xu et al. (2006)

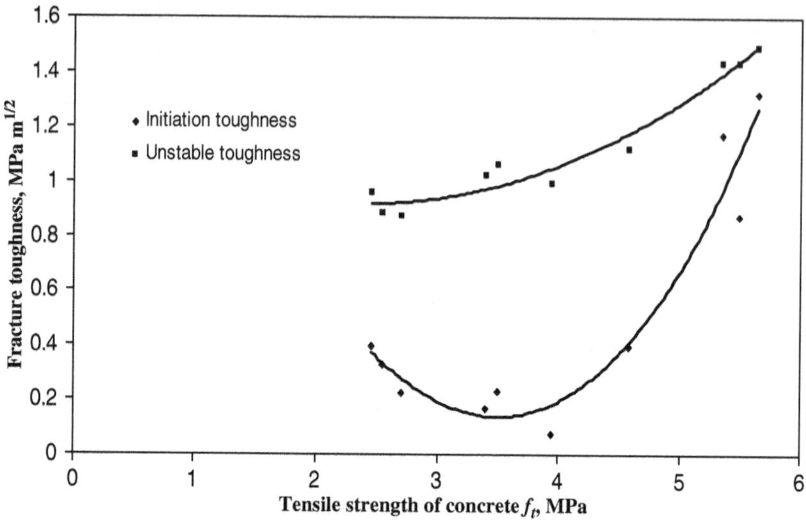

Fig. 4.19 Variation of double-*K* fracture parameters with concrete strength for test series of Kim et al. (1997)

matrix cannot be activated. This results in the failure of high-strength concrete in a more brittle manner. It is difficult to draw some conclusions based on very few test results. However, if the brittleness is caused due to high-strength nature of concrete, then it seems to be true that K_{IC}^{ini} will approach K_{IC}^{un} because the effective crack extension a_c will be approaching the initial notch length a_0. This leads to lower value of K_{IC}^{C} and higher value of K_{IC}^{ini}. The nondimensional brittleness parameter β_B of concrete may then be defined using double-*K* fracture parameters as

$$\beta_B = \frac{K_{IC}^{ini}}{K_{IC}^{un}} = 1 - \frac{K_{IC}^{C}}{K_{IC}^{un}} \tag{4.79}$$

where the nondimensional *brittleness* parameter β_B will vary between 0 and 1. The higher is the value of β_B, the more is the brittleness. However, sufficient experimental results are further needed to verify Eq. (4.79).

4.8 Size-Effect Prediction from the Double-*K* Fracture Model for TPBT

It is clear that a fundamental difference between the fictitious crack model and the double-*K* fracture model is perceptible. In the development of fictitious crack model or cohesive crack model for concrete fracture, no singularity is considered at the crack tip and the crack propagation criterion is based on the cohesive stress in the vicinity of the crack tip. On the contrary, in the double-*K* fracture model, the value of K_{IC}^{un} is equal to the summation of K_{IC}^{ini} and K_{IC}^{C} and hence the cohesive stress does not necessarily abolish the stress singularity condition and the crack propagation is not governed by stress criteria, unlike the concept of fictitious crack model.

From the observations based on the experimental studies (Xu and Reinhardt 1999a, b), it was reported that the double-K fracture parameters were almost independent of the specimen size. Though the size-effect study of double-K fracture parameters based on experimental results is available in the literature, no such numerical study has been attempted in the past. Furthermore, it is pointed that the principles of the development of fictitious crack model and double-K fracture model are contrary to each other. In view of the foregoing observations, a relationship between fictitious crack model and double-K fracture model has been attempted to establish in the present investigation for the TPBT in the manner similar to the previous studies (Planas and Elices 1990, Karihaloo and Nallathambi 1990).

4.8.1 Size-Effect Law for Unstable Fracture Toughness K_{IC}^{un}

A similar size-effect law, employed for effective crack model (Karihaloo and Nallathambi 1990), is used for the size-effect prediction of $K_{\mathrm{IC}}^{\mathrm{un}}$ in double-K fracture model. For instance, the following expression can be written for the TPBT using Eq. (3.51):

$$\left(\frac{K_{\mathrm{IC}}^{\mathrm{un}}}{K_{\mathrm{INu}}}\right)^2 = \left(\frac{\alpha_c}{\alpha_o}\right)\left[\frac{k(\alpha_c)}{k(\alpha_o)}\right]^2 \tag{4.80}$$

in which $k(\alpha_o)$ or $k(\alpha_c)$ can be obtained using Eq. (3.52). The right-hand side of Eq. (4.80) is expanded using Taylor's series in the power of $\Delta\alpha_c = (a_c - a_o)/D$ and only up to fourth-order terms are retained in the series such that $K_{\mathrm{IC}}^{\mathrm{un}}$ and K_{INu} will yield the same result for infinitely large structures ($D \to \infty$):

$$\left(\frac{K_{\mathrm{IC}}^{\mathrm{un}}}{K_{\mathrm{INu}}}\right)^2 = \left[1 + \lambda_1\Delta\alpha_c + \lambda_2\Delta\alpha_c^2 + \lambda_3\Delta\alpha_c^3 + \lambda_4\Delta\alpha_c^4\right] \tag{4.81}$$

where λ_i are the dimensionless coefficients depending upon α_o such that

$$\lambda_1 = \frac{2k_o'}{k_o} + \frac{1}{\alpha_o} = 2D_1 + \frac{1}{\alpha_o}$$

$$\lambda_2 = \frac{k_o''}{k_o} + \left(\frac{k_o'}{k_o}\right)^2 + \frac{2}{\alpha_o}\frac{k_o'}{k_o} = 2D_2 + D_1^2 + \frac{2D_1}{\alpha_o}$$

$$\lambda_3 = \frac{k_o'''}{3k_o} + \frac{k_o'k_o''}{(k_o)^2} + \frac{1}{\alpha_o}\left[\left(\frac{k_o'}{k_o}\right)^2 + \frac{k_o''}{k_o}\right] = 2\left(D_3 + D_1D_2\right) + \frac{D_1^2 + 2D_2}{\alpha_o}$$

$$\lambda_4 = \frac{k_o^{iv}}{12k_o} + \frac{k_o^{'}k_o^{'''}}{3(k_o)^2} + \frac{1}{4}\left(\frac{k_o^{''}}{k_o}\right)^2 + \frac{1}{\alpha_o}\left[\frac{k_o^{'}k_o^{'''}}{(k_o)^2} + \frac{k_o^{'''}}{3k_o}\right]$$
$$= 2D_4 + 2D_1D_3 + D_2^2 + \frac{2\,(D_3 + D_1D_2)}{\alpha_o} \tag{4.82}$$

in which

$$k_o = k(\alpha_o), k_o^{'} = k^{'}(\alpha_o), k_o^{''} = k^{''}(\alpha_o), k_o^{'''} = k^{'''}(\alpha_o), k_o^{iv} = k^{iv}(\alpha_o);$$

where superscripts denote the nth derivative of $k(\alpha)$ with respect to α and

$$\begin{aligned} D_1 &= \frac{B_1 + B_oA_1}{B_o} \\ D_2 &= \frac{B_2 + B_1A_1 + B_oA_2}{B_o} \\ D_3 &= \frac{B_3 + B_2A_1 + B_1A_2 + B_oA_3}{B_o} \\ D_4 &= \frac{B_4 + B_3A_1 + B_2A_2 + B_1A_3 + B_oA_4}{B_o} \end{aligned} \tag{4.83}$$

and

$$\begin{aligned} B_o &= 1.99 - \alpha_o(1-\alpha_o)(2.15 - 3.93\alpha_o + 2.7\alpha_o^2) \\ B_1 &= \alpha_o(1-\alpha_o)(3.93 - 5.4\alpha_o) \\ &\quad - (1-2\alpha_o)(2.15 - 3.93\alpha_o + 2.7\alpha_o^2) \\ B_2 &= -2.7\alpha_o(1-\alpha_o) + (1-2\alpha_o)(3.93 - 5.4\alpha_o) \\ &\quad + (2.15 - 3.93\alpha_o + 2.7\alpha_o^2) \\ B_3 &= -2.7(1-2\alpha_o) - (3.93 - 5.4\alpha_o) \\ B_4 &= 2.7 \end{aligned} \tag{4.84a}$$

$$\begin{aligned} A_1 &= -\frac{2}{(1+2\alpha_o)} + \frac{3}{2(1-\alpha_o)} \\ A_2 &= \frac{4}{(1+2\alpha_o)^2} - \frac{3}{(1+2\alpha_o)(1-\alpha_o)} + \frac{15}{8(1-\alpha_o)^2} \\ A_3 &= -\frac{8}{(1+2\alpha_o)^3} + \frac{6}{(1+2\alpha_o)^2(1-\alpha_o)} - \frac{15}{4(1+2\alpha_o)(1-\alpha_o)^2} \\ &\quad + \frac{35}{16(1-\alpha_o)^3} \\ A_4 &= \frac{16}{(1+2\alpha_o)^4} - \frac{12}{(1+2\alpha_o)^3(1-\alpha_o)} + \frac{15}{2(1+2\alpha_o)^2(1-\alpha_o)^2} \\ &\quad - \frac{35}{8(1+2\alpha_o)(1-\alpha_o)^3} + \frac{315}{128(1-\alpha_o)^4} \end{aligned} \tag{4.84b}$$

The for B_2, B_3, and B_4 appear to be slightly different from those presented elsewhere (Karihaloo and Nallathambi 1990). Assuming the equivalent fracture energy termed as unstable fracture energy G_F^{un} and a material constant corresponding to K_{IC}^{un} which can be expressed as

$$G_F^{un} = \frac{(K_{IC}^{un})^2}{E} \tag{4.85}$$

Equation (4.81) can be recast in the following form using Eq. (4.85) and taking $G_F = F_{FC}$:

$$\frac{EG_{FC}}{K_{INu}^2} = \frac{G_{FC}}{G_F^{un}}\left[1 + \gamma_1\left(\frac{l_{ch}}{D}\right) + \gamma_2\left(\frac{l_{ch}}{D}\right)^2 + \gamma_3\left(\frac{l_{ch}}{D}\right)^3 + \gamma_4\left(\frac{l_{ch}}{D}\right)^4\right] \tag{4.86}$$

where

$$\gamma_1 = \lambda_1\frac{\Delta a_c}{l_{ch}},\ \gamma_2 = \lambda_2\left(\frac{\Delta a_c}{l_{ch}}\right)^2,\ \gamma_3 = \lambda_3\left(\frac{\Delta a_c}{l_{ch}}\right)^3,\ \gamma_4 = \lambda_4\left(\frac{\Delta a_c}{l_{ch}}\right)^4 \tag{4.87}$$

Equation (4.86) gives the required size-effect curve for K_{IC}^{un} which can be plotted on the *X*–*Y* coordinate similar to the size-effect equation for the FCM.

4.8.2 Size-Effect Law for Initiation Fracture Toughness K_{IC}^{ini}

It is clear that both the values K_{IC}^{ini} and K_{INu} depend on the relative size of the initial notch length α_o and therefore, it may be possible to express that

$$\left(\frac{K_{IC}^{ini}}{K_{INu}}\right)^2 = \frac{F(P_{ini}, \alpha_o)}{F(P_u, \alpha_o)} = \left(\frac{P_{ini}}{P_u}\right)^2 \tag{4.88}$$

Alternatively, Eq. (4.88) may also be expressed in the following form using Eq. (4.42):

$$\left(\frac{K_{IC}^{ini}}{K_{INu}}\right)^2 = \left(\frac{K_{IC}^{un} - K_{IC}^{C}}{K_{INu}}\right)^2 \tag{4.89}$$

The right-hand side (RHS) of Eq. (4.88) becomes indeterminate since the value of P_{ini} is an unknown quantity before predicting the size-effect law for K_{IC}^{ini}. Also the RHS terms of Eq. (4.89) containing K_{IC}^{C} is a function of cohesive stress distribution $\sigma(w), f_t$, and Δa_c, therefore, it is difficult to arrive at a closed-form size-effect law. Assuming that a definite relationship exists between K_{IC}^{ini} and K_{IC}^{un} with regard to the size of the structure, an empirical relationship similar to Eq. (4.86) may be arrived at by considering an equivalent fracture energy at crack initiation G_F^{ini} such that

$$G_{\mathrm{F}}^{\mathrm{ini}} = \frac{\left(K_{\mathrm{IC}}^{\mathrm{ini}}\right)^2}{E} \tag{4.90}$$

and

$$\frac{EG_{\mathrm{FC}}}{K_{\mathrm{INu}}^2} = \frac{G_{\mathrm{FC}}}{G_{\mathrm{F}}^{\mathrm{ini}}}\left[\kappa_{\mathrm{o}} + \kappa_1\left(\frac{l_{\mathrm{ch}}}{D}\right) + \kappa_2\left(\frac{l_{\mathrm{ch}}}{D}\right)^2 + \kappa_3\left(\frac{l_{\mathrm{ch}}}{D}\right)^3 + \kappa_4\left(\frac{l_{\mathrm{ch}}}{D}\right)^4\right] \tag{4.91}$$

in which κ_i are dimensionless coefficients depending on the geometry and material properties. For infinitely large-size structures ($D \to \infty$), left-hand side (LHS) of Eq. (4.91) tends to be unity and $G_{\mathrm{F}}^{\mathrm{ini}}/G_{\mathrm{FC}} \to \kappa_{\mathrm{o}}$, which is a constant quantity and does not depend on geometry or size of the structure. Equation (4.91) can be plotted on X–Y coordinates similar to previous size-effect laws for the unstable fracture toughness.

4.8.3 Prediction of Size Effects from FCM and DKFM

For computing the mathematical coefficients, the fracture peak load and the corresponding CMOD are determined using fictitious crack model. The concrete mix for which the material properties are $f_t = 3.21$ MPa, $E = 30$ GPa, and $G_{\mathrm{FC}} = 103$ N/m (Planas and Elices 1990) is taken into account for the present investigation. The value of ν is assumed to be 0.18. For TPBT specimen with $B = 100$ mm, size range $100 \leq D \leq 400$ mm, and $S/D = 4$, half of the beam is discretized having 40 numbers of equal four-node isoparametric elements along the depth of the beam. The number of fracture nodes along the fracture line is considered to be 38. The quasi-exponential softening relation (Eq. (3.39)) is used in FCM and DKFM.

The fracture parameters, viz. K_{INu}, $K_{\mathrm{IC}}^{\mathrm{ini}}$, $K_{\mathrm{IC}}^{\mathrm{un}}$ $G_{\mathrm{F}}^{\mathrm{ini}}$, and $G_{\mathrm{F}}^{\mathrm{un}}$, are computed according to the steps mentioned in the preceding sections for which five-term weight function method is employed. The calculated values of the material fracture parameters are reported in Table 4.11. The difference in the values of the peak load obtained between present calculation and those reported elsewhere (Karihaloo and Nallathambi 1990) is due to the effect of self-weight of the specimens. In the present study, fictitious crack model (Carpinteri 1989a–c) gives the peak load excluding the self-weight of the beam specimen. The mean values of $G_{\mathrm{F}}^{\mathrm{un}}$ (Table 4.12) and $G_{\mathrm{F}}^{\mathrm{ini}}$ (Table 4.13) are taken as the material properties of the concrete mix for the size range $100 \leq D \leq 400$ mm. The size-effect parameters for $K_{\mathrm{IC}}^{\mathrm{un}}$ are computed in the similar manner of the previous work (Karihaloo and Nallathambi 1990). Hence the values of $\Delta a_{\mathrm{c}}/l_{\mathrm{ch}}$ for each specimen are calculated by satisfying size-effect conditions between fictitious crack and double-K fracture models and solving fourth-order equations (4.86) and (4.87) in terms of $\Delta a_{\mathrm{c}}/l_{\mathrm{ch}}$. The gained size-effect parameters λ_1, λ_2, λ_3, λ_4 for each geometry and $\Delta a_{\mathrm{c}}/l_{\mathrm{ch}}$ for each specimen are presented in Table 4.12. Also, a comparison of unstable fracture load calculated according to fictitious crack model and the present size-effect law for unstable fracture toughness is presented in the same table. The size-effect law predicted by $K_{\mathrm{IC}}^{\mathrm{un}}$ of double-K

Table 4.11 Specimen dimensions and double-K fracture parameters

D	a_o/D	P_u (N)	$CMOD_c$ (μm)	a_c/D	K_{INu} (MPa mm$^{1/2}$)	K_{IC}^{ini}(MPa mm$^{1/2}$)	K_{IC}^{un}(MPa mm$^{1/2}$)	G_F^{ini} (N/m)	G_F^{un} (N/m)
400	0.2	13548	60.3	0.3144	33.64	12.22	45.20	4.97	68.10
300	0.2	11248	56.8	0.3386	31.69	12.64	45.34	5.32	68.52
200	0.2	8486.4	43.8	0.3428	28.84	13.08	41.71	5.70	58.00
100	0.2	5065.5	34.5	0.3961	24.03	13.19	40.10	5.80	53.59
400	0.3	10369	80.6	0.4212	33.88	12.68	46.84	5.36	73.15
300	0.3	8654	69.6	0.4277	31.92	13.42	44.95	6.00	67.34
200	0.3	6558.1	59.4	0.45	29.04	13.86	43.63	6.40	63.45
100	0.3	3932.4	39.5	0.4691	24.22	13.56	38.53	6.13	49.48
400	0.4	7646.5	96	0.5095	33.35	13.12	46.20	5.73	71.13
300	0.4	6429.7	83.6	0.5158	31.40	13.83	44.38	6.38	65.66
200	0.4	4898.2	72.7	0.5389	28.53	14.27	43.54	6.79	63.21
100	0.4	2946.4	49.7	0.5606	23.74	13.95	39.07	6.49	50.87
400	0.5	5300.3	126.8	0.6169	32.32	13.12	48.91	5.74	79.74
300	0.5	4513.3	100	0.6049	30.41	13.84	43.91	6.39	64.26
200	0.5	3467.8	79.6	0.6105	27.56	14.06	40.67	6.59	55.13
100	0.5	2094.2	55.9	0.6335	22.81	13.60	36.96	6.16	45.53
							Mean	5.997	62.322

Table 4.12 Size-effect parameters for K_{IC}^{un} of double-K fracture model

D	a_o/D	a_c/D[a]	G_F^{un*}	G_F^{un}/G_{FC}	$\Delta a_c/l_{ch}$	λ_1	λ_2	λ_3	λ_4	Max. diff. in critical fracture loads (%)
400	0.2	0.29727	62.36	0.605	0.12975	5.41	9.57	36.42	23.70	0
300	0.2	0.32074	62.44		0.12079					
200	0.2	0.35768	62.73		0.10516					
100	0.2	0.42869	64.37		0.07626					
400	0.3	0.39261	62.36	0.605	0.12354	5.12	13.98	36.90	67.72	0
300	0.3	0.41434	62.44		0.11438					
200	0.3	0.44803	62.73		0.098724					
100	0.3	0.51119	64.35		0.070424					
400	0.4	0.48885	62.39	0.605	0.11852	5.53	18.39	51.44	125.15	0
300	0.4	0.50814	62.51		0.10818					
200	0.4	0.53814	62.90		0.092129					
100	0.4	0.59398	64.97		0.064687					
400	0.5	0.58525	62.46	0.605	0.11371	6.35	25.32	84.05	251.53	0
300	0.5	0.60165	62.64		0.10169					
200	0.5	0.62748	63.22		0.085023					
100	0.5	0.67568	66.11		0.058583					
∞	All			0.605						−22.22

[a]Predictions from size-effect study

Table 4.13 Size-effect parameters for K_{IC}^{ini} of double-K fracture model

D	a_o/D	G_F^{ini}/G_{FC}	κ_0	κ_1	κ_2	κ_3	κ_4	Max. diff. in crack initiation loads (%)
400	0.2	0.0582	0.09	0.101	−0.0137	0.0019	-1.04×10^{-4}	0.044
300	0.2							0.003
200	0.2							−0.065
100	0.2							−0.174
400	0.3	0.0582	0.09	0.0967	−0.0114	0.0012	-4×10^{-5}	0.017
300	0.3							0.019
200	0.3							−0.003
100	0.3							−0.005
400	0.4	0.0582	0.09	0.1064	−0.0163	0.0026	-1.7×10^{-4}	0.008
300	0.4							−0.043
200	0.4							0.023
100	0.4							0.048
400	0.5	0.0582	0.09	0.125	−0.0247	0.0049	-3.7×10^{-4}	0.132
300	0.5							−0.076
200	0.5							0.046
100	0.5							0.083
∞	All	0.0582						−19.58

fracture model for two particular geometries ($a_o/D = 0.3$ and 0.5) is shown in Fig. 4.20. It is seen from Table 4.12 and Fig. 4.20 that size-effect predictions from K_{IC}^{un} of double-K fracture model and fictitious crack model are in very close agreement for laboratory-size-range specimens and the difference of prediction is not by more than 22.22% for asymptotic large size ($D \to \infty$). The ratio a_c/D obtained from LEFM equations (4.22) and (4.23) and size-effect equations (4.86) and (4.87) for l_{ch}/D ratios 0.75 and 3 is plotted in Fig. 4.21 for the sake of comparison between LEFM formulae and size-effect formulae. It is clear from the figure that the values of $\Delta a_c/l_{ch}$, and therefore a_c/D, so obtained are in good agreement with a_c/D determined using standard procedures with LEFM expressions for the size range taken into study. The variation of $\Delta a_c/l_{ch}$ calculated using size-effect law with D/l_{ch} is shown in Fig. 4.22 for different geometries (0.2–0.5). It is observed from the plot that $\Delta a_c/l_{ch}$ is not a material property for a given geometry of up to nearly 500 mm beam depth and it increases asymptotically with increase in D/l_{ch}. Beyond 500 mm depth, these values seem to be constant and may be considered as a material property for a particular geometry.

Since the size-effect law for initiation toughness K_{IC}^{ini} as introduced in previous section seems to be more complex and is based on the assumption that ratio of K_{IC}^{un}/K_{IC}^{ini} maintains some definite relationship with respect to the size of structure. Therefore, the variation between K_{IC}^{un}/K_{IC}^{ini} and l_{ch}/D is plotted in Fig. 4.23 for three different geometries (0.3–0.5). It is observed from the figure that the

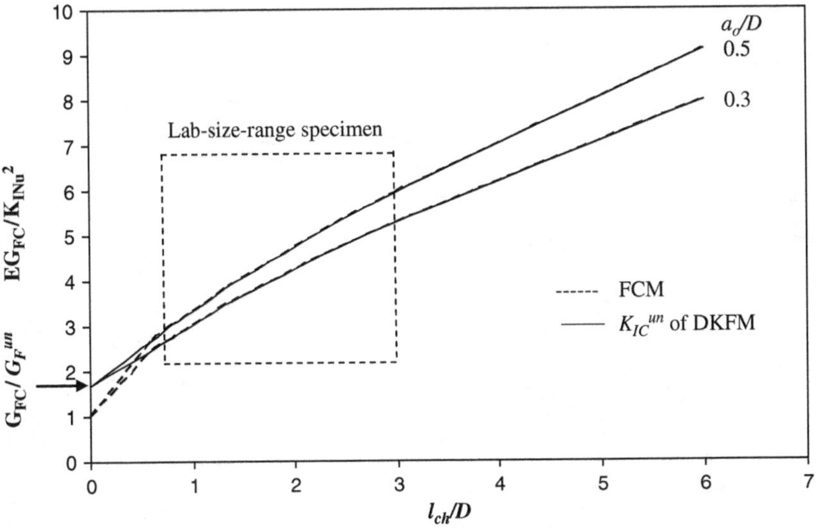

Fig. 4.20 Size-effect curves for FCM and K_{IC}^{un} of DKFM for TPBT specimens

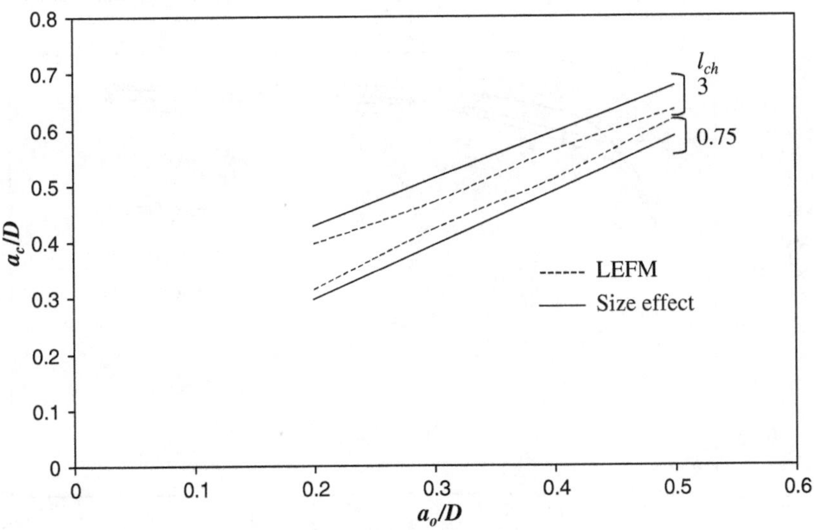

Fig. 4.21 Comparison of effective crack extension to depth (a_c/D) ratio calculated using LEFM (Eqs. (4.22) and (4.23)) and size effect (Eqs. (4.86) and (4.87))

assumption regarding deriving size-effect law for K_{IC}^{ini} turns out to be true. The coefficients $\kappa_i(\kappa_o - \kappa_4)$ are then determined from least square fitting after satisfying the size-effect law using Eq. (4.91) completely. The values of κ_o such determined are slightly different for each geometry for instances 0.0869, 0.0887, 0.0921, and

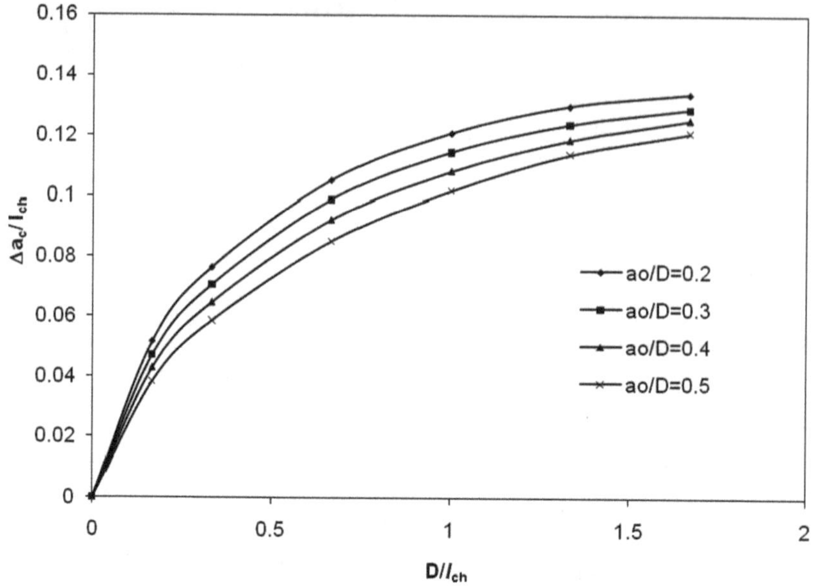

Fig. 4.22 Relationship between $\Delta a_c/l_{ch}$ from size-effect calculation and D/l_{ch} from FCM

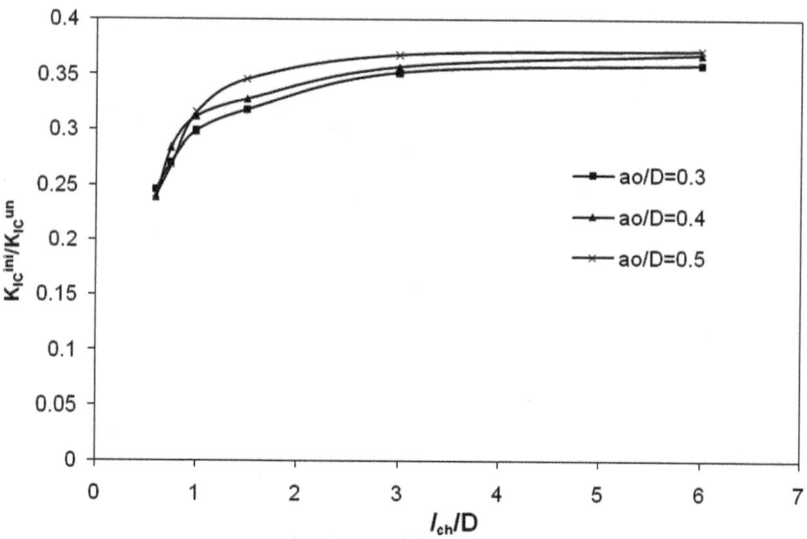

Fig. 4.23 Variation of K_{IC}^{ini}/K_{IC}^{un} ratio with l_{ch}/D for various specimen sizes

0.0908 for $a_o/D = 0.2$, 0.3, 0.4, and 0.5, respectively. Therefore, finally mean of all, $\kappa_o \approx 0.09$, is adopted as material property, which is geometry independent. Keeping this value to be constant, the coefficients $\kappa_1 - \kappa_4$ are determined for different geometries using least square fitting after satisfying Eq. (4.91). The different

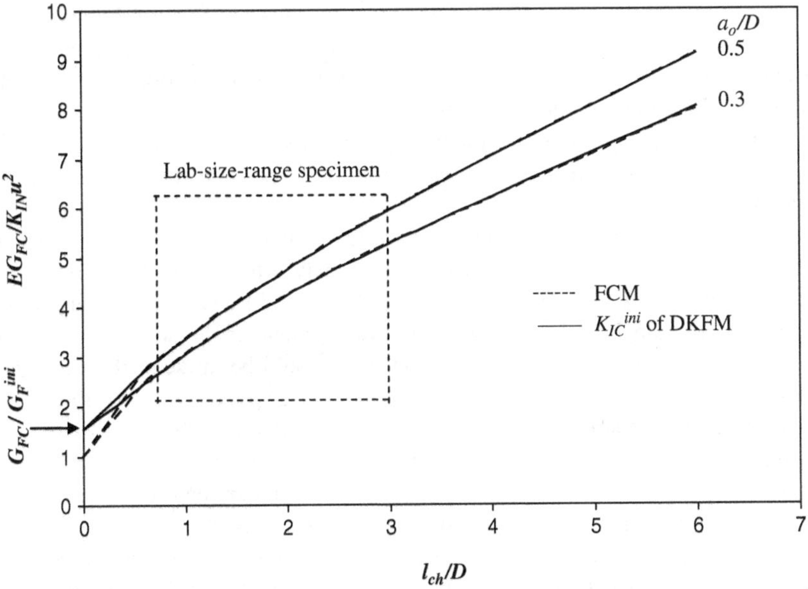

Fig. 4.24 Size-effect curves for FCM and $K_{\mathrm{IC}}^{\mathrm{ini}}$ of DKFM for TPBT specimens

parameters of size-effect law for $K_{\mathrm{IC}}^{\mathrm{ini}}$ are presented in Table 4.13 and the corresponding size-effect law predicted with respect to fictitious crack model for two particular geometries ($a_0/D = 0.3$ and 0.5) is shown in Fig. 4.24.

From Table 4.13 and Fig. 4.24, it is clear that the size-effect curves are in very good agreement and are dependent on the geometrical parameters, a_0/D ratios. It is noticed that the maximum difference in the predicted crack initiation load gained from the two models for laboratory-size-range specimens are less than 0.2% (Table 4.13). Further, it is also clear that the predictions of the two models differed by about 19.58% for asymptotic size ($D \to \infty$), i.e., size-effect law for $K_{\mathrm{IC}}^{\mathrm{ini}}$ of double-*K* fracture model seems to be more conservative by about 20%.

For notched three-point bending laboratory-size specimens, it was shown that the predictions of fictitious crack model and double-*K* fracture parameters ($K_{\mathrm{IC}}^{\mathrm{ini}}$ and $K_{\mathrm{IC}}^{\mathrm{un}}$) are almost the same, similar to the cases between fictitious crack model and two-parameter fracture model, fictitious crack model, and size-effect model (Planas and Elices 1990) and fictitious crack model and effective crack model (Karihaloo and Nallathambi 1990). In the asymptotic limit of infinite size, the predictions of fracture loads between fictitious crack model and the $K_{\mathrm{IC}}^{\mathrm{ini}}$ and fictitious crack model and the $K_{\mathrm{IC}}^{\mathrm{un}}$ are significantly different. This divergence is slightly more than the agreement between fictitious crack model and effective crack model and less than that between fictitious crack model and size-effect model and between fictitious crack model and two-parameter fracture model.

The size-effect equations (4.86) and (4.87) for unstable fracture toughness K_{IC}^{un} and Eq. (4.91) for initial cracking toughness K_{IC}^{ini} are derived making use of linear elastic fracture mechanics formulae for the three-point bending test geometry. Therefore, the equations presented herein the chapter is valid only for the TPBT geometry and for any other specimen geometry, a similar approach can be easily followed for the size-effect predictions using the respective linear elastic fracture mechanics equations.

Since the shape of the softening function is the main ingredient of the cohesive crack method, the size-effect curves are obviously affected by the choice of the softening function. It was reported in the study (Zi and Bažant 2003) that the size-effect curves could be affected by the initial segment of the softening curve for the specimens of positive geometry, while those could be influenced by the entire softening curve for the negative–positive specimen geometry. Furthermore, it was concluded that the shape of the softening function influenced the size-effect curves to a significant extent for large-size specimens. In addition, it is also obvious that the softening function is used for determining the fracture parameters of the double-K fracture model; its shape does affect the size-effect curves of the double-K fracture criterion. As a result, a separate analysis needs to be carried out in order to accurately determine the mathematical coefficients of the size-effect laws for each type of the softening curve.

4.9 Numerical Study on Double-*K* Fracture Parameters

A precise comparative study on determination of double-K fracture parameters using existing *analytical method*, *simplified approach,* and the present *weight function approach* is carried out on TPBT and CT specimen geometries at different a_0/D ratios ranging 0.15–0.80 for specimen size 200 mm. The influence of specimen geometry, loading condition, and softening function on the fracture parameters for the size range 100–600 mm is also investigated. The input data such as the peak load P_u and the corresponding critical value of $CMOD_c$ for TPBT and FPBT or COD_c for CT specimen are obtained using FCM.

4.9.1 Material Properties

The concrete mix for which $f_t = 3.21$ MPa, $E = 30$ GPa, $G_F = 103$ N/m (Planas and Elices 1990), $d_a = 16$ mm, and $\nu = 0.18$ along with quasi-exponential softening function is taken into account for the present investigation. For standard TPBT, CT, and FPBT specimens with $B = 100$ mm having size range $D = 100-600$ mm, the finite element analysis is carried out for which half of the specimens are discretized due to symmetry as shown in Fig. 4.25 and 80 numbers of equal isoparametric plane elements are considered along the dimension D. The same quasi-exponential softening function is employed for determining the double-K fracture parameters.

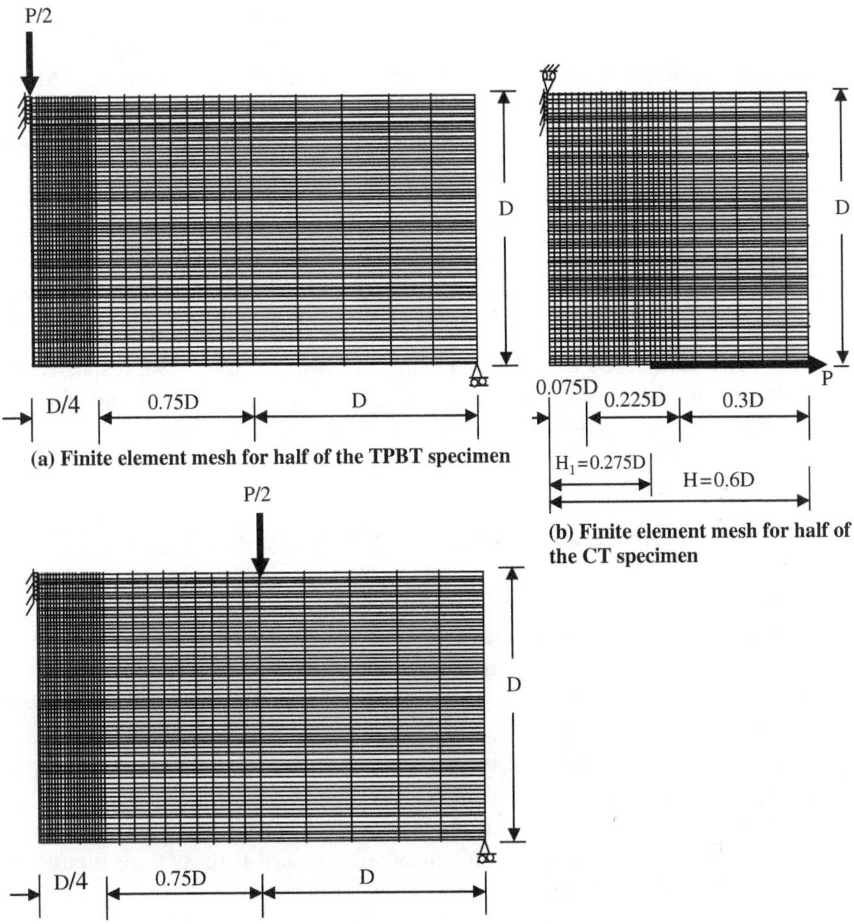

(a) Finite element mesh for half of the TPBT specimen

(b) Finite element mesh for half of the CT specimen

(c) Finite element discretization for half of FPBT specimen

Fig. 4.25 Finite element discretization of TPBT, CT, and FPBT specimens

For the precise numerical investigation, the effect of self-weight in TPBT and FPBT geometries can be accounted for at all computational stages, whereas this effect cannot be obviously considered for the CT specimen. In the TPBT and FPBT specimens, the input data are gained using the developed FCM, which accounts for the effect of self-weight of the beam. Therefore, the FCM yields the external load P without inclusion of the self-weight, whereas the CMOD consists of the contribution of the self-weight. Hence, in addition to the external load P acting on the TPBT and FPBT, the influence due to a concentrated load equal to $w_g.S/2$ and acting at the midspan, is also effective during all stages of computations. For FPBT, Eq. (4.41) is employed to obtain SIF. This yields precise comparison amongst the fracture parameters between the two specimen geometries (TPBT and CT) and the two loading conditions (TPBT and FPBT) considered in the study. In addition, the value of P in Eq. (4.22) is modified as $(P + w_{g}.S/2)$ during the computations of effective crack extension in the case of TPBT. A similar contribution in the CMOD due to TPBT action for the self-weight effect is also taken into account while calculating the equivalent effective crack extension in the case of FPBT specimen. Hence considering the effect of self-weight of the beam, the total CMOD for the FPBT specimen can be written as

$$\mathrm{CMOD} = \frac{3a}{BD^2E}\left[4PLV_1(\beta) + w_g S^2 V_2(\beta)\right] \tag{4.92}$$

in which $V_1(\beta)$ and $V_2(\beta)$ are obtained using Eqs. (4.23) and (4.29), respectively. Two well-known relations of FCM, i.e., $l_{ch} = EG_F/f_t^2$ and the critical value of stress intensity factor $K_C = \sqrt{(G_F E)}$ are also used in the numerical comparison.

4.9.2 Comparison Between Methods of Determination for DKFM

The P-CMOD and P-COD curves, for a_o/D ratio ranging 0.15–0.80 and D = 200 mm obtained using FCM for the TPBT and CT specimen geometries, are presented in Figs. 4.26 and 4.27, respectively. Further, the double-K fracture parameters are computed using the *analytical method*, *simplified approach*, and *weight function approach* having four and five terms. A trial-and-error method is used for calculation of a_c for the analytical method and weight function approach for all the three types of specimens. The direct empirical equations are used to determine the value of a_c for simplified approach for the TPBT and CT specimen geometries.

For systematic comparison, the dimensionless numerical results are plotted in Figs. 4.28, 4.29, 4.30, 4.31, and 4.32.

- The nondimensional parameter K_C/K_{IC}^{un} determined using LEFM equations and simplified empirical equations for the TPBT and CT specimen geometries having $D = 200$ mm is plotted with a_o/D ratio in Fig. 4.28. Up to $a_o/D = 0.75$ the maximum difference in the results of K_{IC}^{un} between simplified approach and analytical method for TPBT geometry is 3.25% and it is slightly more at $a_o/D = 0.8$, that is, 5.34%. This difference calculated on the mean values of K_{IC}^{un} at different a_o/D ratios is 0.13%. For CT test specimen, this maximum difference up to $a_o/D = 0.80$ is 0.31% and the difference obtained on the mean values of K_{IC}^{un} is 0.26%. It is also seen from the figure that results are not greatly affected by the shape of the specimen. The value of K_{IC}^{un} remains almost constant up to a_o/D ratio of 0.6 and beyond this, a marginal decrease is observed for both the specimen geometries. The difference in the average values of K_{IC}^{un} between these two geometries is about 1.76%.

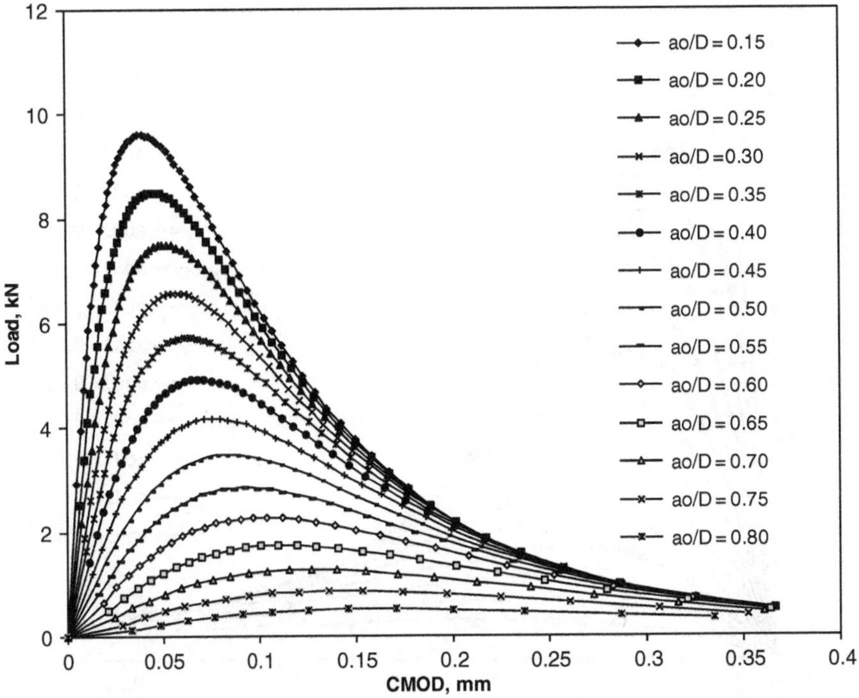

Fig. 4.26 P-CMOD curves for TPBT specimen having $D = 200\,\text{mm}$

- The comparison of nondimensional values K_C/K_{IC}^C with a_o/D ratio obtained using various methods for TPBT specimen $D = 200\,\text{mm}$ is shown in Fig. 4.29. WF4 and WF5 represent the values obtained using four-term and five-term weight function approach, respectively, in the figure. The variation of K_C/K_{IC}^C is similar to that of K_C/K_{IC}^C with respect to a_o/D ratios. The comparison of results obtained using weight function and simplified approach is made with regard to that determined using standard analytical method. It is observed that simplified approach yields relatively more difference at higher a_o/D ratios. However, the percentage difference in the average values of K_{IC}^C obtained using weight function with four-term, five–term, and simplified approach for specimen size 200 mm is 0.85, 0.07, and 0.78%, respectively. These differences in calculated values of K_{IC}^{ini} are 1.62, 0.13, and 1.86%, respectively, as seen from Fig. 4.30. A similar pattern can be observed in Figs. 4.31 and 4.32 for the CT test specimen with $D = 200\,\text{mm}$. It is found from Fig. 4.31 that the percentage difference in mean values of K_{IC}^C using weight function with four-term, five–term, and simplified approach as compared with analytical method is 0.89, 0.07, and 0.78%, respectively. In addition, these differences in the mean values of K_{IC}^{ini} using weight function with four-term, five–term, and simplified approach (from Fig. 4.32) are found to be 1.61, 0.13, and 1.52%, respectively.

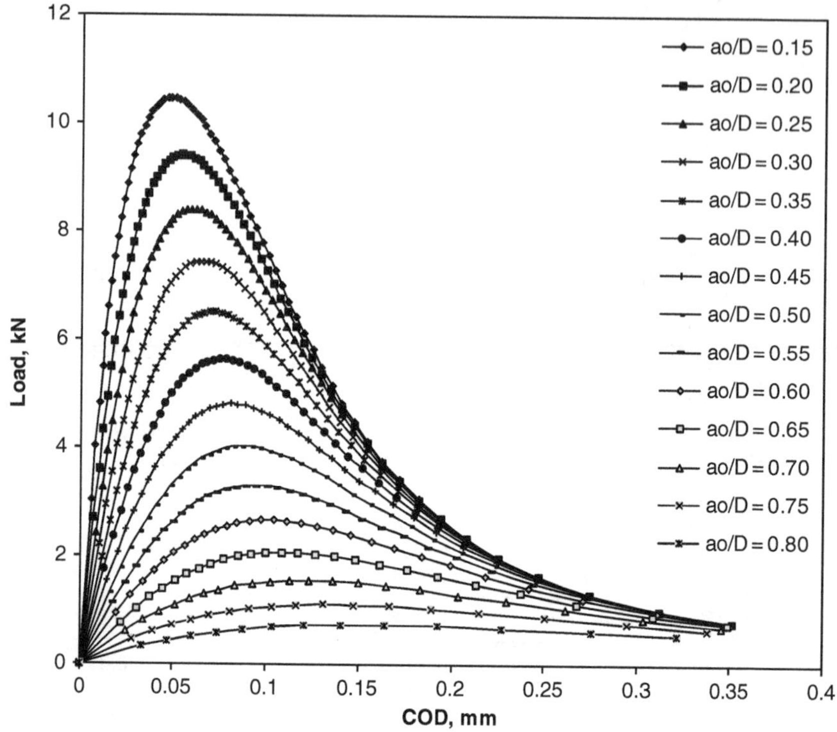

Fig. 4.27 P-COD curves for CT specimen having $D = 200\,\text{mm}$

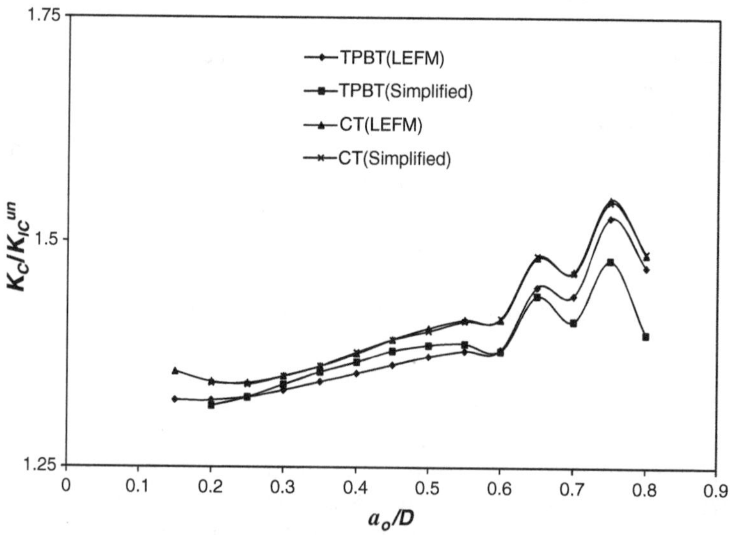

Fig. 4.28 Variation of $K_{\mathrm{IC}}^{\mathrm{un}}$ obtained using LEFM equations and simplified empirical equations for TPBT and CT specimen geometries, $D = 200\,\text{mm}$

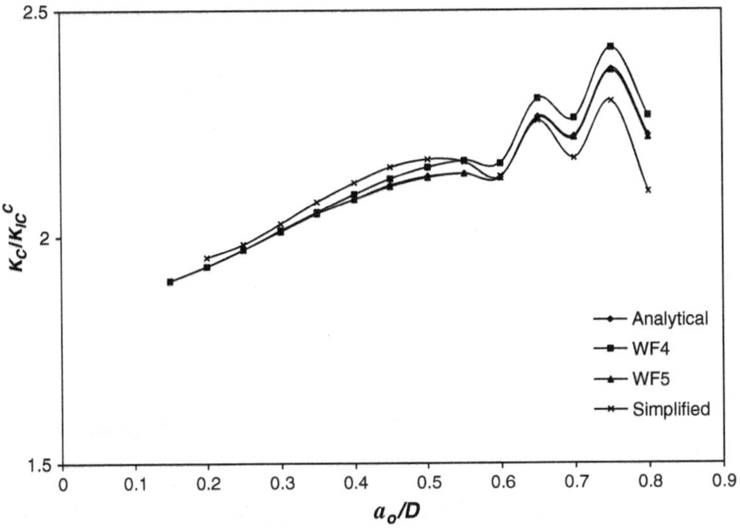

Fig. 4.29 Variation of K_{IC}^{C} obtained using analytical method, weight function approach, and simplified approach for TPBT specimen, $D = 200$ mm

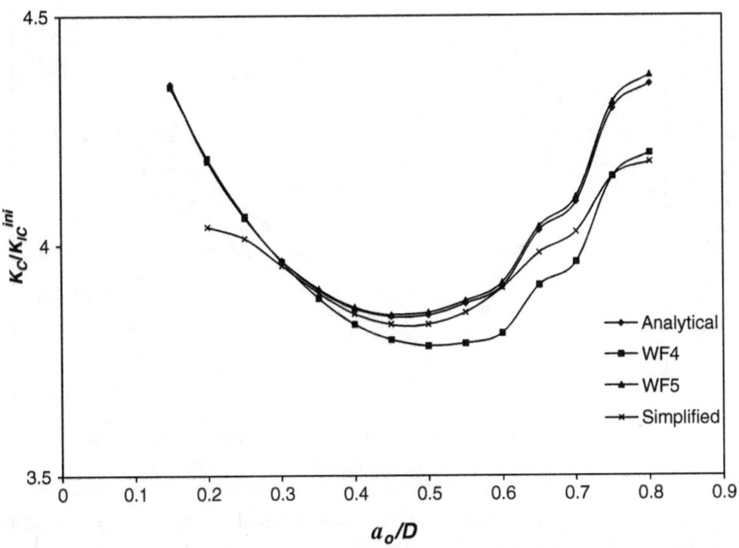

Fig. 4.30 Variation of K_{IC}^{ini} obtained using analytical method, weight function approach, and simplified approach for TPBT specimen, $D = 200$ mm

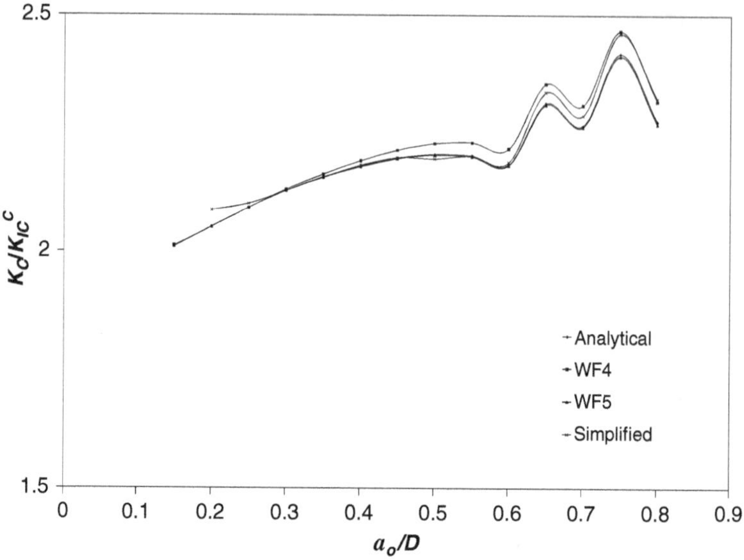

Fig. 4.31 Variation of K_{IC}^{C} obtained using analytical method, weight function approach, and simplified approach for CT specimen, $D = 200$ mm

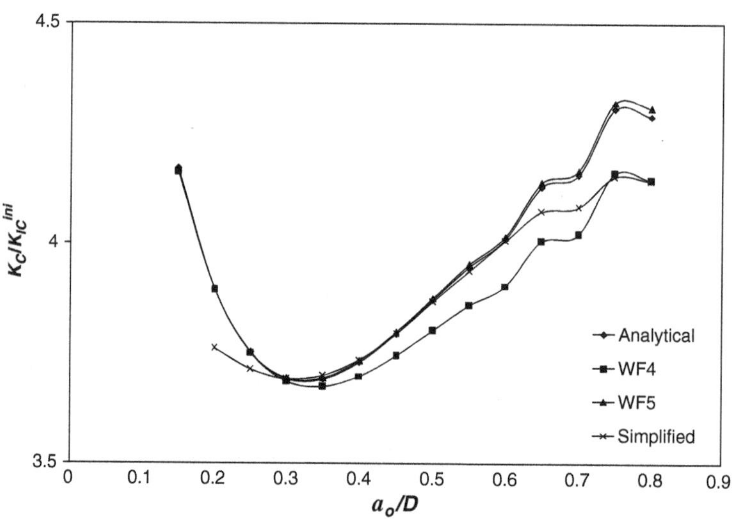

Fig. 4.32 Variation of K_{IC}^{ini} obtained using analytical method, weight function approach, and simplified approach for CT specimen, $D = 200$ mm

4.9.3 Influence of Specimen Geometry

For this investigation, the gained fracture parameters for TPBT and CT specimen geometries are plotted in dimensionless forms in Figs. 4.33, 4.34, 4.35, and 4.36. The following observations can be made from these figures:

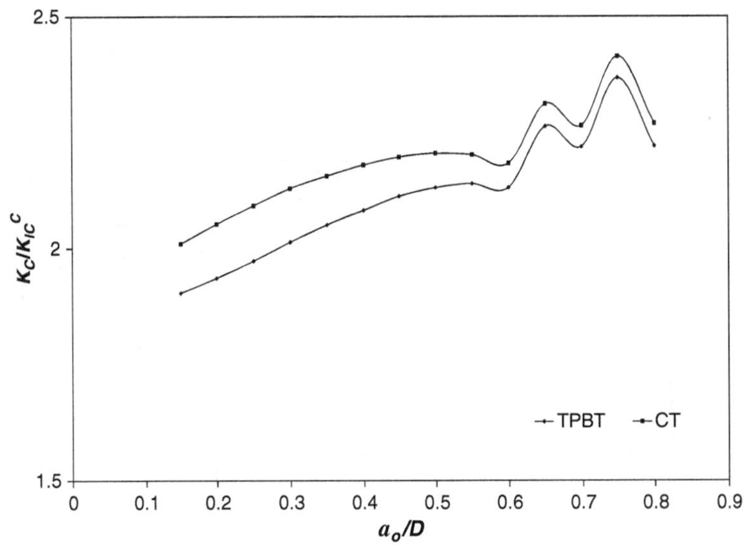

Fig. 4.33 Comparison of $K_{\mathrm{IC}}^{\mathrm{C}}$ obtained using weight function approach with five terms for TPBT and CT specimens, $D = 200\,\mathrm{mm}$

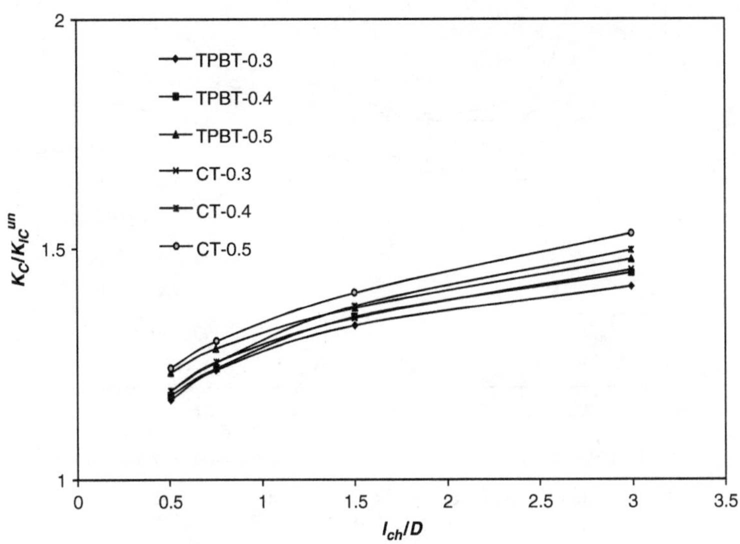

Fig. 4.34 Effect of specimen geometries on $K_{\mathrm{IC}}^{\mathrm{un}}$ for different specimen sizes

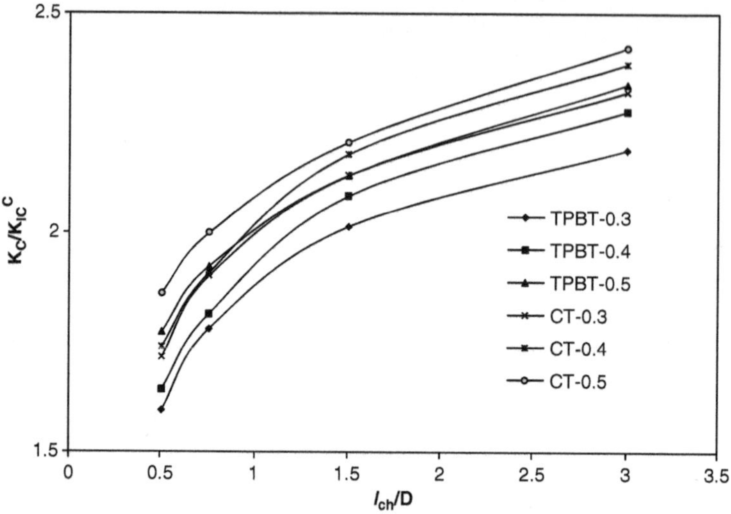

Fig. 4.35 Effect of specimen geometries on K_{IC}^{C} for different specimen sizes

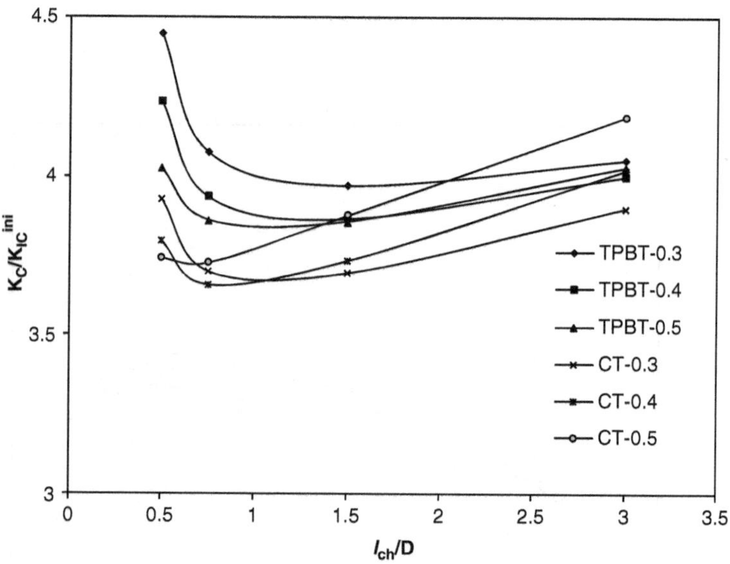

Fig. 4.36 Effect of specimen geometries on K_{IC}^{ini} for different specimen sizes

- Figure 4.33 presents the effect of specimen geometry on numerical results of K_{IC}^{C} for $D = 200$ mm. The value of K_{IC}^{C} was obtained using weight function method with five terms. It is seen that almost a constant value of cohesion toughness can be obtained up to $a_o/D = 0.6$ and between a_o/D values of 0.6–0.8, a marginal decrease in the value of cohesion toughness is observed. The percentage difference in average values of K_{IC}^{C} for TPBT and CT test specimen geometries is about 2.12%.
- The values of K_{IC}^{un}, K_{IC}^{C}, and K_{IC}^{ini} are calculated using five-term weight function approach at sizes ranging 100–600 mm for TPBT and CT test specimen geometries for a_o/D ratios 0.3–0.5. Variations of nondimensional parameters K_C/K_{IC}^{un}, K_C/K_{IC}^{C} and K_C/K_{IC}^{ini} with l_{ch}/D for both the specimen shapes are plotted in Figs. 4.34, 4.35 and 4.36, respectively. Legends in these figures are denoted by the type of specimen geometry and a_o/D ratio (such as TPBT-0.3 in Fig. 4.34). It is observed from the figures that values of fracture parameters are marginally dependent on geometrical parameter (a_o/D ratio) and the type of specimen geometries. Both the parameters K_{IC}^{un} and K_{IC}^{C} increase with increase in specimen size, whereas the values of K_{IC}^{ini} are relatively less dependent on the sizes ranging 100–400 mm; however, beyond the size range 400 mm, a decrease in the value of K_{IC}^{ini} is observed. Neglecting the effect of a_o/D ratio, the mean values of K_{IC}^{un}, K_{IC}^{C} and K_{IC}^{ini} for TPBT and CT test specimens of size range $100 \le D \le 600$ mm are computed and it is found that the values of K_{IC}^{un} for CT specimen are less than those for TPBT specimen by about 2.99, 1.69, 1.20 and 1.16% for sizes 100, 200, 400 and 600 mm, respectively. The values of cohesion toughness determined show that the values of K_{IC}^{C} for CT specimen are less than those for TPBT geometry by about 4.59, 4.43, 5.01 and 5.82% for specimen sizes 100, 200, 400 and 600 mm, respectively. Consequently the average values of K_{IC}^{ini} determined as such for TPBT geometry are less than those for CT specimen by nearly 0, 3.33, 6.55 and 9.63% for specimen sizes 100, 200, 400 and 600 mm, respectively.
- The variation of $CTOD_c$ at the onset of unstable crack propagation in nondimensional form ($l_{ch}/CTOD_c$) with a_o/D ratio is plotted for different specimen shapes and sizes (100 and 400 mm) as shown in Fig. 4.37. The legend in the figure shows the type of specimen geometry and its size in millimeters. From Fig. 4.37 it observed that the $CTOD_c$ varies with the size and shape of the test specimens. Xu and Reinhardt (1999b) concluded a similar observation based on the experimental tests (Refai and Swartz 1987) that the evaluated $CTOD_c$ seems to be size dependent.

4.9.4 Influence of Loading Condition

The gained values of K_{IC}^{un}, K_{IC}^{C} and K_{IC}^{ini} using five-term weight function approach at different specimen sizes ranging 100–600 mm for TPBT and FPBT loading conditions for a_o/D ratios 0.3–0.5 are plotted in Figs. 4.38, 4.39, and 4.40, respectively.

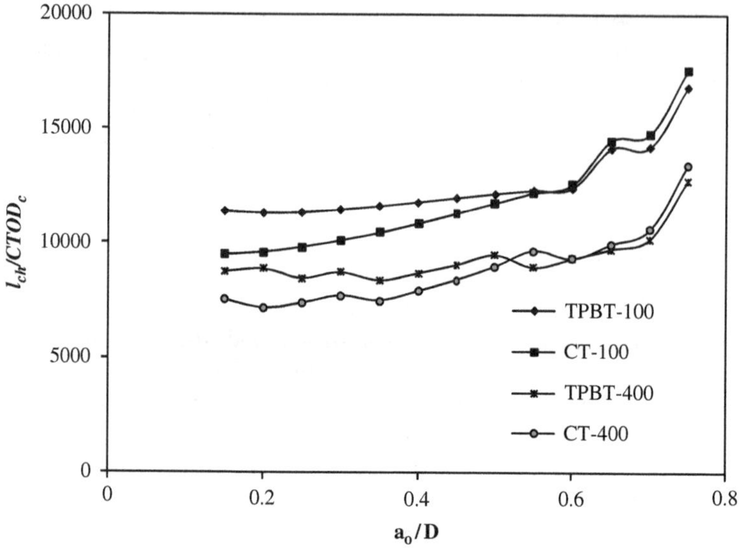

Fig. 4.37 Variation of $CTOD_c$ for TPBT and CT specimen geometries for $D = 100$ and 400 mm

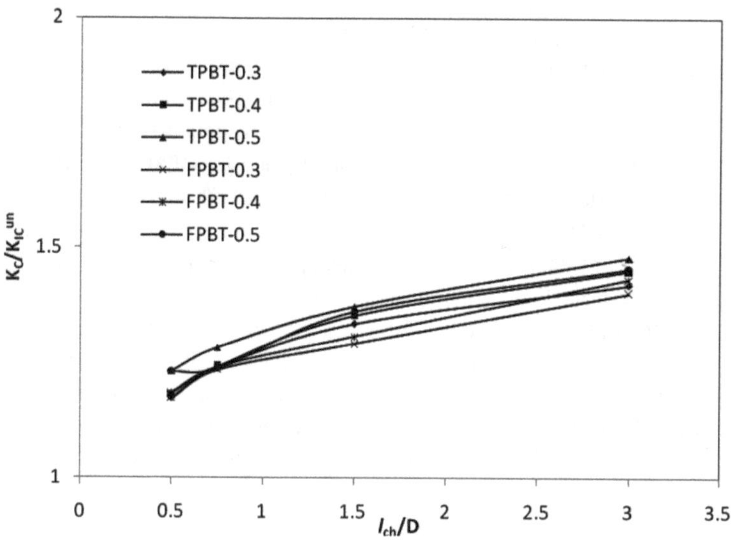

Fig. 4.38 Effect of loading condition on K_{IC}^{un} for different specimen sizes

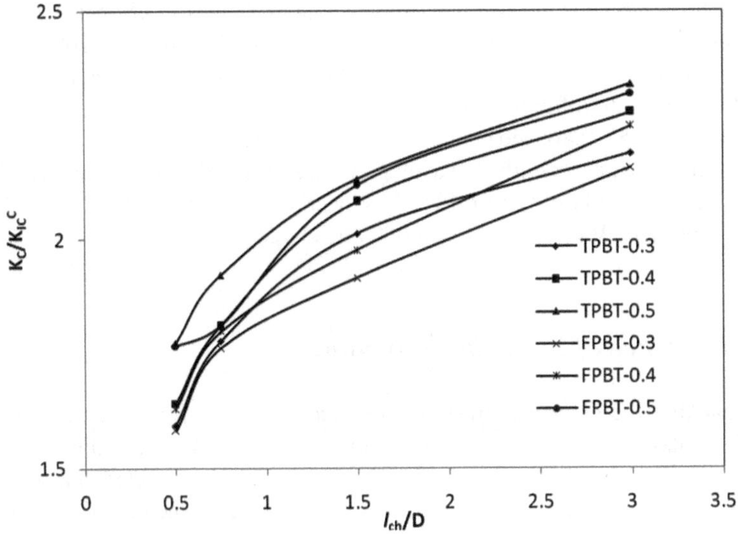

Fig. 4.39 Effect of loading condition on K_{IC}^{C} for different specimen sizes

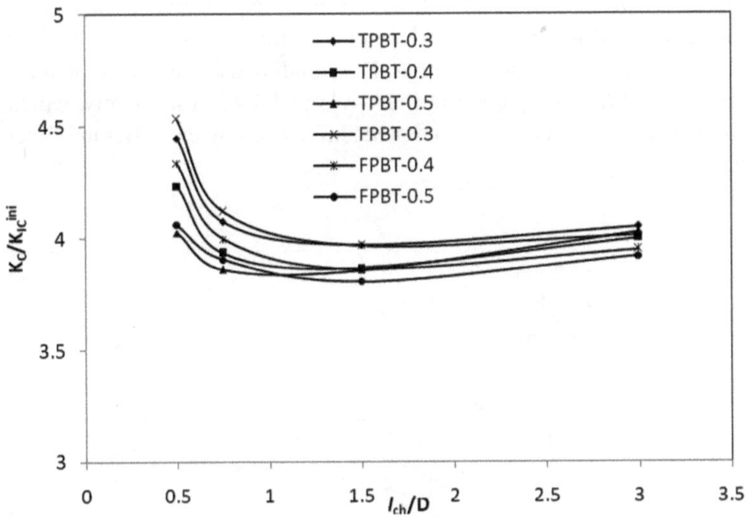

Fig. 4.40 Effect of loading condition on K_{IC}^{ini} for different specimen sizes

1. It is observed from the figures that values of fracture parameters are marginally dependent on geometrical parameter (a_o/D ratio) and the types of loading conditions. Similar to the previous investigation, the mean values of K_{IC}^{un}, K_{IC}^{C}, and K_{IC}^{ini} for TPBT and FPBT specimens are computed neglecting the effect of a_o/D

ratio. It is found that the absolute difference in the values of K_{IC}^{un} between the two loading conditions is about 1.36, 2.53, 1.30, and 0.03% for sizes 100, 200, 400, and 600 mm, respectively. The values of cohesion toughness determined show that the absolute difference in the values of K_{IC}^{C} for these loading conditions is about 1.09, 3.57, 2.46, and 0.62% for specimen sizes 100, 200, 400, and 600 mm, respectively. Consequently, the average values of K_{IC}^{ini} determined as such for the two loading conditions are differed by nearly 1.65, 0.51, 1.28, and 1.69% for specimen sizes 100, 200, 400, and 600 mm, respectively.

4.9.5 Influence of Softening Function

It is clear that the cohesive stress distribution is not involved in computation of K_{IC}^{un} of double-K fracture criterion; hence these values are unaffected by the choice of the softening functions of concrete. Also, it has been reported that the influence of shapes of softening relations on the calculation of the cohesion toughness is not observable (Xu and Reinhardt 1999b). In the present study, some of the commonly used softening functions such as Petersson's (1981) bilinear, Wittmann et al.'s (1988) bilinear, modified bilinear (Xu and Reinhardt 1999b), nonlinear softening functions (Reinhardt et al. 1986), and quasi-exponential (Planas and Elices 1990) are employed to obtain the K_{IC}^{C} and K_{IC}^{ini} for the TPBT specimen at a_o/D ratio 0.3 and size range 100–600 mm. The nondimensional forms of the gained results of K_{IC}^{C} and K_{IC}^{ini} are plotted in Figs. 4.41 and 4.42, respectively, which show that these fracture parameters are somewhat influenced by the softening function of concrete.

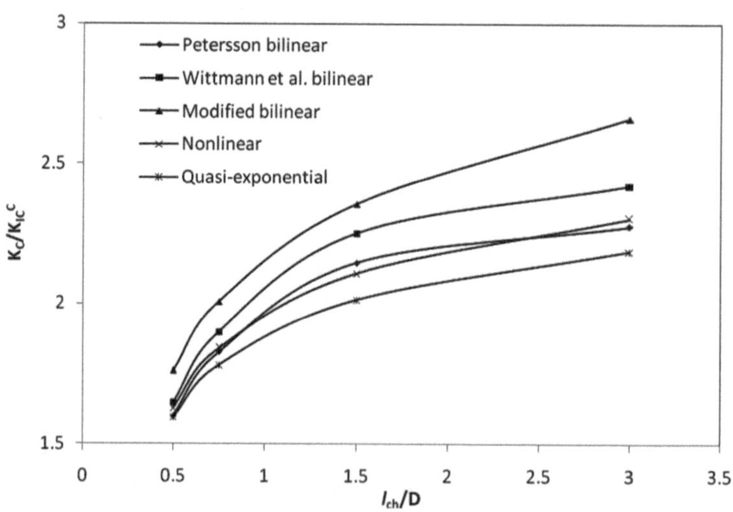

Fig. 4.41 Effect of softening function on K_{IC}^{C} for different specimen sizes for TPBT

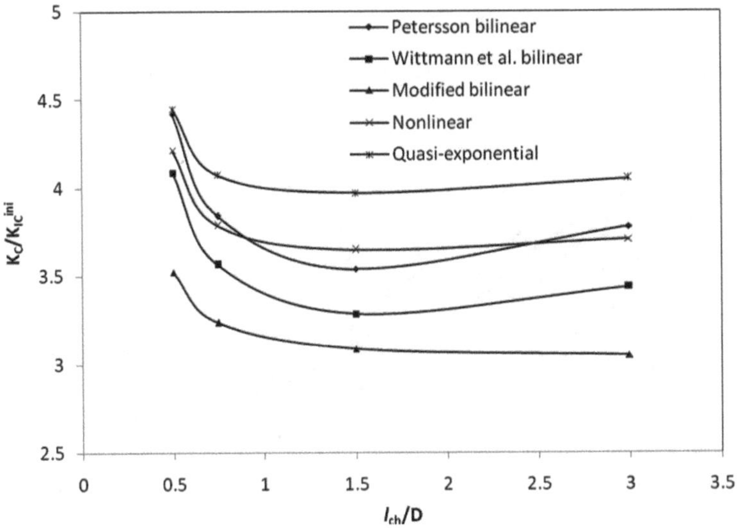

Fig. 4.42 Effect of softening function on K_{IC}^{ini} for different specimen sizes for TPBT

4.10 Equivalence Between Double-*K* and Double-*G* Fracture Parameters

The objective of this study is to present influence of specimen geometry and size effect on the double-*G* fracture parameters and double-*K* fracture parameters and also to investigate the equivalence relationship between the double-*G* fracture criterion and the double-*K* fracture model using numerically obtained input data. Influence of softening function on both the fracture criteria is also reported. For precise comparative study between the two fracture criteria, numerical method has been chosen to obtain the required input data so that no discrepancy can occur due to shape, size, and geometrical parameter of test specimens for a given set of material properties.

4.10.1 Material Properties

The same concrete mix with material properties $f_t = 3.21$ MPa, $E = 30$ GPa, $G_F = 103$ N/m, and $\nu = 0.18$ along with nonlinear softening relation having $c_1 = 3$ and $c_2 = 7$ is considered for this study. This results in the value of $w_c = 167.2\,\mu$m after satisfying Eq. (3.37). With constant specimen thickness, $B = 100$ mm, and variable size range $100 \le D \le 500$ mm, the P-CMOD curves or P-COD curves for a_o/D ratio ranging 0.3–0.4 are obtained using the FCM (Fig. 4.25) for standard TPBT and CT test geometries. These curves are presented in Figs. 4.43 and 4.44. The symbols D_1, D_2, D_3, D_4, and D_5 in the figures correspond to the specimen sizes of 100, 200, 300,

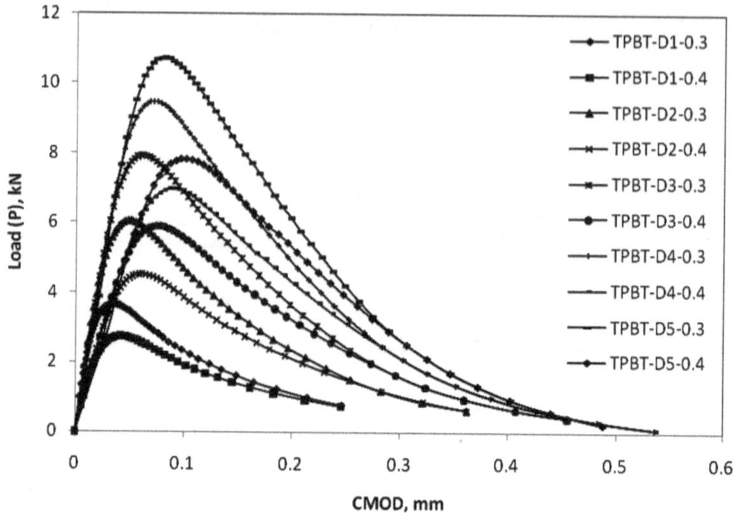

Fig. 4.43 The P-CMOD curves for TPBT specimen geometry

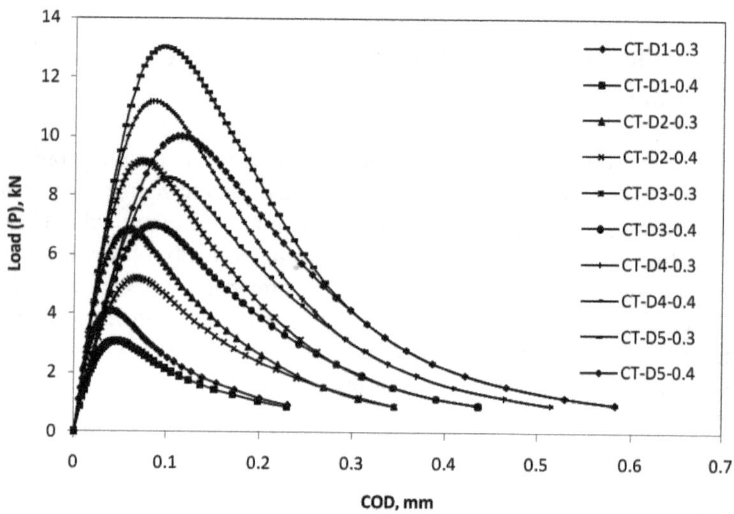

Fig. 4.44 The P-COD curves for CT test specimen geometry

400, and 500 mm, respectively. The legends of the figures are denoted by the type of specimen geometry, specimen size, and the a_o/D ratio. For example, TPBT-D1-0.3 in Fig. 4.43 denotes the TPBT of $D = 100$ mm and $a_o/D = 0.3$.

The effect of self-weight in TPBT is accounted for in the same manner as discussed earlier. The weight function method with five terms is used to calculate double-K fracture parameters. The modified bilinear softening function (Xu and

Reinhardt 1999b) is used to obtain the double-K and double-G fracture parameters. Assuming the maximum size of coarse aggregate d_a to be 16 mm for the considered concrete mix, the values of α_F and w_c are determined for the given values of d_a and G_F. The value of σ_s is then obtained using Eq. (3.32) and finally the bilinear softening function is completely characterized using Eq. (3.26). To determine the double-G fracture criterion, first of all the value G_{IC}^{un} is computed using Eqs. (4.66) and (4.69) for TPBT and CT specimen, respectively, for the given value of P_u and the corresponding computed value of a_c. The Gauss–Legendre numerical integration technique is adopted in the computer program for determining the value of G_{IC}^{C}. Finally, the value of G_{IC}^{ini} is evaluated using Eq. (4.75) for both the test configurations. The effective initiation fracture toughness $\overline{K_{IC}^{ini}}$ and effective unstable fracture toughness $\overline{K_{IC}^{un}}$ are determined using Eq. (4.76) and reported in the subsequent study. The following investigations are carried out from the gained results.

4.10.2 Effect of Specimen Geometry on Double-K Fracture Parameters

The material fracture parameters K_{IC}^{un} and K_{IC}^{ini} for the considered specimen geometries of size range 100–500 mm are graphically presented in nondimensional forms K_C/K_{IC}^{un} and K_C/K_{IC}^{ini} in Figs. 4.45 and 4.46, respectively.

- It is observed from the figures that the parameter K_{IC}^{un} increases with increase in specimen size, whereas the value of K_{IC}^{ini} shows reverse trend. However, the value of K_{IC}^{ini} is relatively less dependent on the size ranging 100–300 mm as seen from

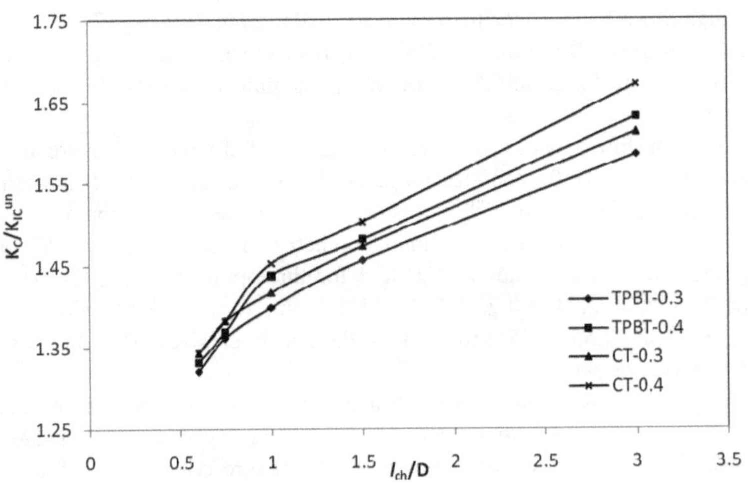

Fig. 4.45 The size-effect curves for K_{IC}^{un} on TPBT and CT test specimens

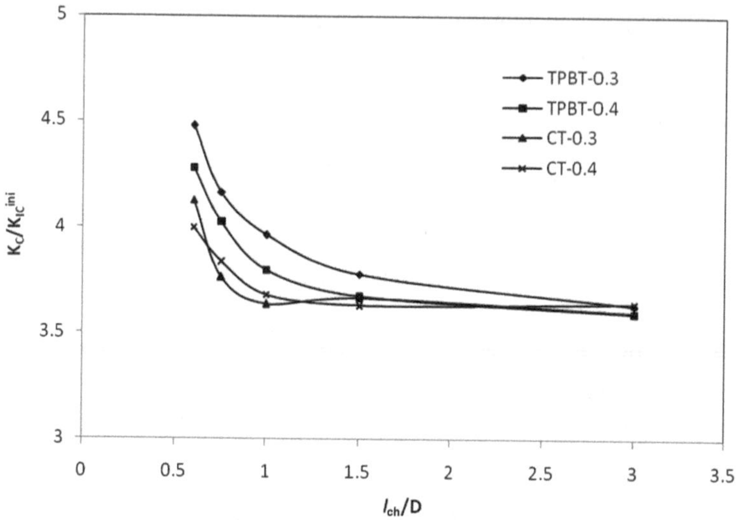

Fig. 4.46 The size-effect curves for K_{IC}^{ini} on TPBT and CT test specimens

Figs. 4.45 and 4.46 and beyond the size range 300 mm; a sharp decrease in the value of K_{IC}^{ini} is observed.

- Neglecting the effect of a_0/D ratio, the mean values of K_{IC}^{un} and K_{IC}^{ini} for TPBT and CT specimens of size range $100 \le D \le 500$ mm are computed. It is found that the values of K_{IC}^{un} for compact tension specimen are less than those for three-point bending specimen by about 2.05, 1.28, 1.15, 1.22, and 1.27% for specimen sizes 100, 200, 300, 400, and 500 mm, respectively.
- On the other hand, the average values of K_{IC}^{ini} determined as such for TPBT are less than those for CT specimen by nearly 2.04, 2.66, 7.22, and 7.19% for specimen sizes 200, 300, 400, and 500 mm, respectively, excepting the specimen size of 100 mm for which the value of K_{IC}^{ini} is almost the same for both the test geometries.
- The relationship between fracture parameters of double-K fracture model is shown in Fig. 4.47 for the size range considered in the study. It is seen from the figure that the ratio of K_{IC}^{un}/K_{IC}^{ini} is dependent on geometrical parameter (a_0/D ratio), specimen size, and specimen geometry. If the effect of a_0/D ratio is neglected, the mean values of K_{IC}^{un}/K_{IC}^{ini} for three-point bending test are found to be 2.25, 2.54, 2.74, 3.0, and 3.30 and the same for compact tension specimen are 2.21, 2.45, 2.55, 2.75, and 3.03 for the specimen sizes 100, 200, 300, 400, and 500 mm, respectively.
- In the present case, result obtained using different combination of softening functions – nonlinear softening used in the cohesive crack model and bilinear softening used in determining double-K fracture criterion – is found to be consistent with those presented earlier.

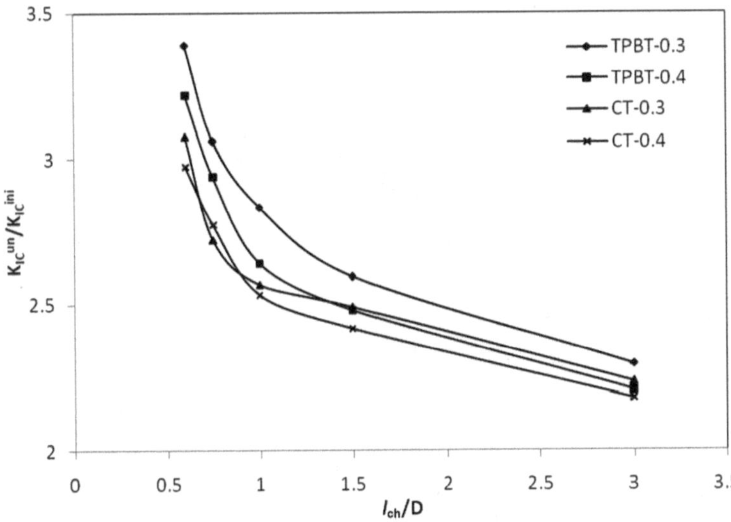

Fig. 4.47 Relationship between the parameters K_{IC}^{un} and K_{IC}^{ini}

4.10.3 Effect of Specimen Geometry on Double-G Fracture Parameters

Influence of specimen geometry and size effect on the effective unstable fracture toughness $\overline{K_{IC}^{un}}$ and effective initiation fracture toughness $\overline{K_{IC}^{ini}}$ of double-*G* fracture criterion is shown in Figs. 4.48 and 4.49, respectively. It is interesting to notice from the figures that similar to the double-*K* fracture parameters, the parameters of double-*G* fracture model are in general dependent on geometrical parameter (a_0/D ratio), specimen size, and specimen geometry. The effective unstable fracture toughness $\overline{K_{IC}^{un}}$ increases with increase in specimen size, while the value of the effective initiation fracture toughness $\overline{K_{IC}^{ini}}$ decreases with the increase in specimen size.

- Similar to the previous study carried out on the double-*K* fracture parameters, the influence of specimen geometry on double-*G* fracture parameters may be observed for a particular specimen size if the fracture parameters are assumed to be independent on a_0/D ratio. Hence after neglecting the effect of a_0/D ratio, the mean values of $\overline{K_{IC}^{un}}$ and $\overline{K_{IC}^{ini}}$ for TPBT and CT specimens of size range $100 \le D \le 500$ mm are evaluated. It is observed that the values of $\overline{K_{IC}^{un}}$ for compact tension specimen are found to be more than those obtained for three-point bending geometry by about 0.79, 1.96, 2.43, 2.55, and 2.61% for specimen sizes 100, 200, 300, 400, and 500 mm, respectively.
- Further, the mean values of $\overline{K_{IC}^{ini}}$ for CT geometry are less than those determined for TPBT by approximately 8.17, 9.03, 8.90, 8.84, and 8.57% for specimen sizes 100, 200, 300, 400, and 500 mm, respectively.

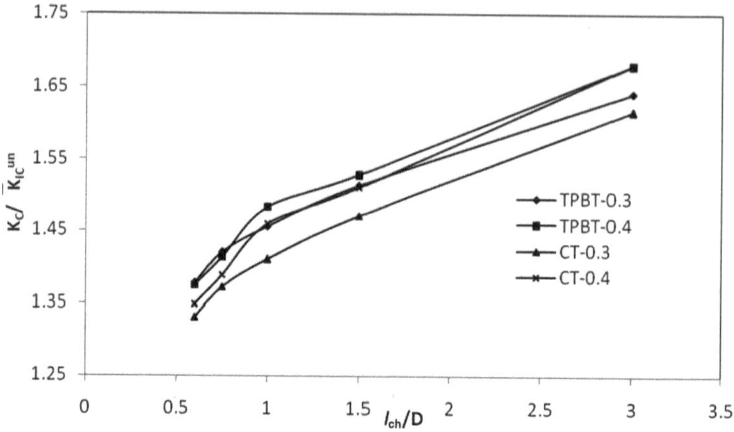

Fig. 4.48 The size-effect curves for $\overline{K_{IC}^{un}}$ on TPBT and CT test specimens

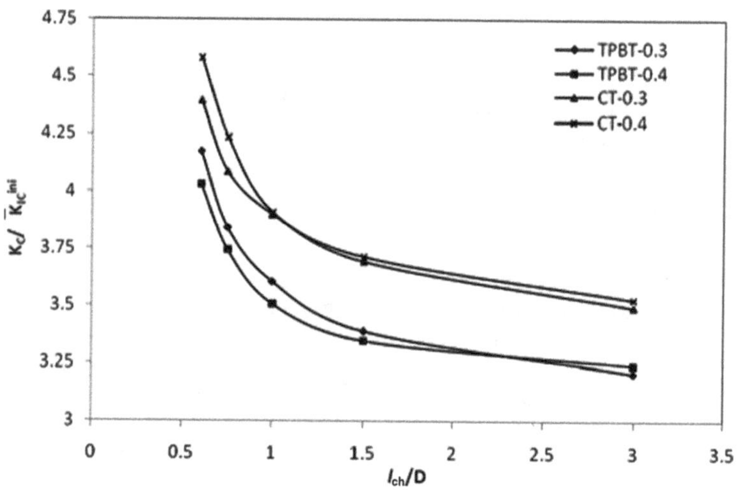

Fig. 4.49 The size-effect curves for $\overline{K_{IC}^{ini}}$ on TPBT and CT test specimens

- Further, the relationship between fracture parameters of double-*G* fracture criterion is presented in Fig. 4.50 for the size range $100 \leq D \leq 500$ mm. It is seen from the figure that the ratio of $\overline{K_{IC}^{un}}/\overline{K_{IC}^{ini}}$ is dependent on geometrical parameter (a_o/D ratio), specimen size, and specimen geometry. Neglecting the influence of a_o/D ratio, the mean values of $\overline{K_{IC}^{un}}/\overline{K_{IC}^{ini}}$ for TPBT are found to be 1.94, 2.22, 2.42, 2.68, and 2.98 and those for CT specimens are 2.13, 2.48, 2.72, 3.01, and 3.35 for the specimen sizes 100, 200, 300, 400, and 500 mm, respectively.

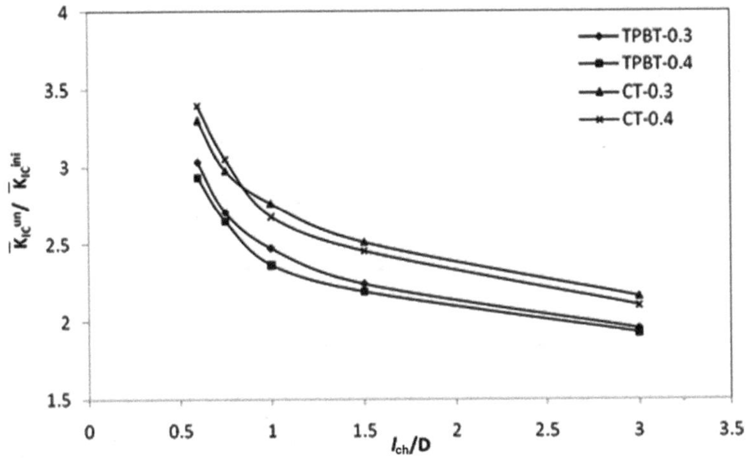

Fig. 4.50 Relationship between the parameters $\overline{K_{IC}^{un}} / \overline{K_{IC}^{ini}}$

4.10.4 Equivalence Between Double-K and Double-G Fracture Parameters

The interaction between the double-K fracture criterion and the double-G fracture model is shown graphically in Figs. 4.51 and 4.52. The figures present variation of the parameters $K_{IC}^{un}/\overline{K_{IC}^{un}}$ and $K_{IC}^{ini}/\overline{K_{IC}^{ini}}$ with l_{ch}/D ratio. It is seen from the figures that both the fracture models – the double-K fracture model based on the stress intensity factor approach and double-G fracture model based on energy release rate approach – are equivalent to each other. From the numerical study, the following observations are made:

- It is observed that the ratios of $K_{IC}^{un}/\overline{K_{IC}^{un}}$ and $K_{IC}^{ini}/\overline{K_{IC}^{ini}}$ are dependent on geometrical parameter (a_0/D ratio), specimen size, and specimen geometry. However, if the effect of a_0/D ratio is neglected, it seen that the mean values of $K_{IC}^{un}/\overline{K_{IC}^{un}}$ for three-point bending specimens are gained to be 1.033, 1.035, 1.037, 1.038, and 1.038 and the values obtained for compact tension specimens are 1.004, 1.002, 1.0, 0.999, and 0.998 for the specimen sizes 100, 200, 300, 400, and 500 mm, respectively.
- Similarly, the mean values of $K_{IC}^{ini}/\overline{K_{IC}^{ini}}$ for three-point bending geometry are calculated as 0.892, 0.904, 0.917, 0.926, and 0.937 and those for compact tension specimens are 0.970, 1.015, 1.0673, 1.095, and 1.105 for the specimen sizes 100, 200, 300, 400, and 500 mm, respectively.
- Based on experimental results of Xu and Zhang (2008), the relationship between fracture parameters of double-K fracture model and double-G fracture model is shown in Figs. 4.53 and 4.54. It is observed from test results of three-point bending geometry of size range 150–500 mm (Fig. 4.53) that the values of $K_{IC}^{ini}/\overline{K_{IC}^{ini}}$

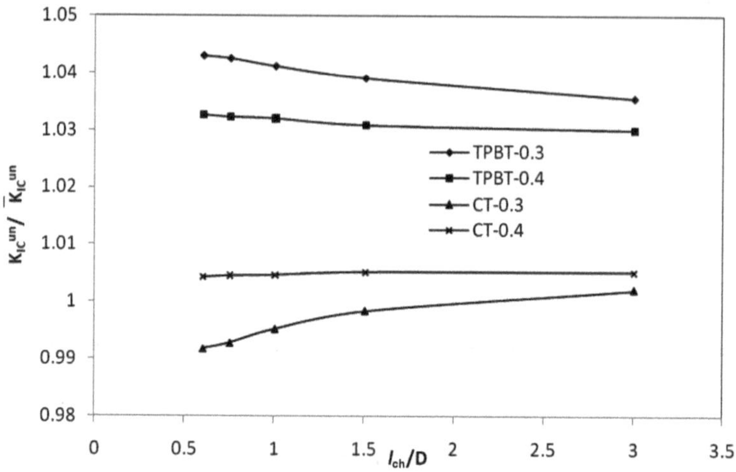

Fig. 4.51 Relationship between K_{IC}^{un} and the $\overline{K_{IC}^{un}}$ for TPBT and CT test specimens

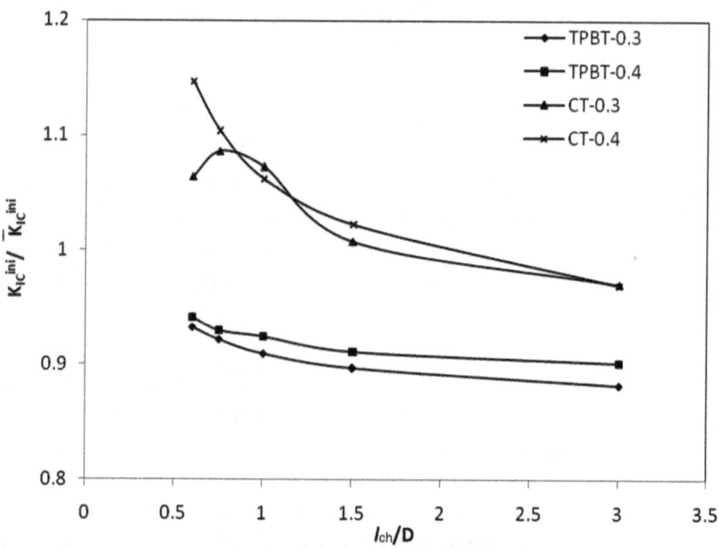

Fig. 4.52 Relationship between K_{IC}^{ini} and $\overline{K_{IC}^{ini}}$ for TPBT and CT test specimens

vary from 0.858 to 0.963, while the values of $K_{IC}^{un}/\overline{K_{IC}^{un}}$ are very close to 1.0. From experimental results of wedge splitting test of size range 200–1000 mm (Fig. 4.54), it is noticed that the values of both $K_{IC}^{ini}/\overline{K_{IC}^{ini}}$ and $K_{IC}^{un}/\overline{K_{IC}^{un}}$ are very close to 1.0.

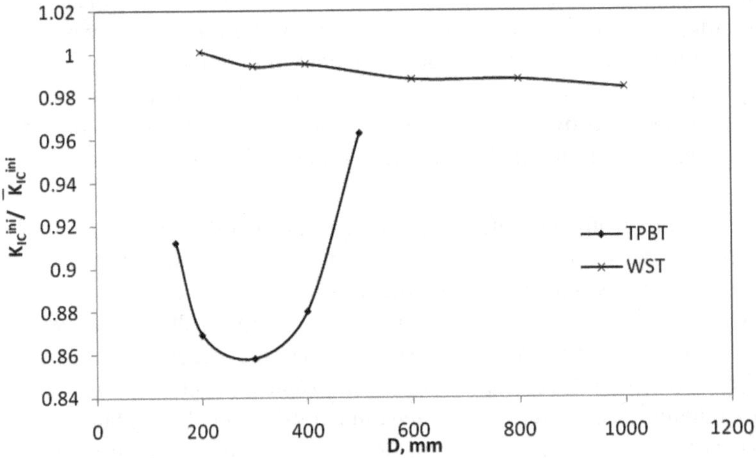

Fig. 4.53 Relationship between $K_{\mathrm{IC}}^{\mathrm{ini}}$ and $\overline{K_{\mathrm{IC}}^{\mathrm{ini}}}$ for TPBT and WST geometries taken from test results of Xu and Zhang (2008)

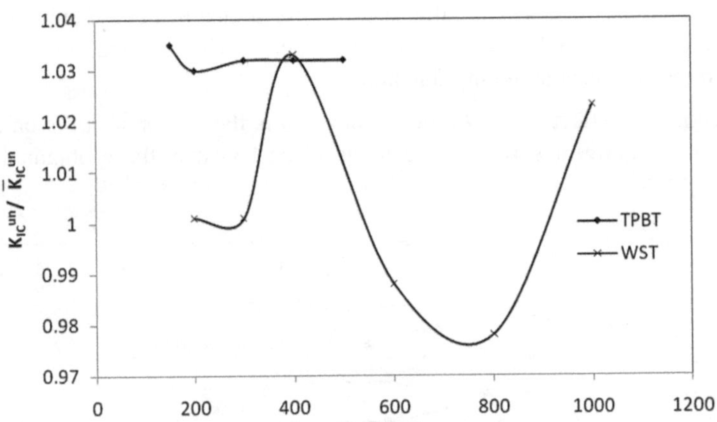

Fig. 4.54 Relationship between $K_{\mathrm{IC}}^{\mathrm{un}}$ and the $\overline{K_{\mathrm{IC}}^{\mathrm{un}}}$ for TPBT and WST geometries taken from test results of Xu and Zhang (2008)

4.10.5 Influence of Softening Function on the Double-K and Double-G Fracture Parameters

The nonlinear softening function of concrete as mentioned in the previous sections is further taken up for the calculation of double-*K* fracture parameters and the double-*G* fracture parameters considering TPBT geometry. This study shows the comparison between the fracture parameters obtained using nonlinear and bilinear softening relations. It is noted that the cohesive stress distribution is not involved in the computation of unstable fracture toughness of double-K fracture criterion or unstable fracture energy release rate of double-G fracture model and hence these values are unaffected by the choice of the softening functions of concrete. The following observations are made from the analysis of the results:

- In the modified bilinear softening function, an important assumption is made between the two parameters: critical value of crack-tip opening displacement and abscissa of kink point of bilinear softening. According to this assumption the value of $CTOD_c$ is equal to the value of w_s which results in relatively lower value of stress level at the initial crack tip than the stress level caused by the nonlinear softening function. Therefore, the value of cohesive toughness in the case of double-K fracture model or the critical value of cohesive breaking energy in the case of double-G fracture criterion is relatively less when bilinear softening function is used in the computation. This results in relatively higher value of the initiation toughness of double-K fracture parameters or effective initiation toughness of double-G fracture parameters.
- The influence of the softening function on the initiation toughness of double-K fracture model is shown in Fig. 4.55. It is observed from the figure that the value crack initiation toughness is influenced by the choice of the softening relation particularly between the bilinear softening function with the condition $CTOD_c = w_s$ and the nonlinear softening function.
- Neglecting the effect of a_0/D ratio in calculation, the values of initiation toughness using nonlinear softening are found to be less than those obtained using

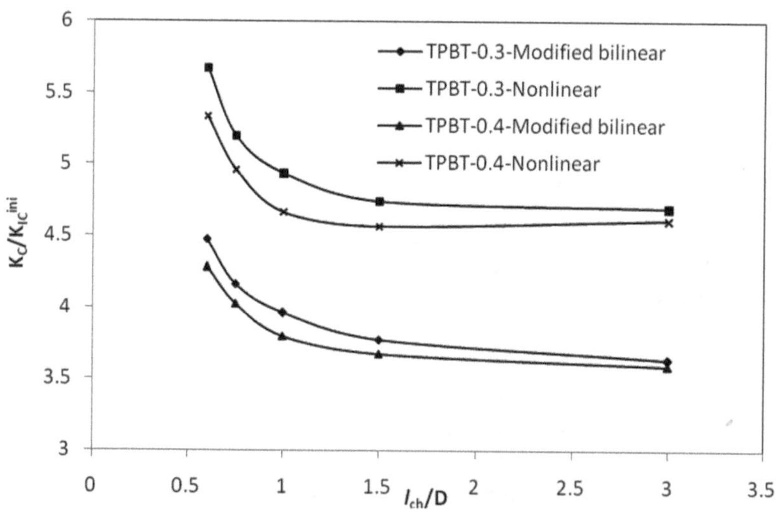

Fig. 4.55 Influence of softening function on K_{IC}^{ini} for TPBT

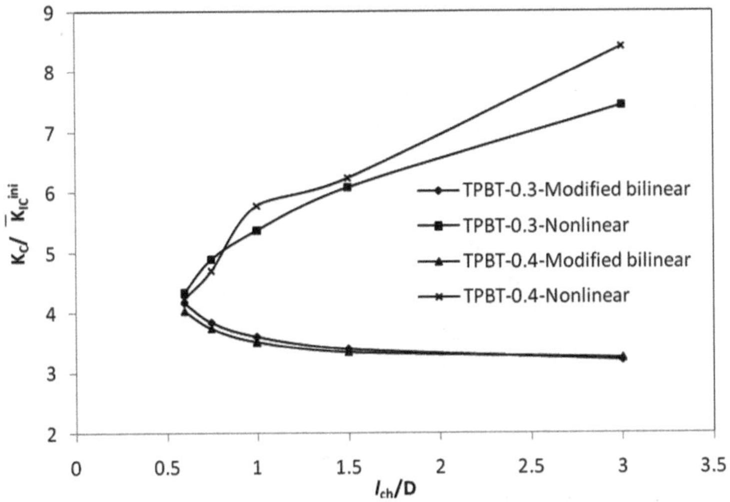

Fig. 4.56 Influence of softening function on $\overline{K_{IC}^{ini}}$ for TPBT

bilinear softening by nearly 22.27, 19.94, 19.17, 19.37, and 20.40% for the specimen sizes 100, 200, 300, 400, and 500 mm, respectively.

- Figure 4.56 shows the dependency of effective crack initiation toughness of double-G fracture parameters on the type of softening function used in computation. It is observed from the figure that the value of critical cohesive energy release is significantly influenced by softening function. At higher depth, i.e., $D = 500$ mm, the result of effective crack initiation toughness determined using either bilinear or nonlinear softening function is almost the same. On the other hand, as compared with the initiation toughness, an opposite trend in the value of effective crack initiation toughness is observed for lower specimen size range between 400 and 100 mm.
- The size effect on the value of effective crack initiation toughness is relatively more pronounced when the nonlinear softening function is used. The approximate difference in the calculated values of effective crack initiation toughness between the use of bilinear softening and nonlinear softening is found to be approximately 4.33% for $D = 500$ mm and 59.18% for $D = 100$ mm. With nonlinear softening function, the value of effective crack initiation toughness for specimen size 100 mm becomes approximately half of that for specimen size 500 mm. In this case, particularly the assumption made while defining the bilinear softening function seems to be more crucial.
- Effect of softening function on the relationship between the double-K fracture model and double-G fracture model in terms of $K_{IC}^{ini}/\overline{K_{IC}^{ini}}$ ratio is presented in Fig. 4.57. It is seen that both the fracture models are equivalent to each other in case bilinear softening function of concrete is taken in calculation. On the

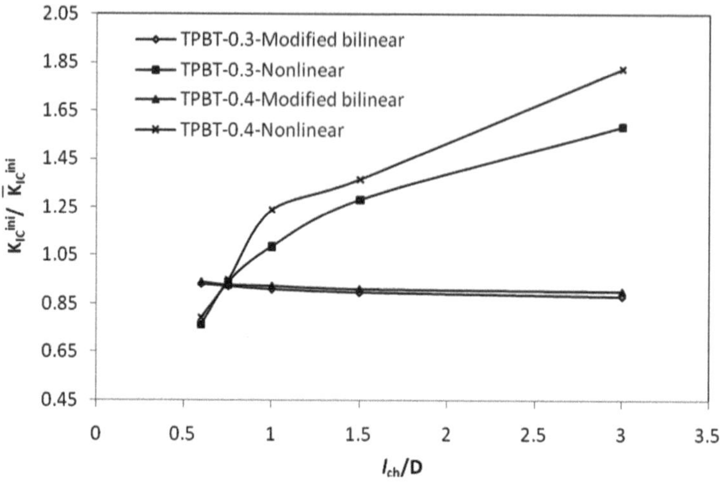

Fig. 4.57 Influence of softening function on $K_{\mathrm{IC}}^{\mathrm{ini}}/\overline{K_{\mathrm{IC}}^{\mathrm{ini}}}$ for TPBT

other hand, this equivalency is violated if the nonlinear softening is used. After neglecting the effect of a_0/D ratio, it is found that the mean values of $K_{\mathrm{IC}}^{\mathrm{ini}}/\overline{K_{\mathrm{IC}}^{\mathrm{ini}}}$ for three-point bending geometry vary approximately from 0.78 to 1.71 for the specimen size ranging from 500 to 100 mm.

4.11 Closing Remarks

The following salient observations can be made from the investigation carried out in the present chapter:

- An analytical method and a simplified approach are available in the literature for determining the double-K fracture parameters. While determining cohesive toughness using the existing analytical method, the solution is done by using numerical integration technique, which requires special treatment because of singularity problem at integral boundary; the simplified approach provides accurate results within acceptable error which employs a correction factor based on empirical relations used in the expression of cohesive toughness. The method introduced in the present work for determination of cohesive toughness using universal weight function may provide a viable solution in closed-form equations. Within the premises of three-point bending test results (Refai and Swartz 1987), as compared with the existing analytical method, the values of double-K fracture parameters can be evaluated with less than 2% error using four-term universal weight function, whereas almost the same results can be obtained using the five terms of the universal weight function. The use of simple forms of weight functions will facilitate a closed-form accurate solution in the analysis of fracture process of concrete structures.

- Trial-and-error or some specialized numerical technique for determining the value of effective crack extension is required in analytical method and weight function approach, while a direct calculation can be done using empirical relations in simplified approach. The double-*K* fracture criterion for crack propagation requires only unstable fracture load (peak load) and corresponding crack opening displacement, thus avoiding the use of closed-loop testing system. Therefore, applying the input data obtained using fictitious crack model, comprehensive comparison of numerical results of double-*K* fracture parameters can be made.
- Calculation of fracture parameters K_{IC}^{ini} and K_{IC}^{un} from the test results of wedge splitting test reported in Xu et al. (2006) showed that these values are not significantly affected by the initial notch length. Since the double-*K* fracture parameters can capture the important stages of concrete fracture, it is possible to define a new nondimensional brittleness parameter using double-*K* fracture parameters. However, this definition needs to be further investigated.
- From precise numerical study, it can be summarized that double-*K* fracture parameters are influenced by many factors such as geometrical parameter, specimen geometry, loading condition, and size effect. The influence of a_o/D ratio, shape of test specimen, and type of loading condition is relatively less than the size effect on the values of fracture parameters.
- From the comparison it is observed that up to a_o/D ratio 0.75, the maximum difference in the results of K_I^{un} between simplified approach and analytical method for TPBT geometry is 3.25% and it is slightly more at a_o/D ratio 0.8, that is, 5.34%. For CT test specimen this maximum difference up to a_o/D ratio 0.80 is 0.31% and the difference obtained on the mean values of K_I^{un} is 0.26%. On an average, the percentage difference in the K_{IC}^{C} for TPBT or CT specimen obtained using weight function with four terms, five terms, and simplified approach for specimen size 200 mm is about 1, 0.1, and 1%, respectively. These differences in calculated values of K_{IC}^{ini} are about 2, 0.2, and 2%, respectively. From various numerical studies, it is observed that as compared to analytical method, the error range in determining the double-*K* fracture parameters using simplified approach and weight function approach with four terms is comparable.
- Further, it is noticed that less than about 3, 6, and 10% difference in results of unstable fracture toughness, cohesion toughness, and crack initiation toughness can be obtained between the TPBT and CT test specimen geometries for specimen size ranging 100–600 mm and a_o/D ratios 0.3–0.5.
- From the investigation of the effect of loading condition on the double-*K* fracture parameters, it is observed that less than about 3, 4, and 2% difference in results of unstable fracture toughness, cohesion toughness, and crack initiation toughness can be obtained between the TPBT and FPBT loading conditions for specimen size range 100–600 mm and a_o/D ratios 0.3–0.5.
- The fracture parameters such as K_{IC}^{C} and K_{IC}^{ini} are somewhat influenced by the softening function of concrete.

- For the specimen size range $100 \le D \le 500\,\text{mm}$, the difference gained in the value of $\overline{K_{IC}^{un}}$ between compact tension and three-point bending test geometries is found to be approximately less than 3% and the difference gained in the value of $\overline{K_{IC}^{ini}}$ between the above test geometries is nearly less than 9%.
- The ratio of the values K_{IC}^{un}/K_{IC}^{ini} and $\overline{K_{IC}^{un}}/\overline{K_{IC}^{ini}}$ is found to be approximately in the range of 2–3.5 depending upon the size and shape of the specimen and initial notch length/depth ratio.
- The double-*K* fracture parameters and the corresponding double-*G* fracture parameters are found almost equivalent to each other in the case when the fracture parameters are determined using bilinear softening function of concrete. This observation agrees well with that of experimental investigation presented by Xu and Zhang (2008).
- Double-*K* fracture parameters and double-*G* fracture parameters are influenced by the softening function of concrete with respect to the bilinear softening and the assumption that $CTOD_c = w_s$. The values of initiation toughness of double-*K* fracture criterion using nonlinear softening are less than those obtained using bilinear softening by nearly 20% for the size range $100 \le D \le 500$. In contrast, the value of the effective initiation toughness of double-*G* fracture model is significantly influenced by the choice of softening function of concrete depending upon size of the specimen.
- The predicted value of effective initiation toughness using nonlinear softening is less than that obtained using bilinear softening by approximately 5% for specimen size 500 mm and 60% for specimen size 100 mm.

References

ASTM International Standard E399-06 (2006) Standard Test Method for Linear-Elastic Method Plane-Strain Fracture Toughness K_{IC} of Metallic Materials. West Conshohocken, PA, Copyright ASTM International: 1–32.

Brühwiler E, Wittmann FH (1990) The wedge splitting test: A method of performing stable fracture mechanics tests. Eng Fract Mech 35: 117–125.

Bueckner HF (1970) A novel principle for the computation of stress intensity factors. Z Angew Math Mech 50: 529–546.

Carpinteri A (1989a) Decrease of apparent tensile and bending strength with specimen size: Two different explanations based on fracture mechanics. Int J Solid Struct 25(4): 407–429.

Carpinteri A (1989b) Cusp catastrophe interpretation of fracture instability. J Mech Phys Solid 37(5): 567–582.

Carpinteri A (1989c) Post-peak and post-bifurcation analysis of cohesive crack propagation. Eng Fract Mech 32: 265–278.

CEB-Comite Euro-International du Beton-EB-FIP Model Code 1990 (1993) Bulletin D'Information No. 2123/214, Lausanne.

Fett T, Mattheck C, Munz D (1987) On calculation of crack opening displacement from the stress intensity factor. Eng Fract Mech 27: 697–715.

Glinka G, Shen G (1991) Universal features of weight functions for cracks in Mode I. Eng Fract Mech 40: 1135–1146.

Jenq YS, Shah SP (1985a) Two parameter fracture model for concrete. J Eng Mech ASCE 111(10): 1227–1241.

Jenq YS, Shah SP (1985b) A fracture toughness criterion for concrete. Eng Fract Mech 21: 1055–1069.

Karihaloo BL, Nallathambi P (1990) Size-effect prediction from effective crack model for plain concrete. Mater Struct 23(3): 178–185.

Karihaloo BL, Nallathambi P (1991) Notched beam test: Mode I fracture toughness. Fracture Mechanics Test Methods for Concrete. Report of RILEM Technical Committee 89-FMT (Edited by SP Shah, A Carpinteri), London, Chapman & Hall: 1–86.

Kim J-K, Lee C-S, Park C-K, Eo S-H (1997) The fracture characteristics of crushed limestone sand concrete. Cement Concr Res 27: 1719–1729.

Kumar S (2010) Behaviour of fracture parameters for crack propagation in concrete. Ph.D. Thesis submitted to Department of Civil Engineering, Indian Institute of Kharagpur, India.

Kumar S, Barai SV (2008a) Influence of specimen geometry and size-effect on the K_R-curve based on the cohesive stress in concrete. Int J Fract 152: 127–148.

Kumar S, Barai SV (2008b) Influence of specimen geometry on determination of double-K fracture parameters of concrete: A comparative study. Int J Fract 149: 47–66.

Kumar S, Barai SV (2008c) Cohesive crack model for the study of nonlinear fracture behaviour of concrete. J Inst Eng (India), CV 89 (Nov.): 7–15.

Kumar S, Barai SV (2009a) Determining double-K fracture parameters of concrete for compact tension and wedge splitting tests using weight function. Eng Fract Mech 76: 935–948.

Kumar S, Barai SV (2009b) Effect of softening function on the cohesive crack fracture parameters of concrete CT specimen. Sadhana-Acad Proc Eng Sci 36(6): 987–1015.

Kumar S, Barai SV (2009c) Equivalence between stress intensity factor and energy approach based fracture parameters of concrete. Eng Fract Mech 76: 1357–1372.

Kumar S, Barai SV (2009d) Size-effect of fracture parameters in concrete: A comparative study. Comput Concr Int J (under review).

Kumar S, Barai SV (2009e) Weight function approach for determining crack extension resistance based on the cohesive stress distribution in concrete. Eng Fract Mech 76: 1131–1148.

Kumar S, Barai SV (2009f) Influence of loading condition and size-effect on the K_R-curve based on the cohesive stress in concrete. Int J Fract 156: 103–110.

Kumar S, Barai SV (2010a) Determining the double-K fracture parameters for three-point bending notched concrete beams using weight function. Fatigue Fract Eng Mater Struct DOI: 10.1111/j.1460-2695.2010.01477.x.

Kumar S, Barai SV (2010b) Size-effect prediction from the double-K fracture model for notched concrete beam. Int J Damage Mech 9: 473–497.

MATLAB Version 7, The MathWorks, Inc., Copyright 1984–2004.

Murakami Y (1987) Stress Intensity Factors Hand Book (Committee on Fracture Mechanics, The Society of Materials Science, Japan), Vol. 1. Oxford, Pergamon Press.

Petersson PE (1981) Crack growth and development of fracture zone in plain concrete and similar materials. Report No. TVBM-100, Lund Institute of Technology.

Planas J, Elices M (1990) Fracture criteria for concrete: Mathematical validations and experimental validation. Eng Fract Mech 35: 87–94.

Refai TME, Swartz SE (1987) Fracture behavior of concrete beams in three-point bending considering the influence of size effects. Report No. 190, Engineering Experiment Station, Kansas State University.

Reinhardt HW, Cornelissen HAW, Hordijk DA (1986) Tensile tests and failure analysis of concrete. J Struct Eng ASCE 112(11): 2462–2477.

Rice JR (1972) Some remarks on elastic crack-tip stress fields. Int J Solid Struct 8: 751–758.

RILEM Draft Recommendations (TC89-FMT) (1990) Determination of fracture parameters K_{IC}^{s} and $CTOD_c$) of plain concrete using three-point bend tests, proposed RILEM draft recommendations. Mater Struct 23(138): 457–460.

Shen G, Glinka G (1991) Determination of weight functions from reference stress intensity factors. Theor Appl Fract Mech 15: 237–245.

Tada H, Paris PC, Irwin G (1985) The Stress Analysis of Cracks Handbook. Hellertown, PA, Del Research Corporation.

Wittmann FH, Rokugo K, Bruhwiller E, Mihashi H, Simopnin P (1988) Fracture energy and strain softening of concrete as determined by compact tension specimens. Mater Struct 21(1): 21–32.

Xu S (1999) Determination of parameters in the bilinear, Reinhardt's non-linear and exponentially non-linear softening curves and their physical meanings. Werkstoffe und Werkstoffprüfung im Bauwesen, Hamburg, Libri BOD: 410–424.

Xu S, Reinhardt HW (1998) Crack extension resistance and fracture properties of quasi-brittle materials like concrete based on the complete process of fracture. Int J Fract 92: 71–99.

Xu S, Reinhardt HW (1999a) Determination of double-K criterion for crack propagation in quasi-brittle materials, Part I: Experimental investigation of crack propagation. Int J Fract 98: 111–149.

Xu S, Reinhardt HW (1999b) Determination of double-K criterion for crack propagation in quasi-brittle materials, Part II: Analytical evaluating and practical measuring methods for three-point bending notched beams. Int J Fract 98: 151–177.

Xu S, Reinhardt HW (1999c) Determination of double-K criterion for crack propagation in quasi-brittle materials, Part III: Compact tension specimens and wedge splitting specimens. Int J Fract 98: 179–193.

Xu S, Reinhardt HW (2000) A simplified method for determining double-K fracture meter parameters for three-point bending tests. Int J Fract 104: 181–209.

Xu S, Zhang X (2008) Determination of fracture parameters for crack propagation in concrete using an energy approach. Eng Fract Mech 75: 4292–4308.

Xu S, Zhao Y, Wu Z (2006) Study on the average fracture energy for crack propagation in concrete. J Mater Civil Eng ASCE 18(6): 817–824.

Zhao G, Jiao H, Xu S (1991) Study on fracture behavior with wedge splitting test method. Fracture Processes in Concrete, Rock and Ceramics (Edited by van Mier et al.), London, E & F.N. Spon: 789–798.

Zhu X (1997) Transport organischer Flüssigkeiten in Betonbauteilen mit Mikro- und Biegerissen. Berlin, Deutscher Ausschuss für Stahlbeton, Heft 475, Beuth Verlag GmbH: S.1–104.

Zi G, Bažant ZP (2003) Eignvalue method for computing size effect of cohesive cracks with residual stress, with application to kink-bands in composites. Int J Eng Sci 41: 1519–1534.

Zi G, Belytschko T (2003) New crack-tip elements for X-FEM and applications to cohesive cracks. Int J Numer Methods Eng 57: 2221–2240.

Chapter 5
Fracture Properties of Concrete Based on the K_R Curve Associated with Cohesive Stress Distribution

5.1 Introduction

This chapter is attributed to the previous works carried out by Kumar and Barai (2008a, b, 2009a–c, 2010) and Kumar (2010) in which mainly the application of *weight function method* is introduced to determine the K_R curve associated with the cohesive stress distribution in the FPZ. Experimental data (Roesler et al. 2007) available in the literature are used to validate the weight function approach for determining the K_R curve and the results are analyzed to study the instability criteria using the K_R-curve method. Finally, the influence of specimen geometry, loading condition, size effect, and softening function of concrete on the K_R curve and the related fracture parameters is carried out for which the load-crack opening displacement curves are obtained using the FCM.

5.2 The K_R-Curve Method

The K_R-curve method based on cohesive stress distribution in the FPZ introduced by Xu and Reinhardt (1998, 1999) and Reinhardt and Xu (1999) for complete fracture process description of concrete differs from the conventional method of the *R* curve. The distribution of cohesive stress along the FPZ at different stages of loading conditions is taken into account in order to evaluate the K_R curve. In addition, the double-*K* fracture parameters are introduced in the stability analysis using the K_R curve. During crack propagation, the contribution of cohesive toughness increases with the increase in process zone length that depends upon cohesive stress distribution σ which is a function of crack opening displacement *w*, tensile strength of concrete f_t, and the propagating crack length *a*. At the onset of unstable crack propagation, the stress intensity factor *K* at the tip of the propagating crack is expressed as

$$K = K_R(\Delta a) \tag{5.1}$$

where $K_R(\Delta a)$ is the crack extension resistance at crack extension length $\Delta a = a - a_0$. Also, the crack extension resistance $K_R(\Delta a)$ can be expressed with the following relation:

S. Kumar, S.V. Barai, *Concrete Fracture Models and Applications*,
DOI 10.1007/978-3-642-16764-5_5, © Springer-Verlag Berlin Heidelberg 2011

$$K_{\rm R}(\Delta a) = K_{\rm IC}^{\rm ini} + K_{\rm I}^{\rm COH}(\Delta a) \tag{5.2}$$

in which

$$K_{\rm I}^{\rm COH}(\Delta a) = F_1\,(f_{\rm t}, \sigma, \Delta a) \tag{5.3}$$

The instability of a propagating crack in cracked solid structures can be represented using the $K_{\rm R}$ curve as shown in Fig. 5.1, in which the value $K(P_{\rm u}, a_{\rm c})$ is the stress intensity factor at the maximum load $P_{\rm u}$ corresponding to critical effective crack extension $a_{\rm c}$ and $K_{\rm C}^{\rm t}$ is the total crack extension resistance at the point of tangency between K and $K_{\rm R}$ curves corresponding to the onset of unstable crack propagation. The fracture criterion using $K_{\rm R}$ curve can be mathematically expressed by Eq. (2.3). Xu and Reinhardt (1999) observed from experiments that the values of $K_{\rm C}^{\rm t}$ determined by the $K_{\rm R}$ curve are equal to the values of $K_{\rm IC}^{\rm un}$ and the values of the initial cracking parts on the $K_{\rm R}$ curves are equal to those of $K_{\rm IC}^{\rm ini}$. As a result, the double-K fracture parameters are used to describe the stability of a propagating crack in complete fracture process. Hence comparing the K curve and $K_{\rm R}$ curve, the stability analysis of crack propagation for the mode I condition can be expressed as given below:

$$\begin{aligned} &K_{\rm I} < K_{\rm IC}^{\rm ini} \quad \text{No crack propagation appears} \\ &K_{\rm I} = K_{\rm IC}^{\rm ini} \quad \text{The crack starts to propagate steadily} \\ &K_{\rm IC}^{\rm ini} < K_{\rm I} < K_{\rm IC}^{\rm un} \quad \text{The crack propagates steadily} \end{aligned} \tag{5.4}$$

$$K_{\rm I} = K_{\rm IC}^{\rm un} \quad \text{Critical unstable crack propagation begins}$$

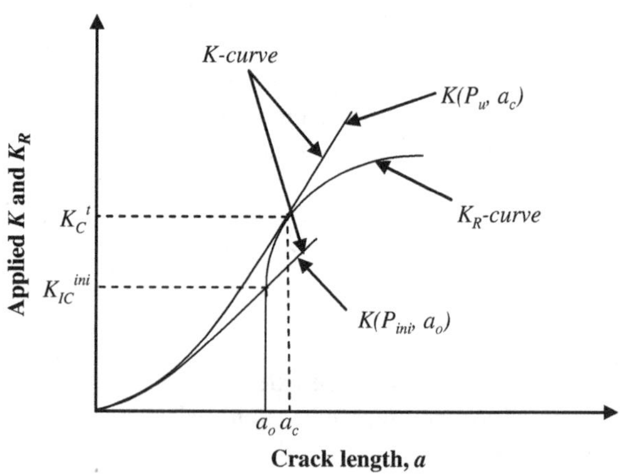

Fig. 5.1 Graphical representation of the $K_{\rm R}$ curve

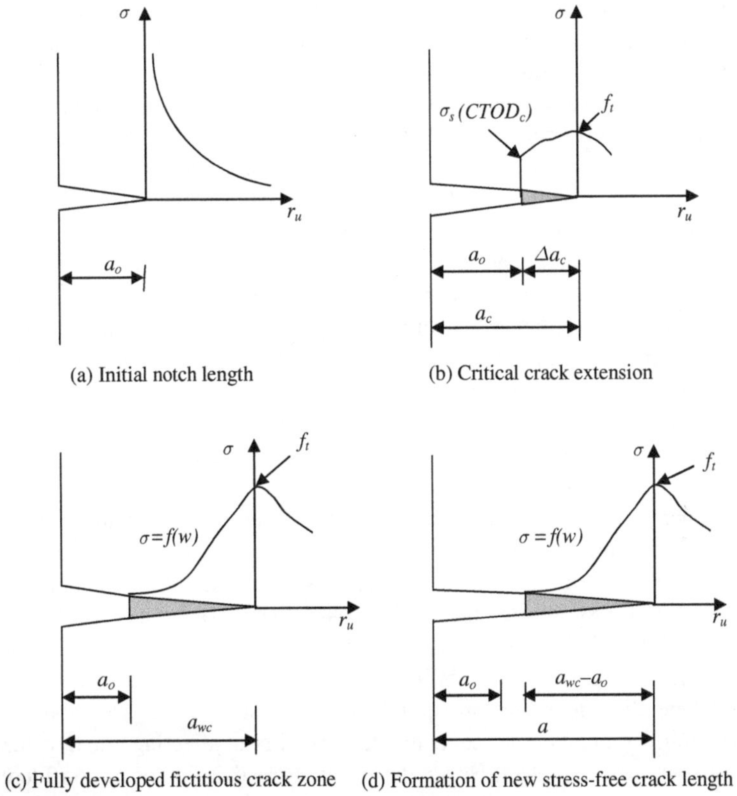

Fig. 5.2 Four stages of crack propagation during the fracture process

$$K_I > K_{IC}^{un} \quad \text{The crack propagates unsteadily}$$

In order to develop the K_R curve for complete fracture process considering the cohesive stress in fictitious fracture zone, it becomes important to determine the value of cohesive toughness K_I^{COH} at every stage of loading on the structure. During the crack propagation, four different loading conditions are considered using three characteristic crack lengths (a_o, a_c, and a_{wc}) as illustrated in Fig. 5.2.

The figure shows the development of cohesive stress distribution during crack initiation and crack growth in fracture process of concrete. With the increase in the fictitious fracture zone from Fig. 5.2a–d, the undamaged length of the ligament r_u decreases. Figure 5.2a represents the first stage of loading in which the external loading on the structure is still less than P_{ini} and the material ahead of the initial crack tip still behaves in a linear elastic manner. In this case, the stress near the crack tip tends to be singular and the classical form of linear elastic fracture mechanics is applicable. As the external loading is equal to the value of P_{ini}, the stable crack begins to form. Further, when the load increases, the stable crack propagation begins to occur and the cohesive stress will start acting across the new crack surfaces. At critical condition, the external load becomes equal to peak load P_u, the crack opening displacement at the initial crack tip w_t reaches to the value of critical value of crack-tip opening displacement $CTOD_c$ and the total crack extension reaches to its critical value a_c. This situation of loading stage is shown in Fig. 5.2b. In the subsequent loading condition as shown in Fig. 5.2c, the crack extension crosses the value of a_c and finally it reaches to the value of a_{wc} at which the value of w_t equals to critical value of crack opening displacement w_c. It means that the cohesive stress at the initial crack tip reduces to zero and a fully developed fracture process zone $(a_{wc} - a_o)$ appears. This is an important phenomenon which is captured by the K_R curve similar to the nonlinear fracture models attributed to numerical methods. In the fourth stage of loading as shown in Fig. 5.2d, when the crack extension a is more than a_{wc}, a new stress-free crack is formed and the subsequent loading step results in the growth of stress-free crack propagation. From the foregoing discussion, it is clear that the K_R curve based on cohesive force has the ability to capture the mechanism of crack initiation and crack growth in the large fracture process zone of concrete. It is a known fact that the actual shape of the softening behavior in the fictitious fracture zone follows a nonlinear pattern. However, the bilinear softening law is often used for all practical purposes in the case of normal concrete. Xu and Reinhardt (1998) reported that the shapes of the distribution of cohesive stress ahead of the crack tip were not sensitive to the overall value of stress intensity factor at the crack tip and linear or bilinear shape of cohesive stress distribution was assumed depending upon the corresponding loading stages. This assumption might lead to simplification in numerical computations without affecting much on the final values of fracture parameters. Hence the linear stress distribution is considered for loading stage shown in Fig. 5.2b and the bilinear stress distribution is assumed for the loading stages shown in Fig. 5.2c, d. Moreover, the above assumption seems to be reasonable because in the second stage of loading (Fig. 5.2b), the process zone length is relatively small and mainly the initial portion of the softening branch is used. In other subsequent loading stages, the process zone length becomes larger in which a larger portion of the softening branch is utilized.

5.3 Analytical Method for Evaluation of the K_R Curve

The general expression for the crack extension resistance for complete fracture associated with cohesive stress distribution in the fictitious fracture zone for three-point bending test geometry (Xu and Reinhardt 1998, Reinhardt and Xu 1999) is given as below:

$$K_R(\Delta a) = K_{IC}^{ini} + \int_{a_o}^{a} \frac{2}{\sqrt{\pi a}} \sigma(x) F\left(\frac{x}{a}, \frac{a}{D}\right) \mathrm{d}x \tag{5.5}$$

in which *F*(*x/a*, *a/D*) is the Green's function given in Eq. (4.12). The crack extension resistance can be determined using Eq. (5.5) according to the situations arisen during crack propagation as shown in Fig. 5.2. In each stage of loading condition, the length of propagating crack, cohesive stress distribution and limits of integral equation (5.5) are different which can be outlined in the following four cases.

5.3.1 Case 1: When $a = a_o$

In this loading stage, there is no advancement in the initial notch length and the body remains in elastic condition, subjected to small load (up to P_{ini}) without any slow crack growth. Consequently, cohesive stress $\sigma(x) = 0$ and the crack growth resistance remains equal to initiation toughness of the material. The value of $K_R(\Delta a)$ is written using Eq. (5.5) as

$$K_R(\Delta a) = K_{IC}^{\text{ini}} \tag{5.6}$$

5.3.2 Case 2: When $a_o \leq a_o \leq a_c$

During this stage of loading, the structure shows the stable slow crack extension until the effective crack extension a_c corresponding to the maximum load P_u is achieved. In the process zone, the cohesive force starts acting across the fictitious fracture zone resulting in the increase of crack extension resistance. At critical condition, the CTOD will be equal to the CTOD_c as shown in Fig. 5.3. During this loading condition, i.e., $a_o \leq a \leq a_c$ or $0 \leq \text{CTOD} \leq \text{CTOD}_c$, the distribution of cohesive stress is approximated to be linear. The variation of cohesive stress along the fictitious fracture zone is expressed as

$$\sigma(x) = \sigma(w_t) + \frac{x - a_o}{a - a_o}[f_t - \sigma(w_t)], \quad \text{for } a_o \leq x \leq a \tag{5.7}$$

where $\sigma(w_t)$ and w_t are the values of cohesive stress and crack opening displacement at the tip of initial notch, respectively. The value of $\sigma(w_t)$ is determined using softening function of concrete. In nonlinear fracture models attributed to numerical

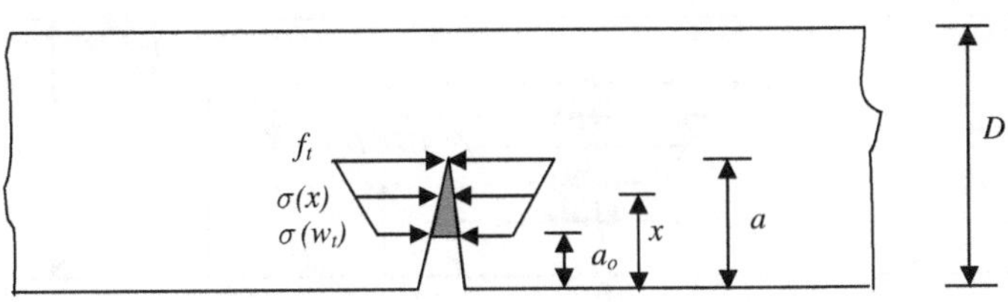

Fig. 5.3 Distribution of cohesive stress in the fictitious fracture zone during crack extension, $a_o \leq a \leq a_c$

methods, the fracture nodes in the process zone are very close and the softening law is compulsorily required to be used at these nodes to obtain the corresponding cohesive stresses. In contrast, according to the development of the K_R-curve method, only two values of cohesive stresses $\sigma(w_t)$ or and $\sigma_s(\mathrm{CTOD_c})$ in the fictitious fracture zone are required to be determined. Any softening function such as bilinear, exponential, and nonlinear can be employed to gain the cohesive stresses.

5.3.3 Case 3: When $a_c \leq a \leq a_{wc}$

During this stage of loading, the applied load P, the value of CTOD, and the effective crack length have increased more than the values of P_u, $\mathrm{CTOD_c}$, and critical effective crack extension a_c, respectively. The cohesive stress distribution for this case takes a bilinear shape as shown in Fig. 5.4 and is written as below:

$$\sigma(x) = \begin{cases} \sigma_1(x) = \sigma(w_t) + \dfrac{x - a_o}{a - \Delta a_c - a_o}[\sigma_s(\mathrm{CTOD_c}) - \sigma(w_t)], \\ \text{for } a_o \leq x \leq (a - \Delta a_c) \\ \sigma_2(x) = \sigma_s(\mathrm{CTOD_c}) + \dfrac{x - a + \Delta a_c}{\Delta a_c}[f_t - \sigma_s(\mathrm{CTOD_c})], \\ \text{for } (a - \Delta a_c) \leq x \leq a \end{cases} \tag{5.8}$$

The crack extension resistance can be obtained using Eq. (5.5) in which the limits of integration should be taken in two steps: $a_o \leq x \leq (a - \Delta a_c)$ for cohesive stress $\sigma_1(x)$ and $(a - \Delta a_c) \leq x \leq a$ for cohesive stress $\sigma_2(x)$. The value of Green's function $F(x/a, a/D)$ for a given effective crack extension is determined using Eq. (4.12).

5.3.4 Case 4: When $a \geq a_{wc}$

This situation arises for the loading condition corresponding to the descending portion of P-CMOD curve. At the effective crack extension a_{wc}, the full shape of cohesive stress distribution is allowed to develop and beyond further extension in

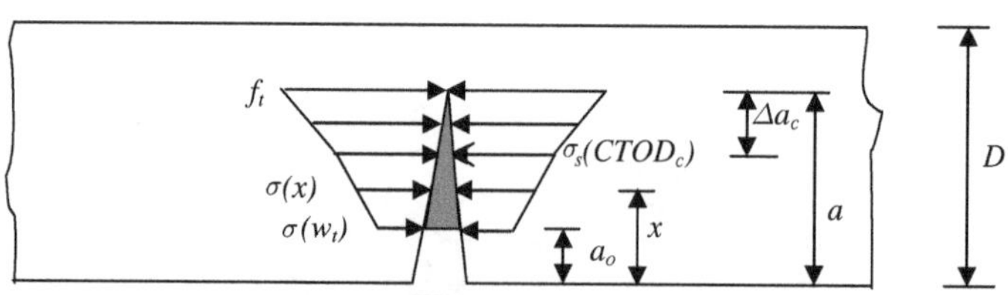

Fig. 5.4 Bilinear distribution of cohesive stress in the fictitious fracture zone during crack extension, $a_c \leq a \leq a_{wc}$

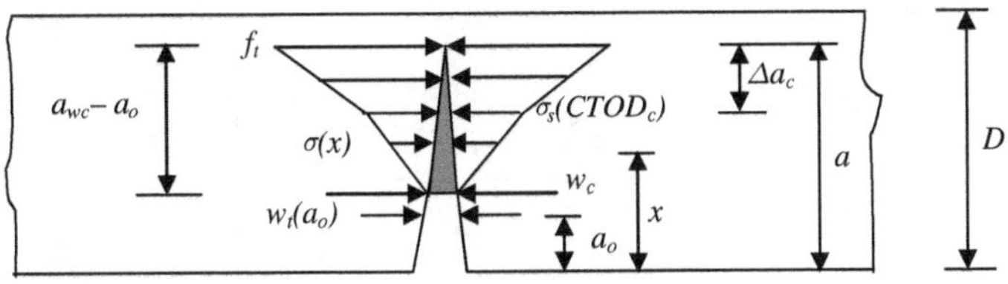

Fig. 5.5 Development of full bilinear cohesive stress in the fictitious fracture zone during new stress-free crack extension, $a \geq a_{wc}$

crack during loading, a new stress-free crack in front of the initial notch tip is formed. In this case, the stress distribution as shown in Fig. 5.5 is expressed with following relations:

$$\sigma(x) = \begin{cases} \sigma_1(x) = 0 \text{ for } a_o \leq x \leq (a - \Delta a_{wc} + a_o) \\ \sigma_2(x) = \dfrac{x + a_{wc} - a_o - a}{a_{wc} - \Delta a_c - a_o}\sigma_s(\text{CTOD}_c), \\ \text{for } (a + a_o - a_{wc}) \leq x \leq (a - \Delta a_c) \\ \sigma_3(x) = \sigma_s(\text{CTOD}_c) + \dfrac{x - a + \Delta a_c}{\Delta a_c}[f_t - \sigma_s(\text{CTOD}_c)], \\ \text{for } (a - \Delta a_c) \leq x \leq a \end{cases} \tag{5.9}$$

Similar to case 3, the crack extension resistance for this case is also evaluated using Eqs. (5.5) and (4.12).

5.4 Weight Function Approach for Evaluation of the K_R Curve

5.4.1 Derivation of Closed-Form Expression for Cohesive Toughness

For calculating the crack extension resistance, a generalized situation of linear stress distribution in the fictitious fracture zone as shown in Fig. 5.6 is considered. It is assumed that the cohesive stress acts between the crack length p and q for a total crack extension a. The magnitude of cohesive stress at the crack length p and q is σ_p and σ_q, respectively. Therefore, at any crack length value x, the distribution of cohesive stress is expressed in the following form:

$$\sigma(x) = \sigma_p + \frac{x - p}{q - p}[\sigma_q - \sigma_p], \text{ for } p \leq x \leq q \tag{5.10}$$

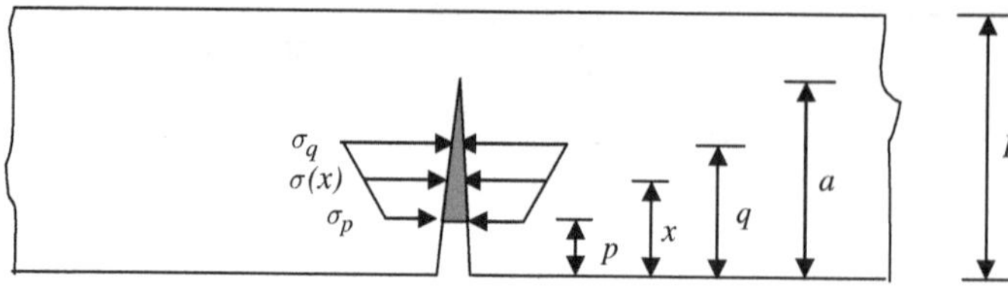

Fig. 5.6 Linear distribution of cohesive stress in the fictitious fracture zone during crack extension, $p \le x \le q$

Let $C_1 = \sigma_p$ and $C_2 = (\sigma_q - \sigma_p)/(q - p)$ in Eq. (5.10); the cohesion toughness due to linear cohesive stress distribution in fictitious fracture zone during crack propagation is computed using Eq. (4.1) in the following expression:

$$K_{\mathrm{I}}^{\mathrm{COH}}(\Delta a) = \int_{p}^{q} [C_1 + C_2(x - p)].m(x, a)\mathrm{d}x \tag{5.11}$$

Using the four-term weight function expressed in Eq. (4.19), the value of $K_{\mathrm{I}}^{\mathrm{COH}}(\Delta a)$ is derived with the help of Eq. (5.11) in a closed-form solution as given below:

$$\begin{aligned} K_{\mathrm{I}}^{\mathrm{COH}}(\Delta a) = {} & kaC_1 \left\{ \begin{array}{l} 2\left(A^{1/2} - B^{1/2}\right) + M_1(A - B) \\ + \dfrac{2M_2}{3}\left(A^{3/2} - B^{3/2}\right) + \dfrac{M_3}{2}\left(A^2 - B^2\right) \end{array} \right\} \\ & + ka^2C_2 \left\{ \begin{array}{l} \dfrac{2}{3}\left[2A^{3/2} - (3A - B)B^{1/2}\right] + \dfrac{M_1}{2}(A - B)^2 \\ + \dfrac{2M_2}{15}\left[B^{3/2}(3B - 5A) + 2A^{5/2}\right] \\ + \dfrac{M_3}{6}\left[\begin{array}{l} 3(2 - A)\left\{(1 - B)^2 - (1 - A)^2\right\} \\ + 6(1 - A)(B - A) + 2\left\{(1 - B)^3 - (1 - A)^3\right\} \end{array} \right] \end{array} \right\} \end{aligned} \tag{5.12}$$

in which $k = \sqrt{\frac{2}{\pi a}}$, $A = 1 - \frac{p}{a}$ and $B = 1 - \frac{q}{a}$.

If five-term weight function is taken into account as expressed in Eq. (4.21), the value of $K_{\mathrm{I}}^{\mathrm{COH}}(\Delta a)$ is determined using Eq. (5.11) in a closed-form solution as expressed in Eq. (5.13):

$$
\begin{aligned}
K_{\mathrm{I}}^{\mathrm{COH}}(\Delta a) = kaC_1 &\left\{ \begin{array}{l} 2\left(A^{1/2} - B^{1/2}\right) + M_1 (A - B) + \dfrac{2M_2}{3}\left(A^{3/2} - B^{3/2}\right) \\ + \dfrac{M_3}{2}\left(A^2 - B^2\right) + \dfrac{2M_4}{5}\left(A^{5/2} - B^{5/2}\right) \end{array} \right\} \\
+ ka^2 C_2 &\left\{ \begin{array}{l} \dfrac{2}{3}\left[2A^{3/2} - (3A - B)\, B^{1/2}\right] + \dfrac{M_1}{2}(A - B)^2 \\ + \dfrac{2M_2}{15}\left[B^{3/2}(3B - 5A) + 2A^{5/2}\right] \end{array} \right. \\
+ \frac{M_3}{6} &\left[\begin{array}{l} 3(2 - A)\left\{(1 - B)^2 - (1 - A)^2\right\} + 6(1 - A)(B - A) \\ + 2\left\{(1 - B)^3 - (1 - A)^3\right\} \end{array} \right] \\
+ \frac{2M_4}{35} &\left.\left[B^{5/2}(5B - 7A) + 2A^{7/2}\right]\right\}
\end{aligned}
\tag{5.13}
$$

where k, A, and B are the same values as mentioned in Eq. (5.12). The weight function approach presented herein is an alternative method for the evaluation of crack extension resistance based on cohesive stress distribution. Instead of using the analytical method currently available in the literature, the authors make use of weight function that does not require specialized numerical integration techniques to obtain the K_R curve. In Chap. 4, (see Sect. 4.3) a similar use of the weight function was presented by the authors to obtain the fracture parameters of double-K fracture criterion in which the closed-form expressions for cohesive toughness were derived for trapezoidal distribution of cohesive stress in the fictitious fracture zone and those could not be applicable to the bilinear variation of cohesive stress. Equations (5.12) and (5.13) are derived in a more generalized manner and they are different from those presented previously. More appropriately, the equations presented in Chap. 4 are the special cases of Eqs. (5.12) and (5.13) and are applicable only to linear variation of cohesive stress acting in the whole length of process zone ($a - a_o$). On the other hand, Eqs. (5.12) and (5.13) are applicable to linear and bilinear distribution of cohesive stresses acting in the partial or whole length of the process zone which are essential requirements for determining the crack extension resistance.

According to the four stages of crack propagation, the crack extension resistance due to cohesive toughness can be determined using either Eq. (5.12) or (5.13). In each stage of propagating crack, length of crack extension a, cohesive stress distribution, i.e., constants C_1 and C_2, and the range of process zone in which the cohesive stress acts along the crack line (the values of p and q) in Eq. (5.12) or (5.13) will be different. These are briefly stated in the following cases.

5.4.2 Case 1: When $a = a_o$

Since cohesive stress distribution $\sigma(x) = 0$, there is no contribution of cohesive toughness in the total crack growth resistance and it remains equal to the value of $K_{\mathrm{IC}}^{\mathrm{ini}}$.

5.4.3 Case 2: When $a_o \le a \le a_c$

For this condition ($a_o \le a \le a_c$ or $0 \le \mathrm{CTOD} \le \mathrm{CTOD_c}$) of loading, the cohesive stress distribution is represented in Eq. (5.7). Referring to Fig. 5.3, the value of $K_I^{COH}(\Delta a)$ is written as

$$K_I^{COH}(\Delta a) = \int_{a_0}^{a} \sigma(x).m(x,a)\mathrm{d}x \tag{5.14}$$

Equation (5.14) can be directly evaluated using Eq. (5.12) or (5.13) in which $C_1 = \sigma(w_t)$, $C_2 = (f_t - \sigma(w_t))\,/\,(a - a_o)$, $p = a_o$, and $q = a$.

5.4.4 Case 3: When $a_c \le a \le a_{wc}$

For this condition of crack propagation (Fig. 5.4), the cohesive stress distribution takes a bilinear shape as expressed in Eq. (5.8). Two parts of cohesive toughness due to stress distribution $\sigma_1(x)$ and $\sigma_2(x)$ are calculated separately and added together for total contribution of cohesion toughness such as

$$K_I^{COH}(\Delta a) = \int_{a_0}^{a-\Delta a_c} \sigma_1(x).m(x,a)\mathrm{d}x + \int_{a-\Delta a_c}^{a} \sigma_2(x).m(x,a)\mathrm{d}x \tag{5.15}$$

Equation (5.15) is evaluated using either Eq. (5.12) or (5.13) in which $C_1 = \sigma(w_t)$, $C_2 = [\sigma_s(\mathrm{CTOD_c}) - \sigma(w_t)]\,/\,(a - \Delta a_c - a_o)$, $p = a_o$, and $q = (a - \Delta a_c)$ associated with cohesive stress $\sigma_1(x)$, whereas these values associated with cohesive stress $\sigma_2(x)$ are $C_1 = \sigma_s(\mathrm{CTOD_c})$, $C_2 = [f_t - \sigma_s(\mathrm{CTOD_c})]/\Delta a_c$, $p = (a - \Delta a_c)$, and $q = a$.

5.4.5 Case 4: When $a \ge a_{wc}$

This case of loading condition is shown in Fig. 5.5 and the cohesive stress distribution is expressed through relation (5.9). Contribution of cohesive toughness in the crack extension resistance is expressed as

$$\begin{aligned} K_I^{COH}(\Delta a) = & \int_{a_0}^{a+a_0-a_{wc}} \sigma_1(x).m(x,a)\mathrm{d}x \\ & + \int_{a+a_0-a_{wc}}^{a-\Delta a_c} \sigma_2(x).m(x,a)\mathrm{d}x + \int_{a-\Delta a_c}^{a} \sigma_3(x).m(x,a)\mathrm{d}x \end{aligned} \tag{5.16}$$

In this case, depending upon the number of terms in the weight function, the cohesion toughness is calculated in three parts using Eq. (5.12) or (5.13). There are the three corresponding stress distribution such as $\sigma_1(x)$, $\sigma_2(x)$, and $\sigma_3(x)$. Since $\sigma_1(x) = 0$, the contribution of toughness due to the cohesive stress $\sigma_1(x) = 0$. For cohesive stress $\sigma_2(x)$, the values of $C_1 = 0$, $C_2 = \sigma_s(\mathrm{CTOD_c})/(a_{wc} - \Delta a_c - a_o)$, $p = (a + a_o - a_{wc})$, and $q = (a - \Delta a_c)$ are taken in Eq. (5.12) or (5.13), whereas those values for cohesive stress $\sigma_3(x)$ are $C_1 = \sigma_s(\mathrm{CTOD_c})$, $C_2 = [f_t - \sigma_s(\mathrm{CTOD_c})]/\Delta a_c$, $p = (a - \Delta a_c)$, and $q = a$. Once the value of $K_I^{COH}(\Delta a)$ is evaluated using either Eq. (5.12) or (5.13), the crack extension resistance $K_R(\Delta a)$ can be computed with the help of Eq. (5.2).

5.5 Computation and Validation of the K_R Curve

The following steps are followed to compute the K_R curve:

1. Effective crack extension is determined employing linear asymptotic superposition assumption and using the LEFM formulae for a particular type of specimen geometry as mentioned in Chap. 4 (see Sects. 4.4 and 4.5).
2. SIF for a particular geometry of the specimen is obtained using the LEFM formulae.
3. Compute the K_{IC}^{C} as mentioned in Chap. 4 (see Sect. 4.5.4) for which crack opening displacement along the fictitious crack line is determined using Eq. (4.45).
4. Determine the initiation toughness K_{IC}^{ini}.
5. Determine K_I^{COH} using analytical method or weight function method.
6. Then compute the K_R curve using Eq. (5.2).

A flowchart for computing the K_R curve is shown in Fig. A.5.

5.5.1 Details of Experimental Results

The primary requirement for evaluation of crack extension resistance for complete fracture process associated with cohesive stress distribution in the fictitious fracture zone is to know the P-CMOD or P–δ curve from the test results. Xu and Reinhardt (1998) have used experimental results of total eight standard notched concrete beams of different sizes and different initial notch length/depth (a_o/D) ratios, in which the complete P-CMOD curves are reported. Out of eight examples, one test beam belongs to Karihaloo and Nallathambi (1991), four results are taken from Refai and Swartz (1987), and three test beams are considered from Jenq and Shah (1985). From the computations, it was shown that the K_R curves were independent of the specimen size and initial notch/depth (a_o/D) ratio. In the present calculation,

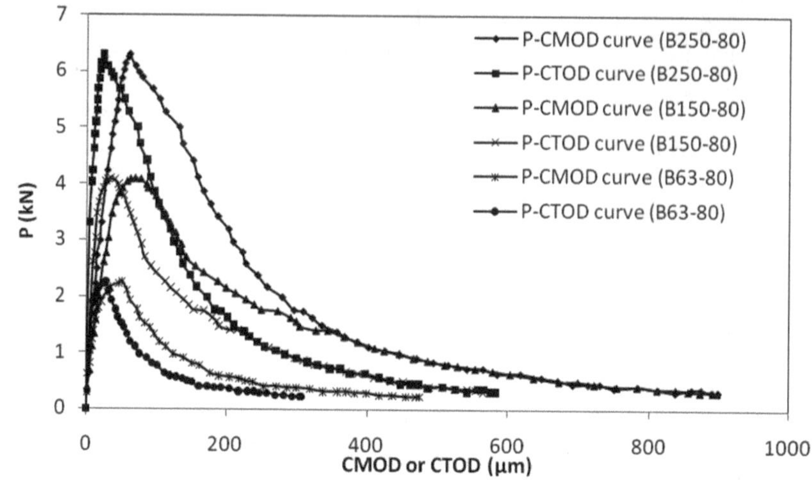

Fig. 5.7 The average experimental P-CMOD envelopes (Roesler et al. 2007) and computed P-CTOD curves for specimen sizes 250, 150, and 63 mm

Table 5.1 Test results of three-point bending geometry for notched concrete beam (E = 32 GPa and f_c = 58.3 MPa)

							K_{IC}^{ini} (MPa m$^{1/2}$)	
Beam No.	D (mm)	B (mm)	S (mm)	a_o (mm)	P_u (kN)	K_{IC}^{un} (MPa m$^{1/2}$)	Using analytical method	Using weight function method
B250-80	250	80	1000	83	6.303	1.427	0.531	0.534
B150-80	150	80	600	50	4.089	1.612	0.458	0.460
B63-80	63	80	250	21	2.264	1.607	0.613	0.616

Roesler et al. (2007)

the primary objective is to validate the *weight function method* with the existing *analytical method* for determining the K_R curves based on cohesive stress distribution. A different set of experimental results recently published by Roesler et al. (2007) and reported in Chap. 3 (see Sect. 3.4) that consists of complete P-CMOD envelopes is taken into account for this study. The designations of specimens B250-80b, B150-80b, and B63-80b correspond to the specimen size of depths 250, 150, and 63 mm, respectively. The same results of the average P-CMOD curves with maximum possible accuracy are reproduced in Fig. 5.7 and the salient features of the test results along with material properties are reported in Table 5.1. Tensile strength of concrete is computed using the relation $f_t = 0.4983\sqrt{f_c'}$ MPa (Karihaloo and Nallathambi 1991).

In real practice, the values of material constants in the nonlinear softening function should be determined from uniaxial tension tests. However, in the absence of test data, the values of c_1, c_2, and w_c of the nonlinear softening (Reinhardt et al. 1986) for normal concrete may be taken as 3, 6.93, and 160 μm, respectively. Also, the values of w_c may be taken in the range of 100–140 μm for concretes with tensile strengths ranging 2.93–4.12 MPa (Phillips and Zhang 1993). Xu and Reinhardt (1998) adopted a trial-and-error approach to determine the constants c_1, c_2, and w_c for three concrete mixes reported in the study. In the computations of K_R curves (Xu and Reinhardt 1998), the parameters c_1, c_2, and w_c were considered to be 3, 7, and 100 μm for the concrete tested in Refai and Swartz (1987); 3, 7, and 140 μm for concrete tested in Jenq and Shah (1985); and 2, 13, and 100 μm for concrete tested in Karihaloo and Nallathambi (1991). In the subsequent study, Reinhardt and Xu (1999) used the material parameters c_1, c_2, and w_c as 3, 10, and 110 μm, respectively, for evaluation of the K_R curves.

5.5.2 Crack Extension Resistance Curves (K_R Curves) and Stability Criterion

A comprehensive computer program in MATLAB environment is developed for the complete mathematical calculations. The test data of CMOD corresponding to load point are carefully read for all the three specimens (B250-80, B150-80, and B63-80) using Fig. 5.7. The existing analytical method (Eq. (5.5)) is applied to calculate reference values of the crack extension resistance and the weight function method is employed to calculate the corresponding values of crack extension resistance for the comparison purpose. Since there is a singularity problem at integral boundary, the Gauss–Chebyshev quadrature numerical technique is used to evaluate

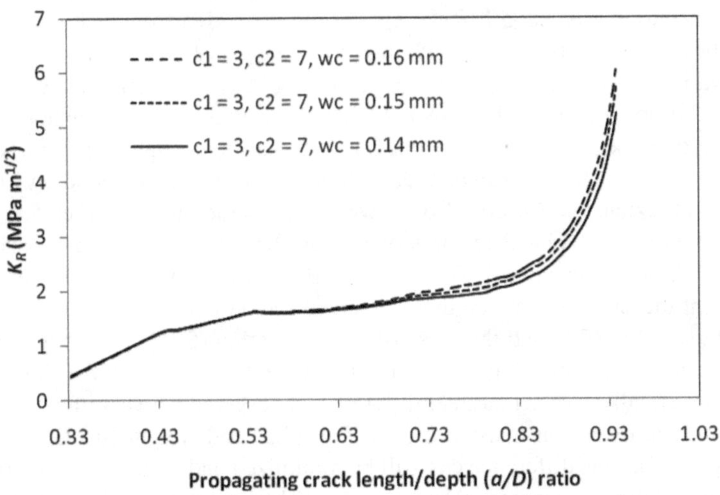

Fig. 5.8 Influence of material constant w_c of the nonlinear softening on the K_R curves for specimen size 150 mm

Eq. (5.5). This problem of using specialized numerical integration method is avoided if the closed-form expression (5.12) or (5.13) employing weight function method is used. Since the double-K fracture parameters can be evaluated with less than 2% error with four-term universal weight function and almost the same results can be gained with the five terms of the universal weight function, for validation of weight function method in present calculation only four-term weight function is taken into account for determination of the crack extension resistance. In this method, first of all the three parameters M_1, M_2, and M_3 of four-term weight function are computed employing Eq. (4.20) and Table 4.1 (see Sect. 4.3). The value of initiation toughness K_{IC}^{ini} is evaluated using the analytical method in case Eq. (5.5) is taken for calculation of the crack extension resistance. On the other hand, the closed-form expression with four-term weight function is used in case the crack extension resistance is calculated using Eq. (5.12). At every stage of loading beyond the peak load, the value of Δa_c along the fictitious crack line is computed satisfying Eq. (4.45).

During computations in the present work, it is realized that the choice of the material constants c_1, c_2, and w_c in the nonlinear softening function does affect the K_R curves. A trial-and-error method is adopted for considering those material constants in which the values of c_1 and c_2 are kept to be constant as 3 and 7, respectively, and the value w_c is varied as 160, 150, and 140 μm. As an example, the influence of the value w_c on the K_R curves gained using weight function method for specimen size 150 mm is shown in Fig. 5.8. It is observed from the figure that the K_R curves are affected by the values of w_c. Especially, the K_R curves for specimen size 63 mm, if compared with the K curve, are found to be unreasonably erroneous for the values of w_c 150 and 160 μm. Therefore, the parameters c_1, c_2, and w_c for the entire computations are taken to be 3, 7, and 140 μm, respectively, which give correct representation of the K_R curves for all the considered specimen sizes. This assumption seems to be reasonable because the computed value of tensile strength for the concrete mix is 3.805 MPa being close to 4.12 MPa for which the value of w_c may be assumed to be 140 μm (Phillips and Zhang 1993).

The variation of crack extension resistance $K_R(\Delta a)$ curve and stress intensity factor at the tip of propagating crack $K(P,a)$ curve with the crack extension Δa for all the three beam specimens using Eq. (5.5) and the weight function method using Eq. (5.12) is shown in Figs. 5.9, 5.10 and 5.11. The corresponding loads P vs. the crack extension Δa are also shown on the same plots. Figures 5.9, 5.10, and 5.11 represent graphical features of fracture behavior for specimen sizes 63, 150, and 250 mm, respectively. The notations B, B′, C, C′, Δa_B, and Δa_c are used to represent the salient points on these graphs. From comparison of the K_R curves obtained using Eq. (5.5) and the closed-form expression (5.12), it is found that no visible difference is noticed at any value of crack extension for all the three specimens. Numerically a very marginal difference is observed between the calculated values of crack extension resistance using Eq. (5.5) and weight function method using Eq. (5.12). This difference can still be minimized and numerically almost the same values of crack extension resistance can be obtained if Eq. (5.13) with the five terms of weight function is considered. Therefore, either Eq. (5.12) or (5.13) can be used to directly calculate the crack extension resistance of the concrete structures without any error. The use of these closed-form expressions also avoids the application of specialized numerical integration technique with a singularity problem at the integral boundary. In addition, implementation of these expressions in computer program is not only simple but also with the help of a calculator, one can calculate the crack extension resistance with improved computational efficiency. It is interesting to point out that the accuracy of fracture parameters using the analytical method depends on the step size of numerical integration. Especially, with the use of bilinear softening function, if the numbers of integration points chosen are relatively less or the integration points located are relatively away from the kink point, the analytical method may lead to some degree of error. On the other side, Eq. (5.12) is a closed-form solution which yields results free from such error.

From Figs. 5.9, 5.10, and 5.11, it is observed that relative comparison between the K_R curves and the K curves can be considered as a fracture criterion to describe the stability condition of a propagating crack in a complete fracture process. In the figures, the values of the K curves are seen lower to those of K_R curves during the ascending portion of P–Δa curve. It means that cracks propagate steadily in ascending portion of P-CMOD curve till the crack extension reaches the value of a_c. In the descending branch of P–Δa curve, the K curves are higher than K_R curves, which means that propagating cracks are unstable. At peak loads, the K_R curves are equal to the K curves corresponding to the onset of unstable crack propagation. Hence these graphical representations are consistent with Eq. (2.3) of fracture criterion using the K_R-curve approach. The beginning point of the K_R curve is the initiation

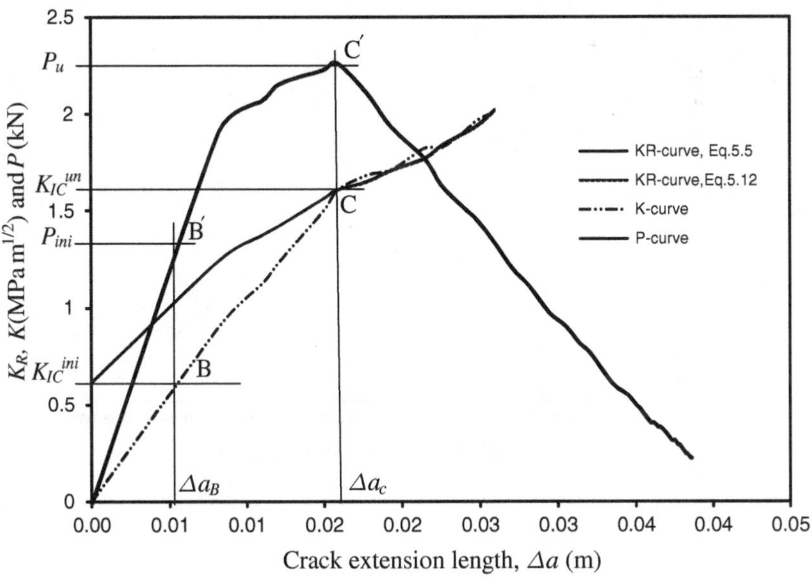

Fig. 5.9 Comparison of K_R curves and stability analysis of crack propagation for specimen size 63 mm

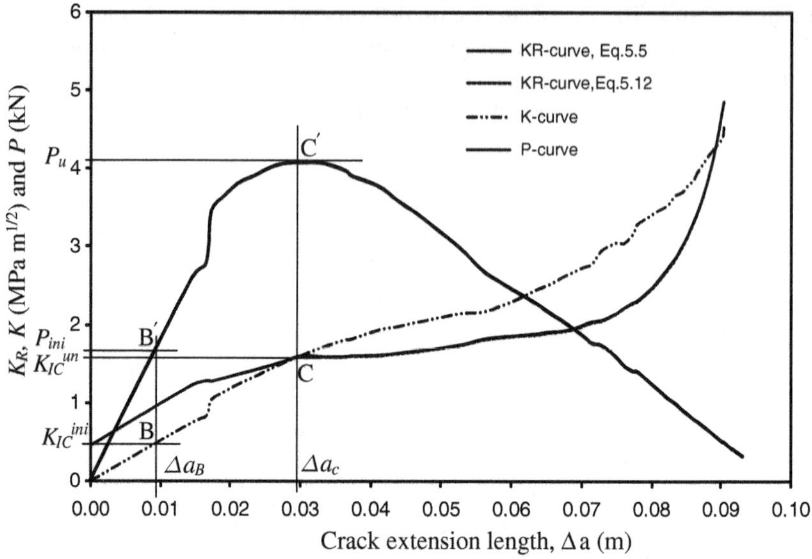

Fig. 5.10 Comparison of K_R curves and stability analysis of crack propagation for specimen size 150 mm

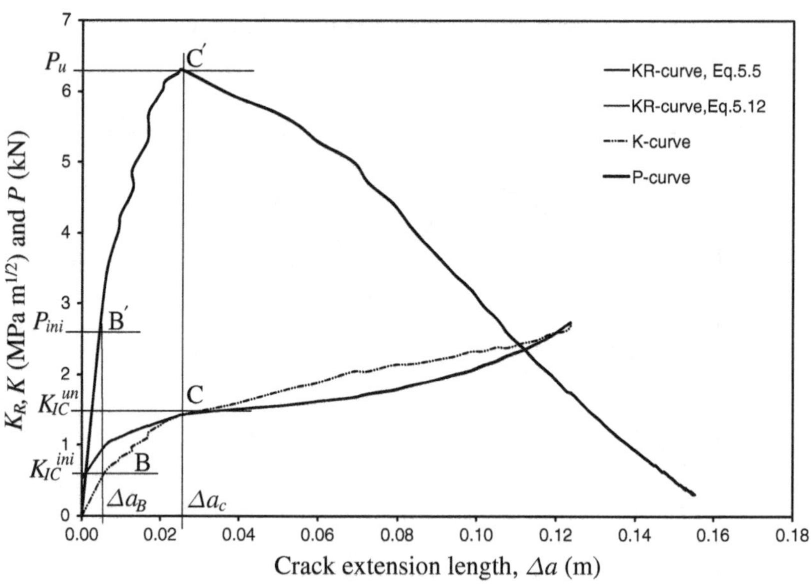

Fig. 5.11 Comparison of K_R curves and stability analysis of crack propagation for specimen size 250 mm

toughness K_{IC}^{ini} of the material and corresponding to the peak load, the toughness of material is the unstable toughness K_{IC}^{un}. Consequently, stability of the crack propagation can be well described using the double-K fracture parameters K_{IC}^{ini} and K_{IC}^{un}. From the K_R-curve analysis, it is clear that K_{IC}^{ini} is the inherent toughness of the material, being the initiation point of the K_R curve at point B. A horizontal line parallel to the Δa-axis is drawn passing through the ordinate K_{IC}^{ini} until it crosses the K curve. From this point, a vertical line is drawn until it cuts the P–Δa curve at point B′. The point B′ gives the crack initiation load P_{ini}, which is difficult to measure from the experiment. The crack extension corresponding to P_{ini} is termed as the length of small micro cracking zone Δa_B. Similarly, a line parallel to Δa-axis is drawn from the intersection point of the K_R curve and the K curve (point C), which cuts the ordinate at a point known as the unstable fracture toughness K_{IC}^{un}. If a vertical line is drawn from the intersection point C until it crosses the P–Δa curve at point C′, the corresponding load point is the peak load. The value of crack extension length Δa_C corresponding to K_{IC}^{un} is the effective crack extension. The value of $(\Delta a_C - \Delta a_B)$ is a macro-crack extension and known as the steady slow crack growth. The complete fracture process in fact can be described using double-K fracture criterion, which is equivalent to the K_R-curve fracture criterion as expressed in Eq. (5.4).

5.5.3 Effect of Specimen Size on the K_R Curves

The size-effect study on the K_R curve was presented by Xu and Reinhardt (1998) for the three-point bending test specimens tested in Refai and Swartz (1987) and Jenq and Shah (1985). It seemed difficult to analyze the size effect quantitatively for the tested specimens in Refai and Swartz (1987) because the values of a_o/D ratio were not the same for the companion specimens. On the other set of test results (Jenq and Shah, 1985), the values of modulus of elasticity taken into calculation for the K_R curves were widely varied for different specimens. However, it was found that the K_R curves were independent of both specimen size and relative size of initial notch length. In the subsequent work, Reinhardt and Xu (1999) performed the size-effect study on the K_R curve for the geometrically similar three-point bending test specimens with the same material properties and a_o/D ratio for the size range $200 \leq D \leq 400$ mm. The input data in the study were gained numerically using nonlinear fracture model based on crack band model (Bažant and Oh 1983). The advantage of choosing the numerical input data might be that it could be free from unseen errors in experiment. In the study (Reinhardt and Xu 1999), the size effect can be quantitatively observed corresponding to the characteristic fracture parameters K_{IC}^{ini} and K_{IC}^{un}. It is found that difference in the values of K_{IC}^{ini} and K_{IC}^{un} calculated between the maximum size (400 mm) and the minimum size (200 mm) is 9.33 and 10.53%, respectively. This observation shows that the K_R curves are dependent on the specimen size.

The gained K_R curves using four-term weight function method (Eq. (5.12)) for the three specimen sizes 63, 150, and 250 mm are shown in Fig. 5.12 and the values

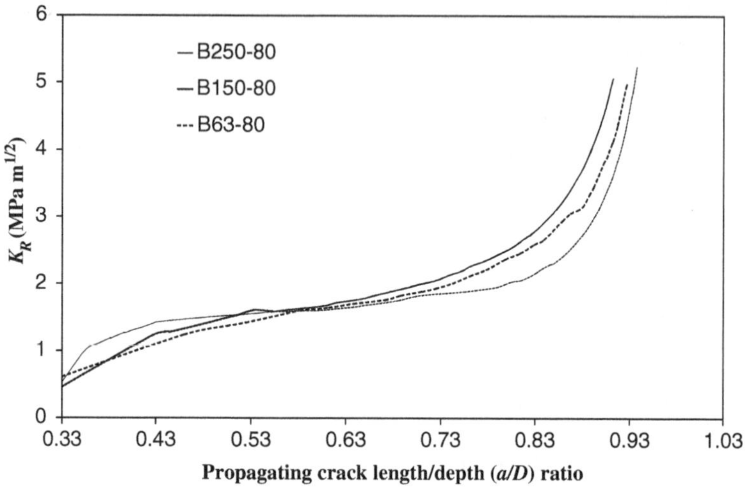

Fig. 5.12 The size-effect analysis on the K_R curves determined using weight function method

of double-K fracture parameters for those specimens are presented in Table 5.1. In present computations it can be seen from Table 5.1 that values of K_{IC}^{ini} and K_{IC}^{un} are different by 10.88 and 11.20% between specimen sizes 250 and 63 mm, respectively, and those differences are 13.86 and 11.48% for specimen sizes 250 and 150 mm, respectively. This size effect can also be observed from Fig. 5.12. It is seen from the figure that the size effect on the K_R curve is more prominent almost after the unstable fracture condition is reached. Especially, the K_R curve for specimen size 150 mm is relatively more scattered than that for specimen sizes 63 and 250 mm.

5.5.4 The P-CTOD Curves

The relation between monotonic increase in load and the corresponding gained values of crack-tip opening displacement for all the three specimens is plotted in Fig. 5.7. In the figure the computed values of P-CTOD curves are merged with the experimentally observed P-CMOD curves to observe the similarity between the experimental data and the computed results. It can be seen from the comparison that the P-CTOD curves for the notched-beam specimens are similar to the experimentally observed P-CMOD curves. This observation further validates the preciseness achieved in the evaluation process of fracture parameters using the K_R-curve method.

5.5.5 The Relationship Between CTOD and Δa

The evaluated results of the CTOD and Δa during the complete fracture process are plotted in Fig. 5.13for all the three notched concrete beam specimens. An

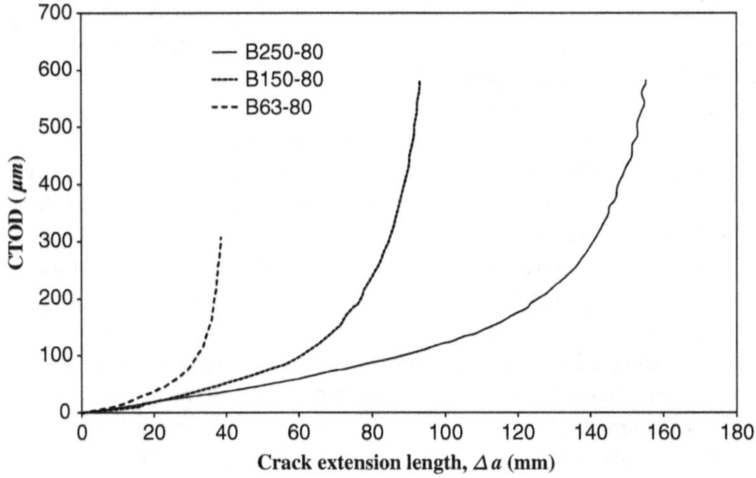

Fig. 5.13 Variation of CTOD with crack extension (Δa) for the beam specimens

exponential relationship exists between the gained values of CTOD and Δa. Du, Kobayashi and Hawkins (1987) experimentally measured the crack opening displacement along the process zone in the three-point bending notched concrete beams using white light moiré interferometry. The dimensions of the notched beams were $S = 162.4\,\text{mm}$, $B = 51\,\text{mm}$, $D = 40.6\,\text{mm}$, and $a_o = 20.3\,\text{mm}$. The increasing COD values at the blunt notch tip were plotted with microcrack-tip extensions for these small concrete beam bend specimens. All the data points fall on the same continuous curves for the four specimens showing an exponential function (Du et al. 1987), which is similar in nature as represented in Fig. 5.13. This further validates the practical use of the K_R curve associated with cohesive force in the fictitious fracture zone.

5.5.6 Analysis and Discussion

The crack extension resistance $K_R(\Delta a)$ based on cohesive stress distribution along the fictitious fracture zone is developed from experimental results of P-CMOD curve on the three-point bending notched concrete beams. Two approaches are mentioned in the chapter. The first approach was proposed by Xu and Reinhardt (1998) in which the cohesive toughness due to cohesive stress distribution is determined using specialized numerical integration technique. This analytical method has the singularity problem at the integral boundary. The second approach is the weight function method proposed by the authors (Kumar and Barai 2008a, 2009a) in which the cohesive toughness due to cohesive stress distribution is determined using the simple form of the universal weight function. In the weight function method, first, the parameters of weight functions are determined using the simple algebraic expressions and then the value of cohesive toughness is evaluated employing derived closed-form solution. It is found from the comparison that the values of crack extension resistance gained using the existing analytical method (Xu and Reinhardt, 1998) and the closed-form expression with four-term weight function are very close. The use of present method does not require any specialized computing technique and hence improves the computational efficiency. Moreover, any possible error due to step size in numerical integration especially with the use of bilinear softening curve for concrete is avoided if the weight function method is used. The double-K fracture criterion is a simple and practically useful tool which can be conveniently applied for the stability analysis of the K_R curve. For describing the complete fracture process of concrete starting from crack initiation to failure, the K_R curves associated with cohesive stress distribution can be used with less computational efforts. Many basic fracture parameters, such as the crack initiation load P_{ini}, length of microcracks, length of macro-cracks which are difficult to be measured experimentally, can be determined easily by making use of the present form of the K_R-curve analysis. In addition, the values of crack-tip opening displacement at every incremental value of crack extension Δa can be evaluated numerically during the complete fracture process.

Experimental test data (Xu and Reinhardt 1998) and numerically obtained data (Reinhardt and Xu 1999) were used to study the size-effect behavior of the K_R curves. A better reflection of size effect on the K_R curves can be captured from the work of Reinhardt and Xu (1999) in which it is observed that the values of K_{IC}^{ini} and K_{IC}^{un} are varied within 11% for the size range $200 \leq D \leq 400$ mm. In the present study, the authors make use of experimental results on the three-point bending configuration to show the size-effect behavior of the K_R curves. It is observed that the maximum difference in the values of K_{IC}^{ini} and K_{IC}^{un} is found to be less than 14% for the size range 63–250 mm and after almost unstable fracture condition, the size effect is more prominent.

It is well known that the R curves depend on the specimen size, the geometry of specimen, the loading condition, the shape of softening function, and the materials properties of concrete. Reinhardt and Xu (1999) carried out systematic studies in this regard showing the effect of compressive strength of concrete, the size effect, and the influence of the shape of the softening law on the K_R curves. The influence of specimen geometry and type of loading condition on the K_R curves is not available in the literature. It is interesting to find out the influence of specimen geometry and loading condition on the K_R curves. The P-CMOD curves are the preliminary requirements for the above studies. For the investigation of influence of specimen geometry, the results of P-CMOD curves tested under the similar conditions for different specimen geometries are needed. Similarly, for the effect of loading condition on the K_R curves, the P-CMOD plots tested under the similar conditions except for different loading conditions are required. Most of the experimentally measured P-CMOD curves available in the literature are not favorable to investigate the influence of specimen geometry and loading condition on the K_R curves. In other way, these investigations can be carried out in the foregoing section with the P-CMOD curves generated using nonlinear fracture models like the cohesive crack model. The advantage of gaining such P-CMOD curves is that they may be free from unseen errors due to experimental measurements.

5.6 Numerical Study on the K_R Curve

5.6.1 Material Properties and Numerical Computations

The concrete mix for which $f_t = 3.21\,\text{MPa}$, $E = 30\,\text{GPa}$, $G_F = 103\,\text{N/m}$ (Planas and Elices 1990), and $\nu = 0.18$ is taken into account for the present investigation. The nonlinear stress–displacement softening relation expressed in Eq. (3.36) (see Sect. 3.3.4) is used for FCM in the study. The values of material constants in nonlinear softening function are taken to be $c_1 = 3$, $c_2 = 5$ for all computations in the present work. This results in the value of $w_c = 108.8\,\mu\text{m}$ after satisfying Eq. (3.37). For standard TPBT, CT, and FPBT specimens with $B = 100\,\text{mm}$ having size range $D = 100 - 400\,\text{mm}$, the finite element analysis is carried out for which half of the specimens are discretized due to symmetry as shown in Fig. 4.25 and 80 numbers of equal isoparametric plane elements are considered along the dimension D (see Sect. 4.9). The P-CMOD curves for standard TPBT and FPBT or P-COD curves for CT test specimen at a_o/D ratios 0.3 and 0.5 are obtained using FCM. The P-CMOD curves for the TPBT and FPBT specimens and P-COD curves for CT test specimens for size range considered in the study are presented in Figs. 5.14, 5.15 and 5.16, respectively. The symbols D_1, D_2, D_3, and D_4 in the figures are the specimen size of 100, 200, 300, and 400 mm, respectively.

The *analytical method* (Eq. (5.5)) is applied to calculate reference values of crack extension resistance and the *weight function method* (Eqs. (5.13) and (5.2))

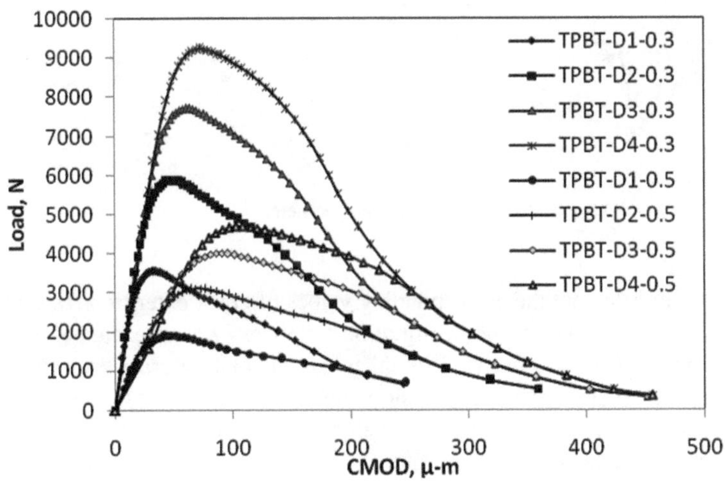

Fig. 5.14 The P-CMOD curves for TPBT specimen geometry

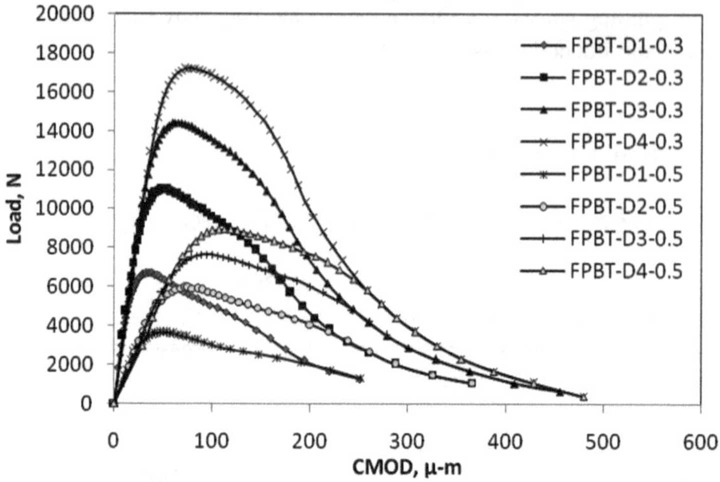

Fig. 5.15 The P-CMOD curves for FPBT loading condition

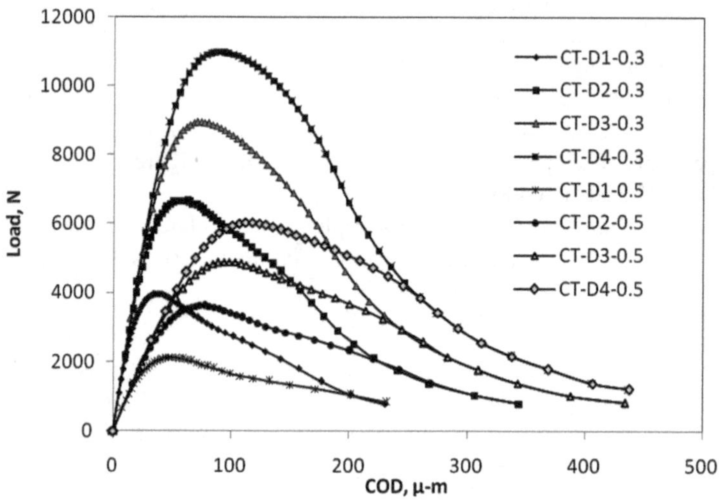

Fig. 5.16 The P-COD curves for CT test specimen geometry

is employed to calculate the corresponding values of crack extension resistance for comparison purpose. The value of initiation toughness K_{IC}^{ini} is evaluated using the *analytical method* in case Eq. (5.5) is taken for calculating crack extension resistance, whereas the closed-form expression with five-term weight function is used in case Eqs. (5.2) and (5.13) are used. The value of Δa_c *is* computed at every stage of loading beyond the peak load. This value is obtained satisfying Eq. (4.45) along the fictitious crack line for locating the value of $CTOD_c$. The same nonlinear softening function is used in FCM, analytical method, and weight function method to maintain

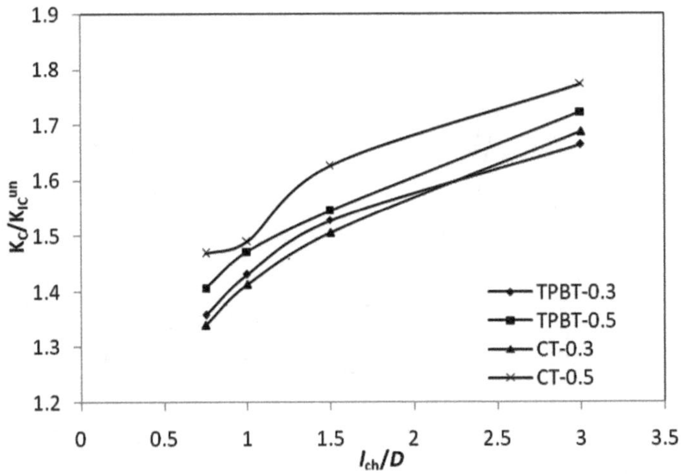

Fig. 5.17 Effect of specimen geometry and size effect on the unstable fracture toughness for TPBT and CT geometries

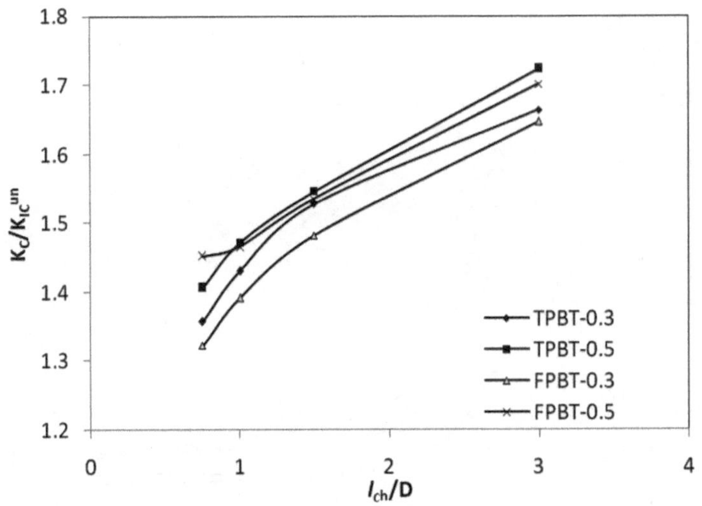

Fig. 5.18 Effect of loading condition and size effect on the unstable fracture toughness for TPBT and FPBT

consistency in the results. For the precise numerical investigation, the effect of self-weight in TPBT and FPBT geometries is accounted for at all computational stages, whereas this effect is not obviously considered for the CT specimen. The fracture material property K_{IC}^{un} for specimen geometries TPBT and CT and loading conditions for TPBT and FPBT for the considered specimens are graphically presented in Figs. 5.17 and 5.18 respectively. The parameter K_{IC}^{ini} for specimen geometries TPBT and CT and loading conditions for TPBT and FPBT for the considered specimens are graphically presented in Figs. 5.19 and 5.20 respectively.

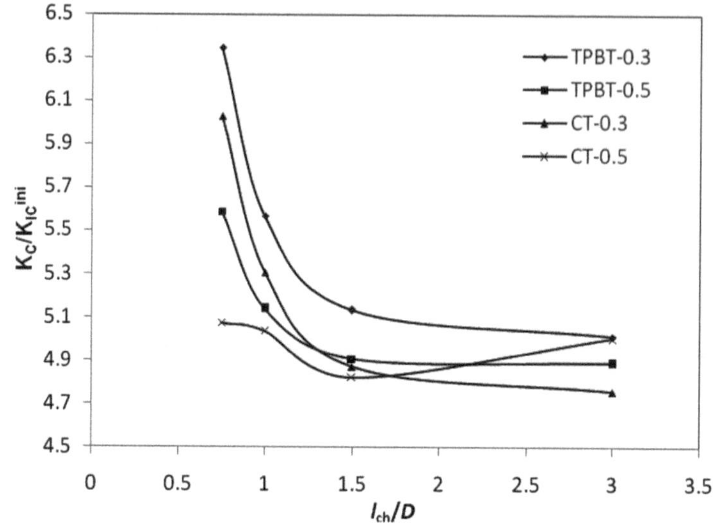

Fig. 5.19 Effect of specimen geometry and size effect on the initial cracking toughness for TPBT and CT geometries

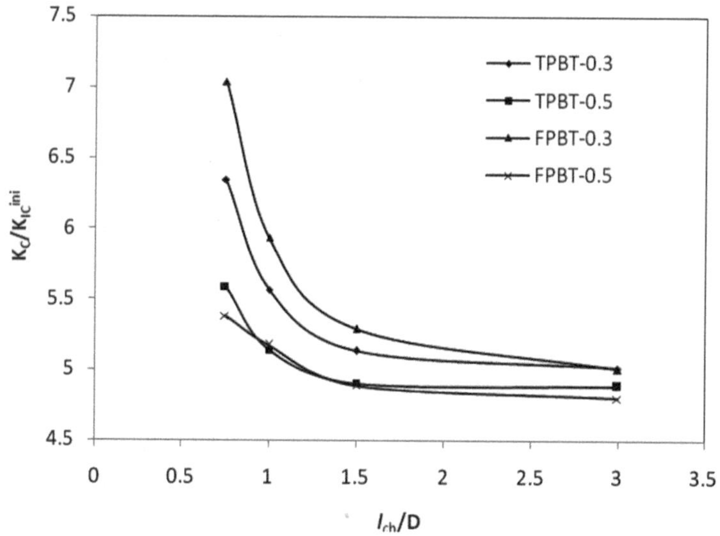

Fig. 5.20 Effect of loading condition and size effect on the initial cracking toughness for TPBT and FPBT

Effect of specimen geometry between TPBT and CT specimens on the double-*K* fracture parameters is similar to the previous studies observed in Chap. 4. Further, it is found that the maximum absolute difference in the values K_{IC}^{un} between the two loading conditions (TPBT and FPBT) is less than 2% and that in the values K_{IC}^{ini} is less than 3.5% for specimen size range 100–400 mm.

5.6.2 Crack Extension Resistance Curves (K_R Curves) and Stability Criterion

Since the K_R curve is considered as a criterion for complete description of crack propagation in structure, it is regarded as the material properties of the complete fracture process. On the other side the stress intensity factor curve (*K* curve) at the crack tip for an arbitrary load *P* and corresponding propagating crack length can be obtained using LEFM formulae. For the demonstration of the complete fracture process, the variation of crack extension resistance curve $K_R(\Delta a)$ and stress intensity factor at the tip of propagating crack *K*(*P*,*a*) curve with crack extension, *a/D* ratio for all TPBT and CT specimens using analytical method (Eq. (5.5)), and the weight function method (Eqs. (5.2) and (5.13)) are shown in Figs. 5.21–5.36. The corresponding load *P* vs. propagating crack extension *a/D* is also shown on the same plot for the 16 specimens (Figs. 5.21–5.36). KR-AN, KR-WF, KI, and *P* denote K_R curve obtained using analytical method, K_R curve obtained using weight function method, stress intensity factor curve, and load curve, respectively. These curves will enable to carry out the stability analysis of the crack propagation for complete fracture process and to compare the results of the K_R curves obtained using analytical method and weight function method.

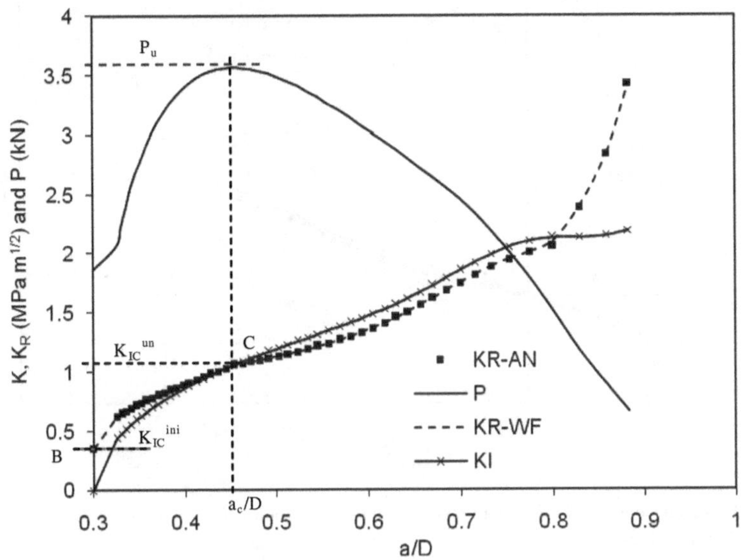

Fig. 5.21 K_R curve and stability analysis of the propagating crack for TPBT specimen of $D = 100$ mm and a_o/D ratio $= 0.3$

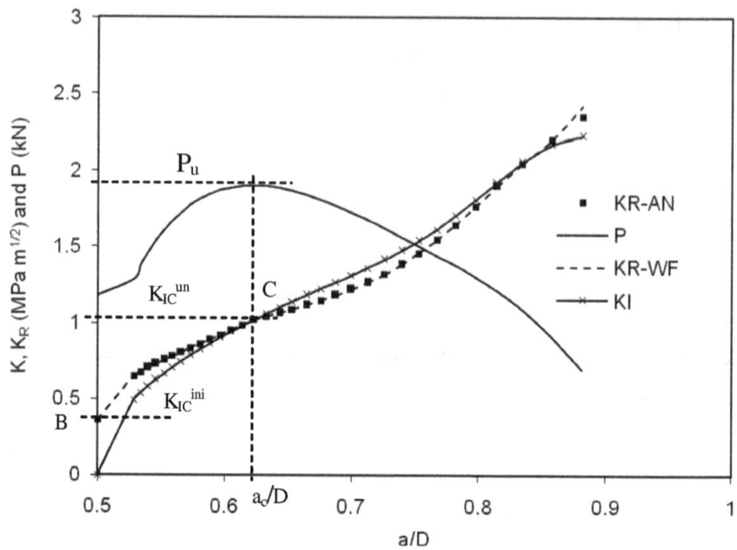

Fig. 5.22 K_R curve and stability analysis of the propagating crack for TPBT specimen of $D = 100\,\text{mm}$ and a_o/D ratio $= 0.5$

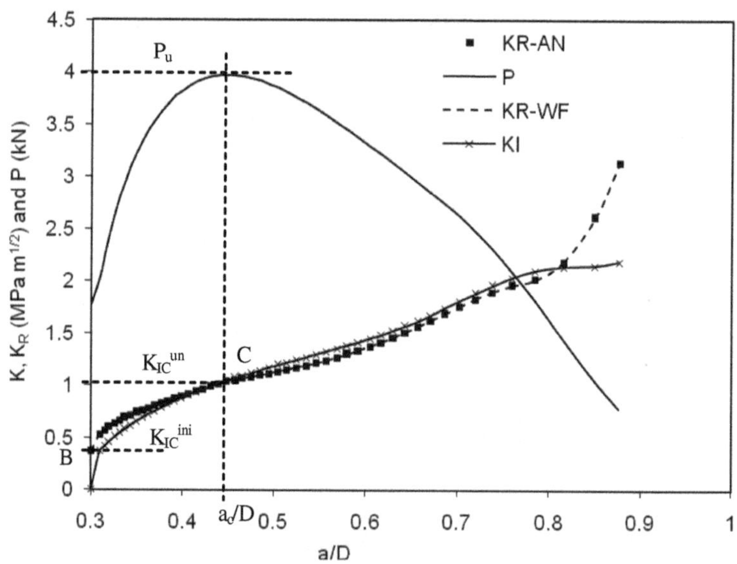

Fig. 5.23 K_R curve and stability analysis of propagating crack for CT specimen of $D = 100\,\text{mm}$ and a_o/D ratio $= 0.3$

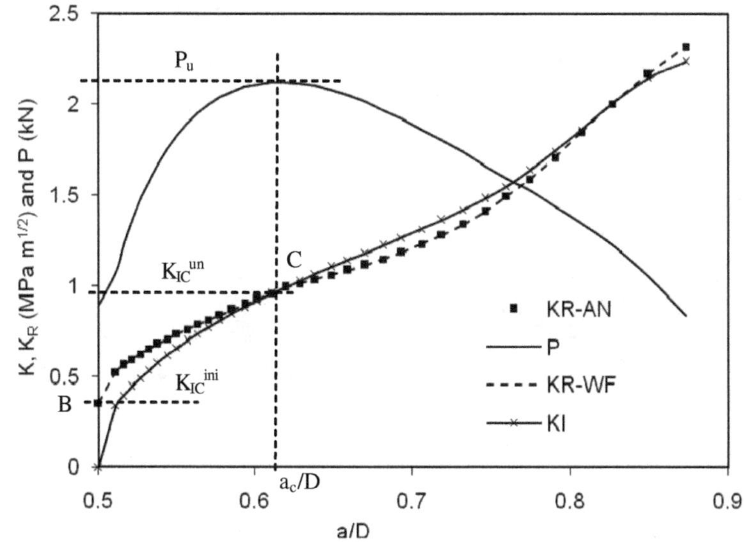

Fig. 5.24 K_R curve and stability analysis of the propagating crack for CT specimen of $D = 100$ mm and a_o/D ratio $= 0.5$

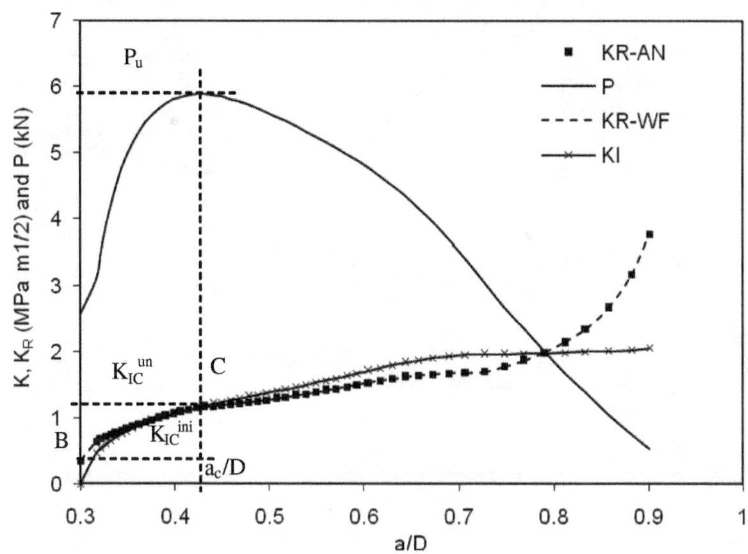

Fig. 5.25 K_R curve and stability analysis of the propagating crack for TPBT specimen of $D = 200$ mm and a_o/D ratio $= 0.3$

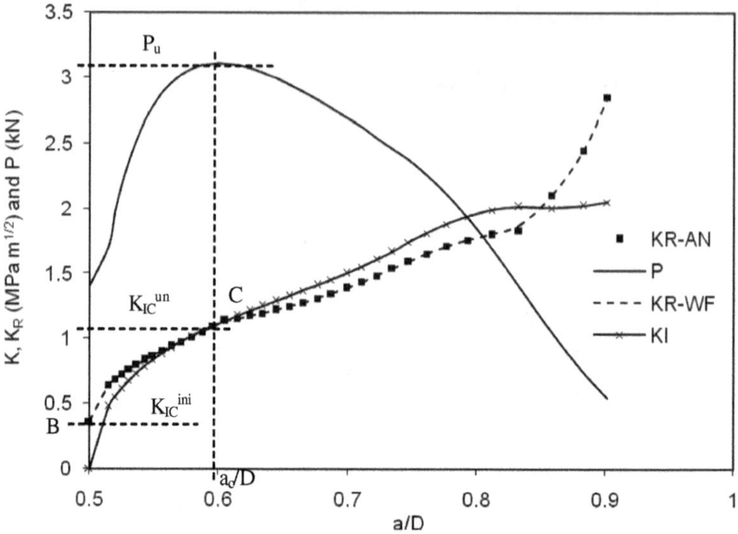

Fig. 5.26 K_R curve and stability analysis of the propagating crack for TPBT specimen of $D = 200\,\text{mm}$ and a_o/D ratio $= 0.5$

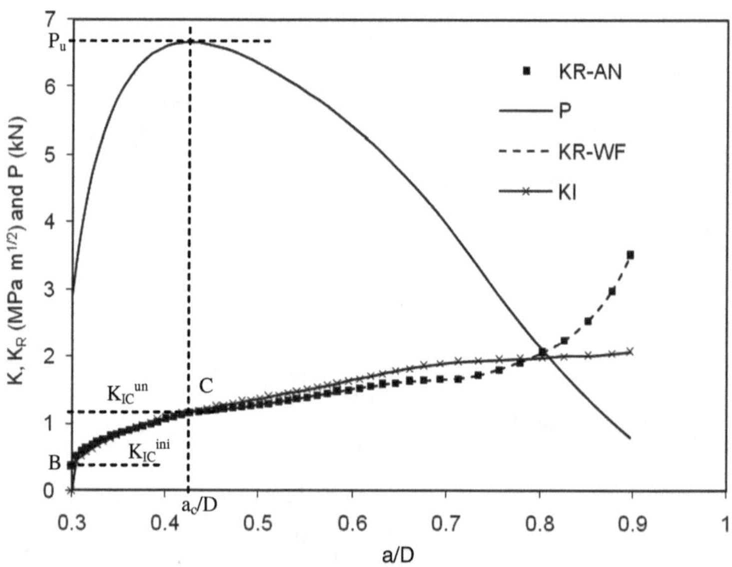

Fig. 5.27 K_R curve and stability analysis of the propagating crack for CT specimen of $D = 200\,\text{mm}$ and a_o/D ratio $= 0.3$

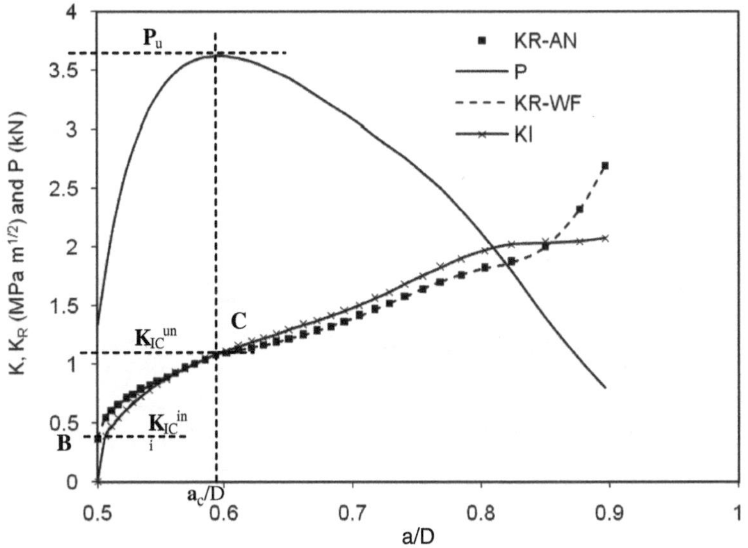

Fig. 5.28 K_R curve and stability analysis of the propagating crack for CT specimen of $D = 200$ mm and a_o/D ratio $= 0.5$

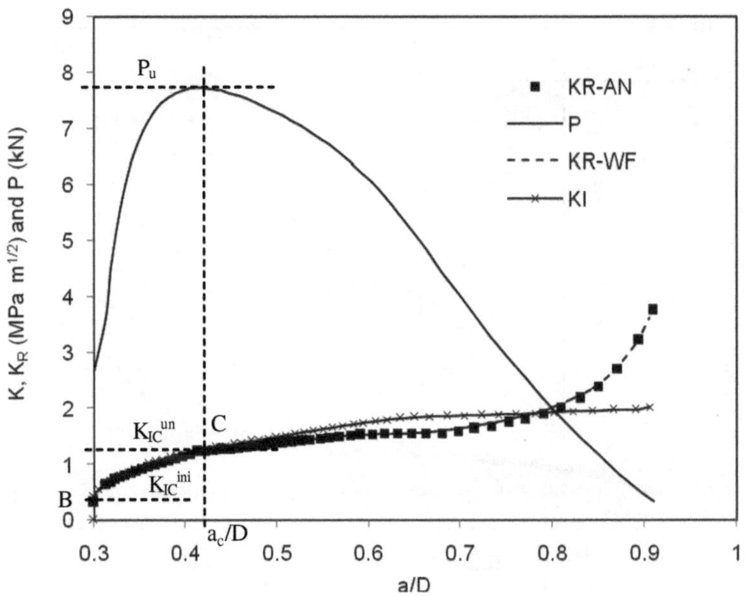

Fig. 5.29 K_R curve and stability analysis of the propagating crack for TPBT specimen of $D = 300$ mm and a_o/D ratio $= 0.3$

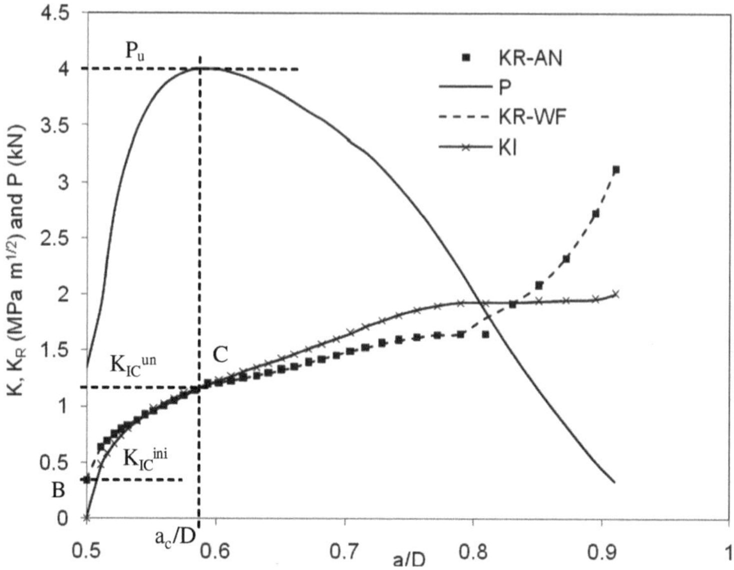

Fig. 5.30 K_R curve and stability analysis of the propagating crack for TPBT specimen of $D = 300\,\text{mm}$ and a_o/D ratio $= 0.5$

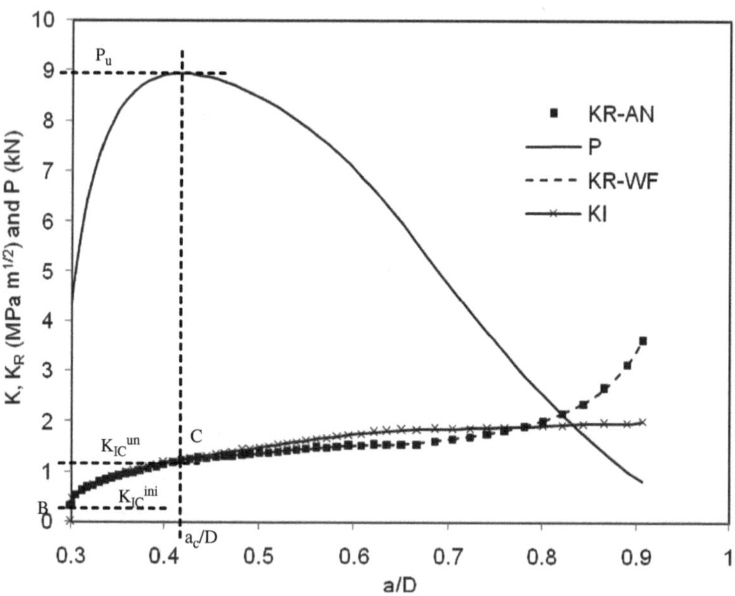

Fig. 5.31 K_R curve and stability analysis of the propagating crack for CT specimen of $D = 300\,\text{mm}$ and a_o/D ratio $= 0.3$

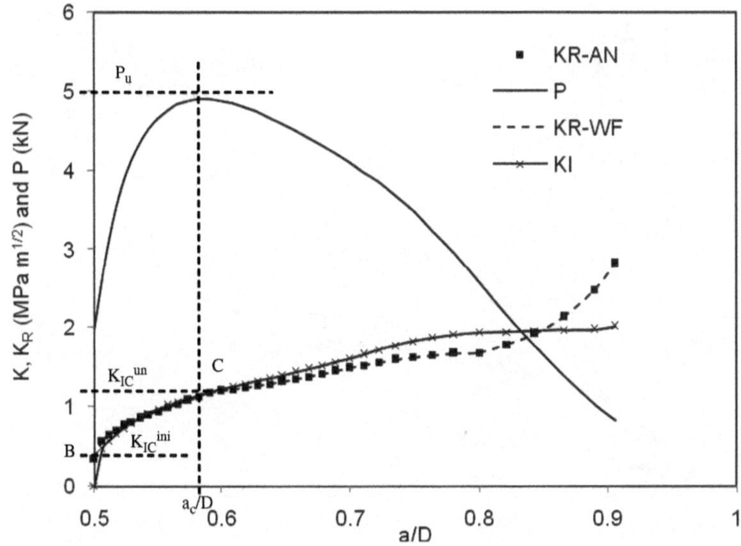

Fig. 5.32 K_R curve and stability analysis of the propagating crack for CT specimen of $D =$ 300 mm and a_o/D ratio $= 0.5$

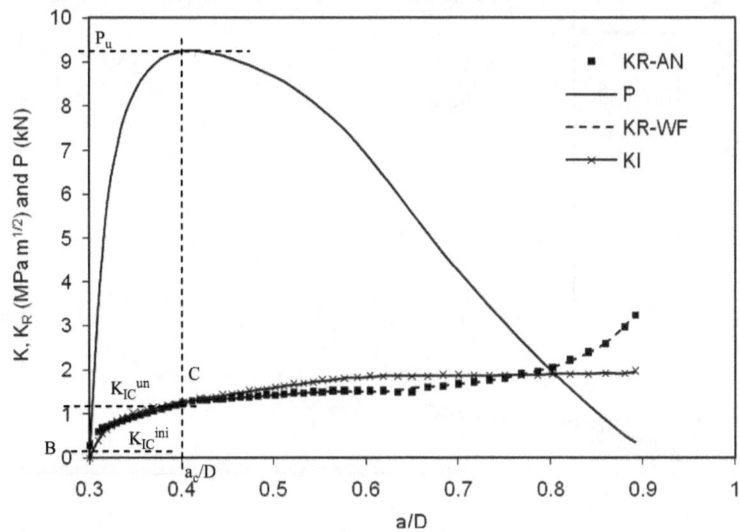

Fig. 5.33 K_R curve and stability analysis of the propagating crack for TPBT specimen of $D =$ 400 mm and a_o/D ratio $= 0.3$

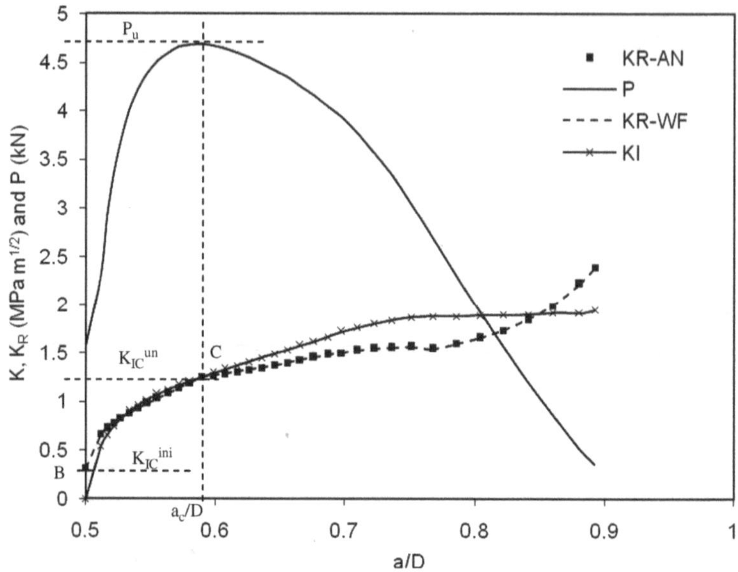

Fig. 5.34 K_R curve and stability analysis of the propagating crack for TPBT specimen of $D = 400$ mm and a_o/D ratio $= 0.5$

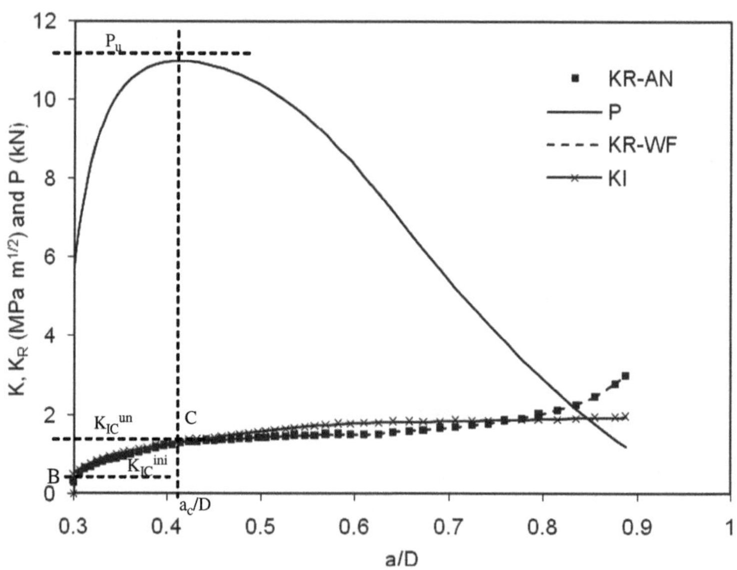

Fig. 5.35 K_R curve and stability analysis of the propagating crack for CT specimen of $D = 400$ mm and a_o/D ratio $= 0.3$

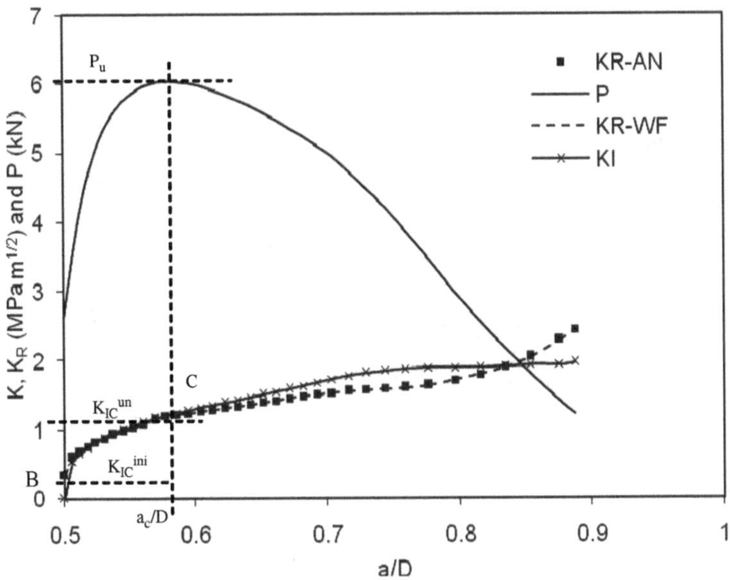

Fig. 5.36 K_R curve and stability analysis of the propagating crack for CT specimen of $D = 400$ mm and a_o/D ratio $= 0.5$

It is observed from Figs. 5.21–5.36 that the K_R curves obtained using weight function method and analytical method for all the cases do not have any distinguished difference. It is also to be pointed out that weight function method does not involve any specialized numerical technique; moreover, straightforward solution can be obtained in closed-form equation resulting in computational efficiency. Hence for the sake of brevity the K_R curves presented in the subsequent part of the chapter are obtained using weight function method.

The K_R curve, K curve, and P curve (Figs. 5.21–5.36) can be considered to investigate the initiation point and the critical unstable point of crack propagation during fracture process in concrete. The toughness of the material increases with increase in crack length from the inherent toughness K_{IC}^{ini} corresponding to point B to the value of the crack extension resistance at the critical unstable point C, denoted as $K_R(\Delta a_c)$. At this condition, a point of inflection is exhibited on the K_R curve and the corresponding load becomes peak load P_u and the crack length achieves its critical value a_c. The difference in the toughness of the material at any propagating crack length is due to the cohesive toughness and the initiation toughness.

The stability analysis of crack propagation for complete fracture process can be performed by comparing the K curves and K_R curves in the Figs. 5.21–5.36. Initially, the K curves yield lower value of stress intensity factor till the crack propagates steadily which are the regions between the points B and C. Further, the K curves coincide with the K_R curves at point C and the crack will start to propagate unsteadily beyond this point. The stability analysis can be expressed mathematically as stated in Eq. (5.4).

5.6.3 Effect of Specimen Geometry and Size Effect on the K_R Curves

The evaluated K_R curves using five-term weight function method for TPBT and CT specimen geometries of laboratory size range ($100 \leq D \leq 400$) at a_o/D ratio of 0.3 and 0.5 are shown in Figs. 5.37 and 5.38, respectively. It is observed from both the figures that at a particular value of specimen size and a_o/D ratio, no distinguished

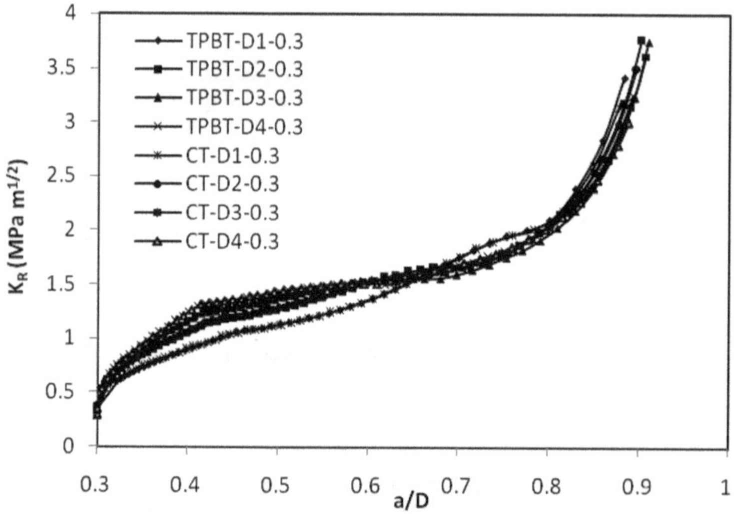

Fig. 5.37 Influence of specimen geometry and size effect on the K_R curves at $a_o/D = 0.3$

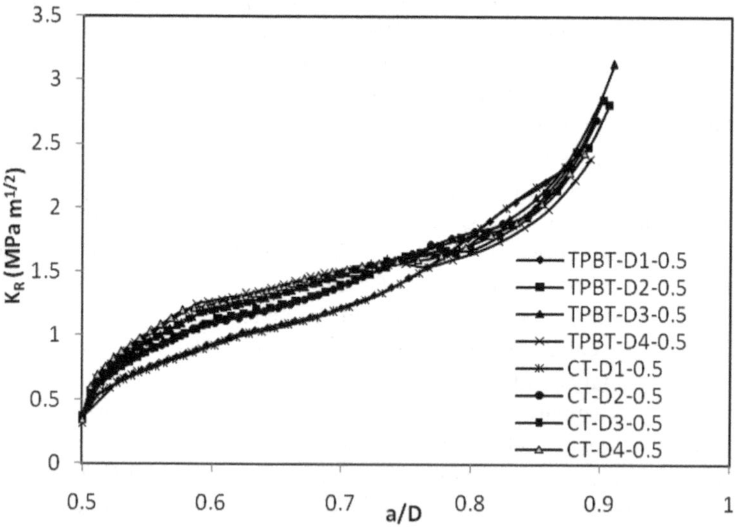

Fig. 5.38 Influence of specimen geometry and size effect on the K_R curves at $a_o/D = 0.5$

difference in the K_R curves gained between TPBT and CT test specimen geometries is noticed. Further, it is noticed from the figures that K_R curves are affected by the size of the specimens; however, it is difficult to quantify graphically the size effect on the gained K_R curves. It is also seen from the figures that the influence of specimen size is more pronounced near the values of K_{IC}^{ini} and K_{IC}^{un} which is clearly discernable in Figs. 5.16 and 5.18.

5.6.4 Effect of Specimen Geometry and Size Effect on the CTOD Curves

The relation between monotonic increments of the evaluated values of crack-tip opening displacement at different loading stages of crack propagation (CTOD curves) for all the TPBT and CT specimens having a_o/D ratio 0.3 and 0.5 is plotted in Figs. 5.39 and 5.40, respectively. It is interesting to note that there is no influence of specimen geometry at a particular value of specimen size and a_o/D ratio. This observation is similar to the effect of specimen geometry on the K_R curves developed between TPBT and CT test specimen geometries.

5.6.5 Effect of Specimen Geometry and Size Effect on the Process Zone Length

The process zone development is an important phenomenon in concrete fracture and the zone length can be accurately evaluated using nonlinear fracture analysis based

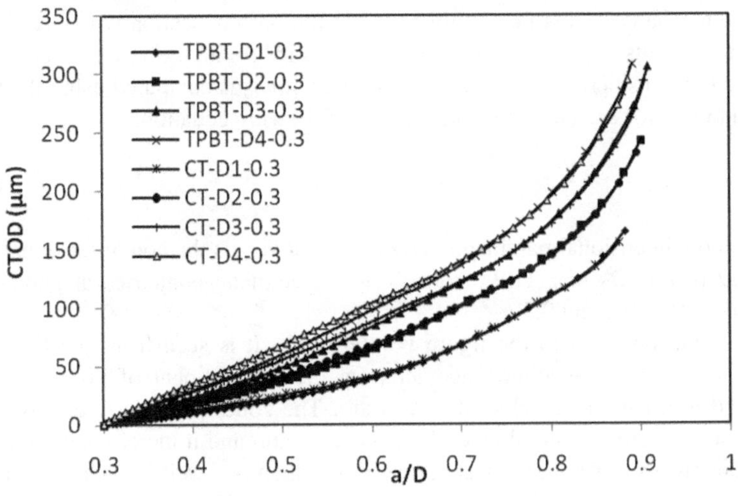

Fig. 5.39 Influence of specimen geometry and size effect on the CTOD curves at $a_o/D = 0.3$

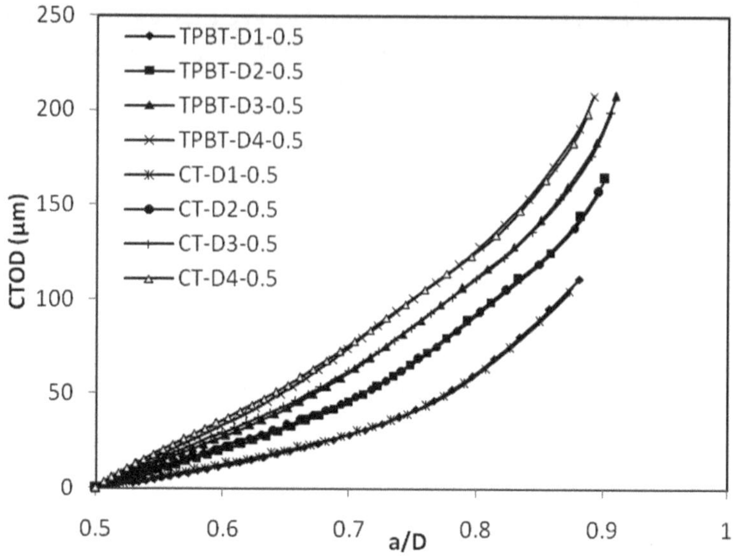

Fig. 5.40 Influence of specimen geometry and size effect on the CTOD curves at $a_o/D = 0.5$

on numerical method like fictitious crack model or crack band model. However, these models are computationally more involved as compared to the modified LEFM models for concrete fracture. Further, modified LEFM models are practically useful due to less computational work for common engineering applications. In those circumstances, the K_R-curve method based on cohesive stress distribution in the fictitious crack zone can give similar information as those obtained from nonlinear fracture mechanics. This fact is illustrated in the present section which reports the numerical results of process zone length determined for all the TPBT and CT specimens under consideration. The process zone length c_p is computed using the values of a_o and a_{wc} for each test specimen and the following equation:

$$c_p = a_{wc} - a_o \tag{5.17}$$

The nondimensional parameter c_p/D is plotted versus the nondimensional value of l_{ch}/D in Fig. 5.41for TPBT and CT test specimen geometries at geometrical factors $a_o/D = 0.3$ and 0.5.

The trend obtained in the figure is interesting. It is seen from the figure that almost no effect of specimen geometry on the development of process zone is observed at a particular value of a_o/D ratio. The zone length is relatively higher for the lower value of initial crack length/depth ratio and it increases with decease in specimen size for both the test geometries, which is a well-known phenomenon in fracture process of concrete. It is further observed that on an average, the difference in the process zone length obtained at different a_o/D ratios (0.3 and 0.5) is

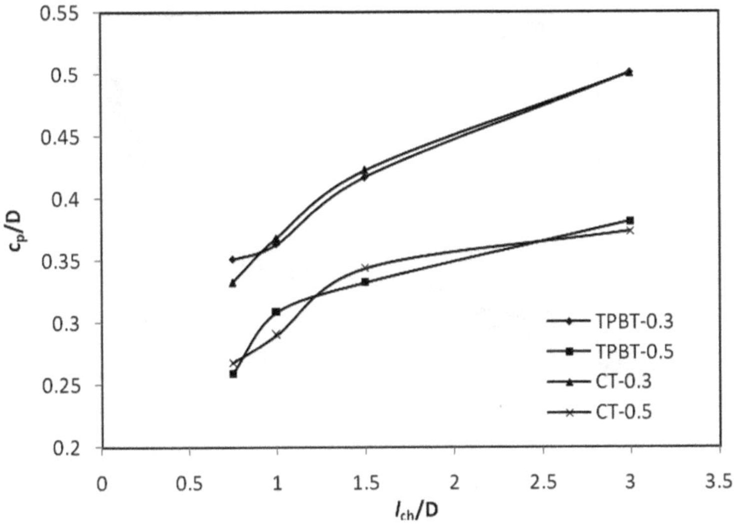

Fig. 5.41 Influence of specimen geometry and size effect on the process zone length

more at lower size specimens and seems to be converging and asymptotic at infinite size specimen for both values of a_0/D ratios. Moreover, the above observations with regard to K_R-curve analysis need more numerical and experimental investigations in future.

5.6.6 Effect of Loading Condition and Size Effect on the K_R Curves

The gained K_R curves for both the loading conditions, i.e., TPBT and FPBT of laboratory size range ($100 \leq D \leq 400$) at a_0/D ratio of 0.3 and 0.5 are shown in Figs. 5.42 and 5.43, respectively.

It is observed from both the figures that at a particular value of specimen size and a_0/D ratio, no distinguished difference in the K_R curves gained between the both loading conditions TPBT and FPBT is noticed. The size-effect behavior is similar to the observation made from Figs. 5.37 and 5.38.

5.6.7 Effect of Loading Condition and Size Effect on the CTOD Curves

The CTOD curves for both the loading conditions (TPBT and FPBT) having a_0/D ratio 0.3 and 0.5 are plotted in Figs. 5.44 and 5.45, respectively.

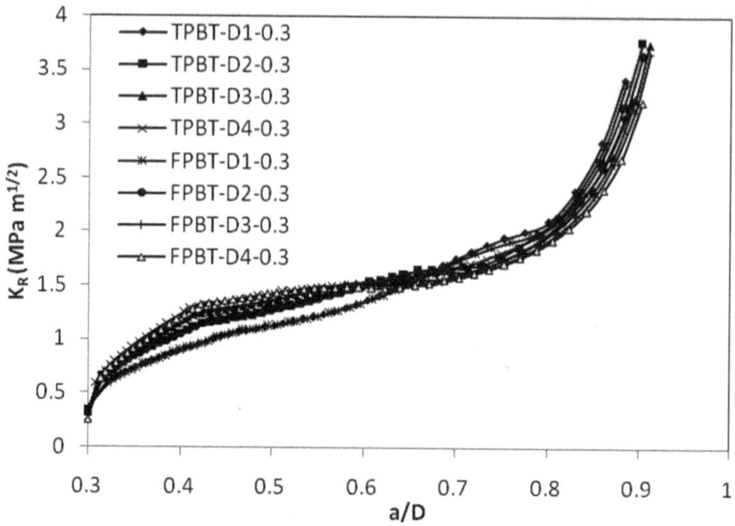

Fig. 5.42 Effect of loading condition and size effect on the K_R curves at $a_o/D = 0.3$

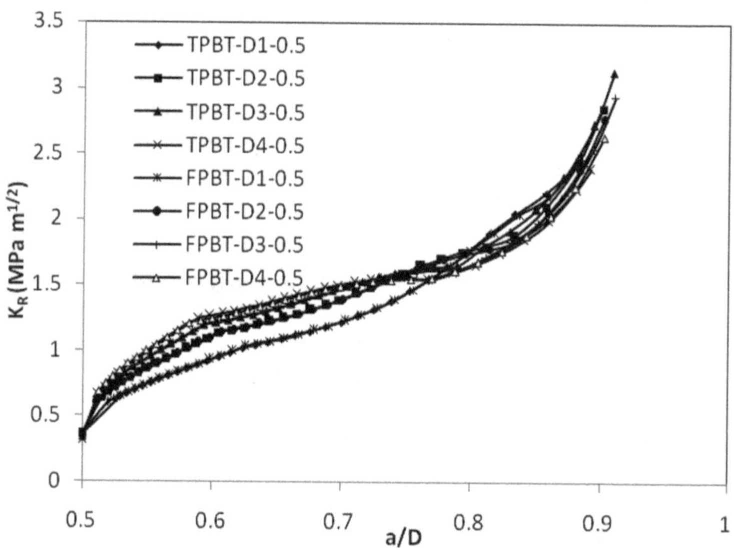

Fig. 5.43 Effect of loading condition and size effect on the K_R curves at $a_o/D = 0.5$

It is interesting to note that there is no influence of loading condition on the CTOD curves at a particular value of specimen size and a_o/D ratio. This observation is similar to the effect of loading condition on the K_R curves developed between TPBT and FPBT specimens.

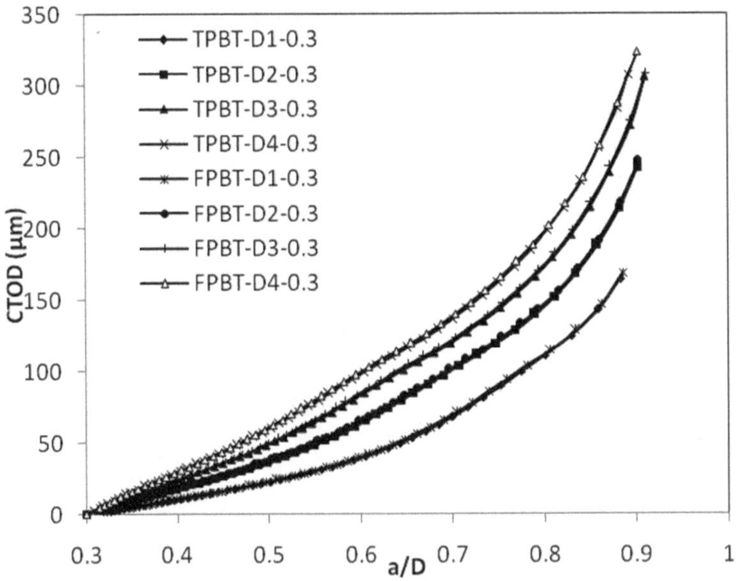

Fig. 5.44 Effect of loading condition and size effect on the CTOD curves at $a_o/D = 0.3$

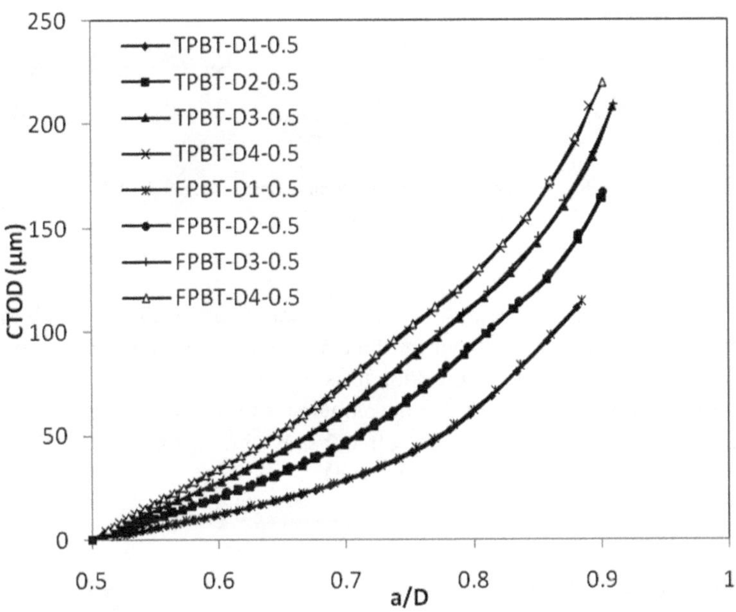

Fig. 5.45 Effect of loading condition and size effect on the CTOD curves at $a_o/D = 0.5$

5.6.8 Effect of Loading Condition and Size Effect on the Process Zone Length

The nondimensional parameter c_p/D is plotted versus the nondimensional value of l_{ch}/D in Fig. 5.46 for TPBT and FPBT specimens at relative initial notch length $a_o/D = 0.3$ and 0.5.

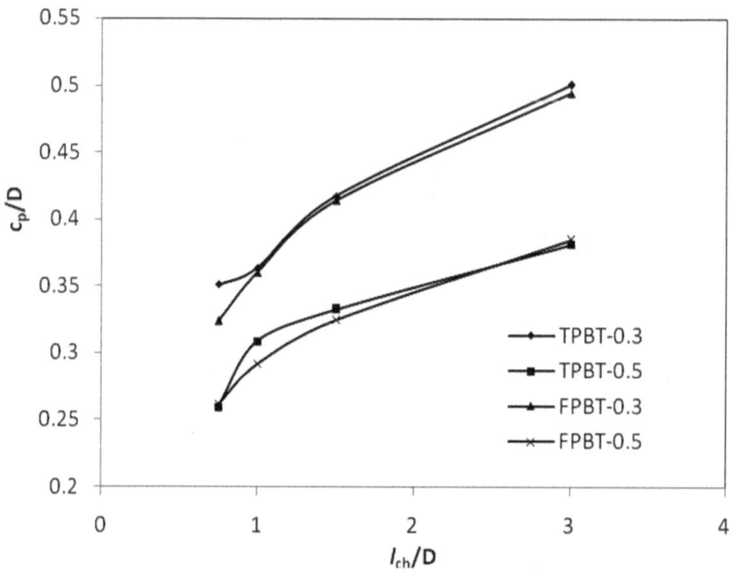

Fig. 5.46 Effect of loading condition and size effect on the process zone length

The trend obtained in the figure is similar to previous results in which the plot between c_p/D and l_{ch}/D was shown for the two different specimen geometries, i.e., three-point bending and compact tension tests. It is seen from Fig. 5.46 that almost no effect of loading condition on the development of process zone is observed at a particular value of a_o/D ratio. The size-effect behavior on the process zone length is similar to the observation made from Fig. 5.41.

5.7 Closing Remarks

In this chapter, weight function approach was introduced to determine the K_R curve based on cohesive stress distribution in the FPZ. In addition, a precise investigation of the various parameters on the K_R curve was carried out. The following salient points can be highlighted from the studies:

- Examples of K_R curves for experimental results on the three-point bending test specimens were evaluated using analytical method and weight function approach with four terms. It was observed that *weight function approach* yields indistinguishable difference in the results of K_R curve as compared to those obtained using the *analytical method*. Therefore, the weight function approach can be used as an alternative tool for determining the K_R curve based on cohesive stress distribution in fictitious crack extension without any appreciable error. This approach improves the computational efficiency over the existing analytical method. The measured K_R curves based on experimental results are influenced by the specimen sizes.
- From numerically obtained input data, many features of the K_R curve are noticeable. The influence of the specimen geometry on the K_R curves and CTOD curves is not observed for the specified specimen size and initial crack length/depth ratio. However, these curves are affected by specimen size. The size effect on the CTOD curves is more pronounced than is on the K_R curves. The size effect on the K_R curve near the zone of initiation toughness and the unstable fracture toughness of the material are found to be observable, whereas in the other parts of the curve this effect is almost absent. The length of the fracture process zone is found to be influenced by the specimen size and the zone length was almost independent of the type of specimen geometry for a given value of relative crack size.
- It was further noticed that the influence of loading condition on the K_R curves and the CTOD curves for a specified specimen size and initial crack length/depth ratio is not observable. The length of fracture process zone is almost independent of the type of loading condition for a given value of relative initial crack size.

References

Bažant ZP, Oh BH (1983) Crack band theory for fracture of concrete. Mater Struct 16(93): 155–177.

Du JJ, Kobayashi AS, Hawkins NM (1987) Fracture process zone of a concrete fracture specimen. SEM/RILEM International Conference on Fracture of Concrete and Rock (Edited by SP Shah, SE Swartz), Springer-Verlag, Houston, TX: 199–204.

Jenq YS, Shah SP (1985) A fracture toughness criterion for concrete. Eng Fract Mech 21: 1055–1069.

Karihaloo BL, Nallathambi P (1991) Notched beam test: Mode I fracture toughness. Fracture Mechanics Test methods for concrete. Report of RILEM Technical Committee 89-FMT (Edited by SP Shah, A Carpinteri), London, Chapman & Hall: 1–86.

Kumar S (2010) Behaviour of fracture parameters for crack propagation in concrete. Ph.D. Thesis submitted to Department of Civil Engineering, Indian Institute of Kharagpur, India.

Kumar S, Barai SV (2008a) Influence of specimen geometry and size-effect on the K_R-curve based on the cohesive stress in concrete. Int J Fract 152: 127–148.

Kumar S, Barai SV (2008b) Influence of specimen geometry on determination of double-*K* fracture parameters of concrete: A comparative study. Int J Fract 149: 47–66.

Kumar S, Barai SV (2009a) Weight function approach for determining crack extension resistance based on the cohesive stress distribution in concrete. Eng Fract Mech 76: 1131–1148.

Kumar S, Barai SV (2009b) Determining double-K fracture parameters of concrete for compact tension and wedge splitting tests using weight function. Eng Fract Mech 76: 935–948.
Kumar S, Barai SV (2009c) Influence of loading condition and size-effect on the K_R-curve based on the cohesive stress in concrete. Int J Fract 156:103–110.
Kumar S, Barai SV (2010) Determining the double-K fracture parameters for three-point bending notched concrete beams using weight function. Fatigue Fract Eng Mater Struct 33(10):645–660. DOI: 10.1111/j.1460-2695.2010.01477.x.
MATLAB Version 7, The MathWorks Inc., Copyright 1984–2004.
Phillips DV, Zhang BS (1993) Direct tension tests on notched and un-notched plain concrete specimens. Magn Concr Res 45(162): 25–35.
Planas J, Elices M (1990) Fracture criteria for concrete: Mathematical validations and experimental validation. Eng Fract Mech 35: 87–94.
Refai TME, Swartz SE (1987) Fracture behavior of concrete beams in three-point bending considering the influence of size effects. Report No. 190, Engineering Experiment Station, Kansas State University.
Reinhardt HW, Cornelissen HAW, Hordijk DA (1986) Tensile tests and failure analysis of concrete. J Struct Eng ASCE 112(11): 2462–2477.
Reinhardt HW, Xu S (1999) Crack extension resistance based on the cohesive force in concrete. Eng Fract Mech 64: 563–587.
Roesler J, Paulino GH, Park K, Gaedicke C (2007) Concrete fracture prediction using bilinear softening. Cement Concr Compos 29: 300–312.
Xu S, Reinhardt HW (1998) Crack extension resistance and fracture properties of quasi-brittle materials like concrete based on the complete process of fracture. Int J Fract 92: 71–99.
Xu S, Reinhardt HW (1999) Determination of double-K criterion for crack propagation in quasi-brittle materials, Part I: Experimental investigation of crack propagation. Int J Fract 98: 111–149.

Chapter 6
Comparison of Fracture Parameters of Concrete Using Nonlinear Fracture Models

6.1 Introduction

From the past studies it is clear that the fracture mechanics can be a useful and powerful tool for the analysis of growth of distributed cracking and its localization in concrete if the softening behavior of the material is taken into account. The presence of FPZ causes size-effect behavior in concrete structures. Various fracture models such as two-parameter fracture model, size-effect model, effective crack model, and double-*K* fracture model (Planas and Elices 1990, Karihaloo and Nallathambi 1991, Kumar and Barai 2010) can also predict this size-effect behavior. The size-effect study of fracture parameters obtained from two-parameter fracture model, effective crack model, double-*K* fracture model, and double-*G* fracture model is presented in this chapter. The studies carried out in subsequent sections are based on the work of Kumar and Barai (2008, 2009a, b, 2010) and Kumar (2010). Results of fictitious crack model for three-point bending test geometry of cracked concrete beam of laboratory size range 100–400 mm are used and the different fracture parameters from size-effect model, effective crack model, double-*K* fracture model, and double-*G* fracture model are evaluated using the input data obtained from FCM. In addition, the fracture parameters of two-parameter fracture model are obtained using the mathematical coefficients available in literature (Planas and Elices 1990). From the study it is concluded that the fracture parameters obtained from various nonlinear fracture models are influenced by the specimen size and they maintain some definite interrelationship depending upon the specimen size and relative size of initial notch length.

6.2 Material Properties and Determination of Fracture Parameters

Fictitious crack model or cohesive crack model for standard specimens of three-point bending test as shown in Fig. 2.11a (see Sect. 2.6) is used in the present study. The discretization as shown in Fig. 4.25a (see Sect. 4.9) is considered along with the quasi-exponential stress–displacement softening relation (Planas and Elices

S. Kumar, S.V. Barai, *Concrete Fracture Models and Applications*,
DOI 10.1007/978-3-642-16764-5_6, © Springer-Verlag Berlin Heidelberg 2011

Table 6.1 Peak load and the corresponding CMOD for standard TPBT obtained using FCM for material properties $f_t' = 3.21$ MPa, $E = 30$ GPa and $G_F = 103$ N/m

	a_o/D							
	0.2		0.3		0.4		0.5	
D (mm)	P_u (N)	$CMOD_c$ (μm)	P_u (N)	$CMOD_c$ (μm)	P_u (N)	$CMOD_c$ (μm)	P_u (N)	$CMOD_c$ (μm)
100	5070.94	32.5	3934.50	41.1	2947.20	48.9	2095.20	58.7
200	8502.80	45.3	6571.40	56.0	4909.70	68.2	3477.20	83.3
300	11276.49	56.1	8672.38	68.8	6447.90	86.8	4529.49	104.5
400	13608.21	62.5	10405.00	79.7	7683.40	99.7	5335.20	118.5

1990) in the FCM. Concrete mix for which $f_t = 3.21$ MPa, $E = 30$ GPa, and $G_F = 103$ N/m is considered as input parameters for FCM. The value of ν is assumed to be 0.18. The fracture peak load and the corresponding CMOD at a_o/D ratios ranging between 0.2 and 0.5 are obtained for TPBT specimen with $B =$ 100 mm, size range $100 \leq D \leq 400$ mm and $S/D = 4$. The peak load P_u and corresponding critical value of CMOD ($CMOD_c$) gained from the numerical model FCM are presented in Table 6.1.

In addition to above, the maximum size of coarse aggregate d_a is taken as 19 mm for all the subsequent computations. Since loading and unloading during test of fracture specimen is required to obtain the fracture parameters of TPFM – K_{IC}^{s} and $CTOD_{cs}$ ($CTOD_c$ of TPFM) according to the procedure outlined in RILEM Draft Recommendations TC89-FMT (1990a) – it is not possible with the available results obtained using FCM to determine the fracture parameters (also see Fig. A.6). Therefore, the parameter K_{IC}^{s} is precisely evaluated with the help of inverse analysis using the expressions and mathematical coefficients presented by Planas and Elices (1990) in which the fracture parameters of TPFM for the same TPBT specimen and material properties were computed. The critical effective crack extension (see Sect. 2.8) for infinite size $\Delta a_{cs\infty}$ is determined using Eq. (2.36) in which $c_f = \Delta a_{cs\infty}$, $G_{FB} = G_{FS}$. Finally, the size-effect equation of TPFM is cast in the following form:

$$\frac{EG_{FC}}{K_{INu}^{2}} = \frac{G_{FC}}{G_{FS}} \left[1 + \frac{\Delta a_{cs\infty}}{l_{ch}} \frac{2\,k'(\alpha_o)}{k(\alpha_o)} \frac{l_{ch}}{D} \right] \tag{6.1}$$

where $\alpha = a/D$, $k'(\alpha)$ is the first derivative of $k(\alpha)$ with respect to α, G_{FC} is equal to G_F, and G_{FS} is the equivalent fracture energy obtained using TPFM. In Eq. (2.36) the mathematical coefficient $\Delta a_{cs\infty}/l_{ch}$ was obtained as 0.0746 (Planas and Elices 1990) for each geometry (a_o/D) ranging between 0.2 and 0.5 within an accuracy level of 3%. The stress intensity factor K_{IN} corresponding to nominal stress σ_N is determined using LEFM formula given in Tada et al. (1985).

The K_{INu} of Eq. (6.1) can be obtained using Eq. (3.51) (see Sect. 3.4.7) in which $K_{IN} = K_{INu}$ for $\sigma_N = \sigma_{Nu}$ (when $P = P_u$) and $\alpha = \alpha_o = a_o/D$. In the present computation, the value of K_{INu} is determined using the value of P_{u} obtained from FCM for a particular TPBT specimen. Then, for a given geometry and material properties, the G_{FS} is determined using Eq. (6.1). Finally, the $\mathrm{CTOD_{cs}}$ is evaluated using Eq. (2.36) and the $K_{\mathrm{IC}}^{\mathrm{s}}$ is calculated using the following LEFM formula:

$$K_{\mathrm{IC}}^{\mathrm{s}} = \sqrt{EG_{\mathrm{FS}}} \tag{6.2}$$

The computed values of both the fracture parameters $K_{\mathrm{IC}}^{\mathrm{s}}$ and $\mathrm{CTOD_c}$ are given in Table 6.2.

For the given peak load and initial notch length, the fracture parameters of SEM, G_{FB}, and c_{f} are determined adopting the procedure (also see Fig. A.2 of Appendix) given in RILEM Draft Recommendations TC89-FMT (1990b) for three-point bending test specimen. Further, the equivalent critical stress intensity factor $K_{\mathrm{IC}}^{\mathrm{b}}$ is obtained using the standard LEFM equation for comparison purpose. These results are presented in Table 6.2.

Fracture parameters critical stress intensity factor $K_{\mathrm{IC}}^{\mathrm{e}}$ and critical effective crack length a_{e} of ECM are obtained using the equations given by Karihaloo and Nallathambi (1990). In this method, first of all a_{e} is obtained by using the regression equation (Karihaloo and Nallathambi 1990) for given material and geometrical properties of a TPBT specimen and then the value of $K_{\mathrm{IC}}^{\mathrm{e}}$ is calculated using LEFM equations. The method of determination of these fracture parameters is briefly outlined in Sect. 2.7.5 and presented in a flowchart of Fig. A.7). Both the fracture parameters determined are presented in Table 6.2 for TPBT specimen at a_{o}/D ratios ranging between 0.2 and 0.5.

In the present work, the double-K fracture parameters are determined using five-term weight function method. Since the softening relation of concrete is also required for determining the parameters of DKFM, modified bilinear softening function of concrete (Eq. (3.32)) is adopted in the present calculation. The flowchart (Fig. A.3) can be seen for determining the fracture parameters from DKFM. The effect of self-weight in the computation of effective crack length and the fracture parameters are taken into consideration. The results of fracture parameters $K_{\mathrm{IC}}^{\mathrm{im}}$ and $K_{\mathrm{IC}}^{\mathrm{un}}$ for the TPBT specimen are presented in Table 6.2.

The *analytical method* (Xu and Zhang 2008, Kumar and Barai 2009a) is used for determining double-G fracture parameters. Therefore, it is convenient to obtain the effective double-K fracture parameters, i.e., effective initiation fracture toughness $\overline{K_{\mathrm{IC}}^{\mathrm{ini}}}$ and effective unstable fracture toughness $\overline{K_{\mathrm{IC}}^{\mathrm{un}}}$ in terms of equivalent stress intensity factors using double-G fracture model. Modified bilinear softening function of concrete (Eq. (3.32)) (see Sect. 3.3.2.) is also used for determining the fracture parameters of DGFM in the present calculation. The flowchart for determining the fracture parameters from DGFM is shown in Fig. A.4. The computed values of double-G fracture parameters are shown in Table 6.2. All calculations are performed with developed computer program using MATLAB (version 7).

Table 6.2 Comparison of various fracture parameters for the material and geometrical properties: $f'_t = 3.21$ MPa, $E = 30$ GPa, $G_F = 103$ N/m, $d_a = 19$ mm, $B = 100$ mm, $S/D = 4$

		Fracture parameters of SEM		Fracture parameters of TPFM			Fracture parameters of ECM		Double-*K* fracture parameters				Double-*G* fracture parameters	
D (mm)	a_o/D	K_{IC}^{b} (MPa-m$^{1/2}$)	c_f (mm)	K_{IC}^{s} (MPa m$^{1/2}$)	CTOD$_{cs}$ (μm)	$a_{cs\infty}$ (mm)	K_{IC}^{e} (MPa m$^{1/2}$)	a_e/D	K_{IC}^{un} (MPa m$^{1/2}$)	K_{IC}^{ini} (MPa m$^{1/2}$)	a_c/D	CTOD$_c$ (μm)	$\overline{K_{IC}^{un}}$ (MPa m$^{1/2}$)	$\overline{K_{IC}^{ini}}$ (MPa m$^{1/2}$)
100	0.2	1.30	36.73	0.795	12.64	22.4	1.227	0.384	1.224	0.553	0.383	20.24	1.171	0.639
200	0.2			0.934	14.87		1.333	0.346	1.328	0.547	0.345	26.53	1.267	0.637
300	0.2			1.020	16.23		1.431	0.337	1.400	0.532	0.329	31.77	1.335	0.634
400	0.2			1.080	17.19		1.512	0.334	1.419	0.520	0.310	33.76	1.352	0.625
100	0.3	1.31	38.52	0.907	14.43	22.4	1.281	0.485	1.238	0.572	0.474	20.94	1.197	0.659
200	0.3			1.008	16.04		1.334	0.438	1.316	0.565	0.433	26.19	1.267	0.644
300	0.3			1.077	17.14		1.419	0.426	1.377	0.554	0.416	30.68	1.323	0.636
400	0.3			1.128	17.96		1.495	0.423	1.420	0.539	0.405	34.31	1.362	0.626
100	0.4	1.30	36.87	0.973	15.48	22.4	1.357	0.586	1.212	0.576	0.555	20.00	1.177	0.649
200	0.4			1.047	16.66		1.332	0.528	1.299	0.576	0.521	25.49	1.260	0.646
300	0.4			1.103	17.55		1.402	0.515	1.383	0.566	0.510	31.37	1.341	0.647
400	0.4			1.146	18.23		1.473	0.511	1.416	0.553	0.498	34.47	1.372	0.637
100	0.5	1.27	33.91	1.014	16.13	22.4	1.548	0.695	1.188	0.575	0.637	19.05	1.148	0.624
200	0.5			1.065	16.95		1.371	0.626	1.281	0.578	0.609	24.70	1.241	0.627
300	0.5			1.110	17.67		1.416	0.609	1.351	0.572	0.597	29.58	1.310	0.627
400	0.5			1.146	18.23		1.480	0.604	1.370	0.562	0.584	31.59	1.329	0.618

6.3 Size-Effect Analysis Using Various Fracture Models

6.3.1 Size Effect of Critical Stress Intensity Factors

In Table 6.2, the K_{IC}^{b} denotes the equivalent critical value of SIF obtained using G_{FB} and LEFM equations. From the table it is clear that the fracture parameters of SEM are independent of specimen size, whereas they are marginally dependent on geometrical factor a_o/D ratio. The reason is obvious. In the SEM, the fracture energy G_{FB} by definition is independent of test specimen size, although this is true only approximately since the size-effect law is not exact. The G_{FB} is also independent of the specimen shape. This becomes clear by realizing that the fracture process zone occupies a negligibly small fraction of the specimen's volume in an infinitely large specimen. Therefore, most of the specimen is elastic, which implies that the fracture process zone at its boundary is exposed to the asymptotic near-tip elastic stress and displacement fields which are known from LEFM and are the same for any specimen geometry. Here, the fracture process zone must be in the same state regardless of the specimen shape. For this reason, the computed fracture parameters K_{IC}^{s} of TPFM, K_{IC}^{e} of ECM, K_{IC}^{un} and K_{IC}^{ini} of DKFM and $\overline{K_{IC}^{un}}$ and $\overline{K_{IC}^{ini}}$ of DGFM at a_o/D ratios 0.2–0.5 are scaled down to K_{IC}^{b} and plotted in Figs. 6.1, 6.2, 6.3 and 6.4, respectively.

From the figures it is observed that all the fracture parameters are influenced by specimen size, hence exhibit size effect. These fracture parameters of various fracture models with reference to K_{IC}^{b} of SEM maintain certain relationship with the nondimensional parameter l_{ch}/D. From Fig. 6.1 it is observed that fracture parameters at critical condition K_{IC}^{e} of ECM, K_{IC}^{un} of DKFM, and $\overline{K_{IC}^{un}}$ of DGFM are close to

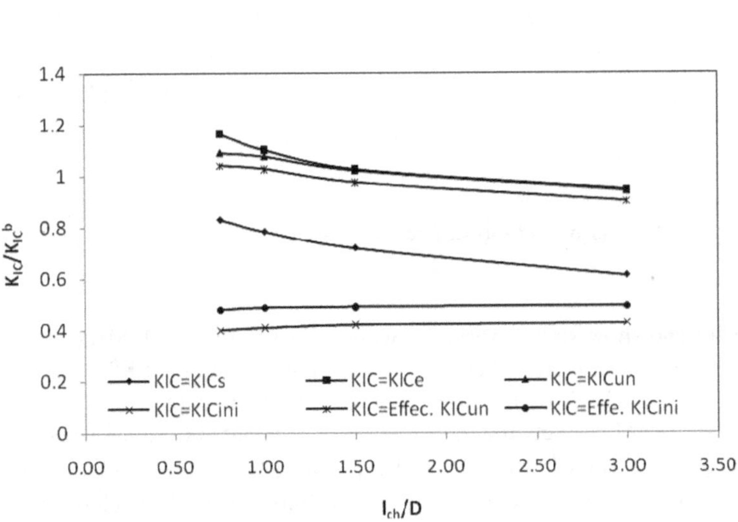

Fig. 6.1 Size-effect behavior of various fracture parameters at a_o/D ratio = 0.2

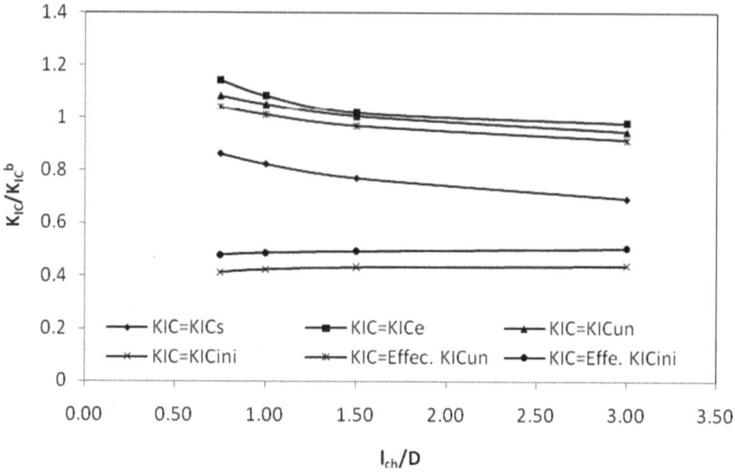

Fig. 6.2 Size-effect behavior of various fracture parameters at a_o/D ratio = 0.3

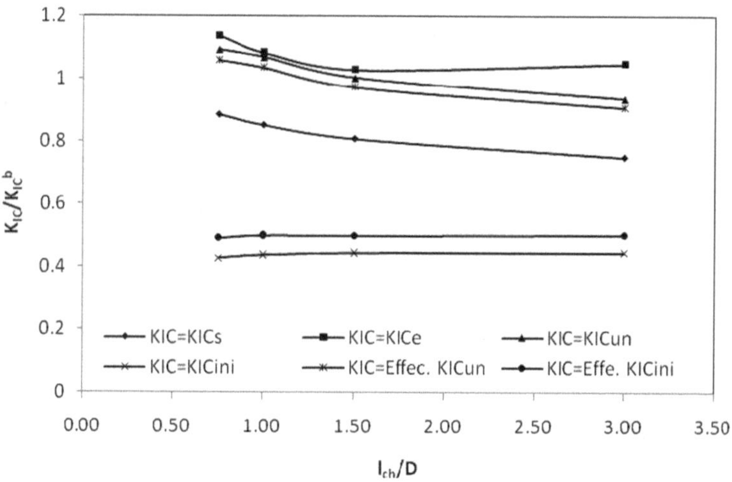

Fig. 6.3 Size-effect behavior of various fracture parameters at a_o/D ratio = 0.4

each other and show similar variation with respect to the l_{ch}/D. Size-effect behavior of K_{IC}^{s} of TPFM at critical condition is similar to that of the K_{IC}^{e}, K_{IC}^{un}, and $\overline{K_{IC}^{un}}$; however, the magnitude of the K_{IC}^{s} is somewhat less than that mentioned above. This means that TPFM predicts the most conservative results of critical stress intensity factor at unstable failure. The K_{IC}^{ini} of DKFM and $\overline{K_{IC}^{ini}}$ of DGFM are found to be very close at initial cracking load and show almost similar size-effect behavior.

Figures 6.2, 6.3, and 6.4 also show the same size-effect behavior as demonstrated in Fig. 6.1 except for the parameter of K_{IC}^{e} of specimen size 100 mm at a_o/D ratio

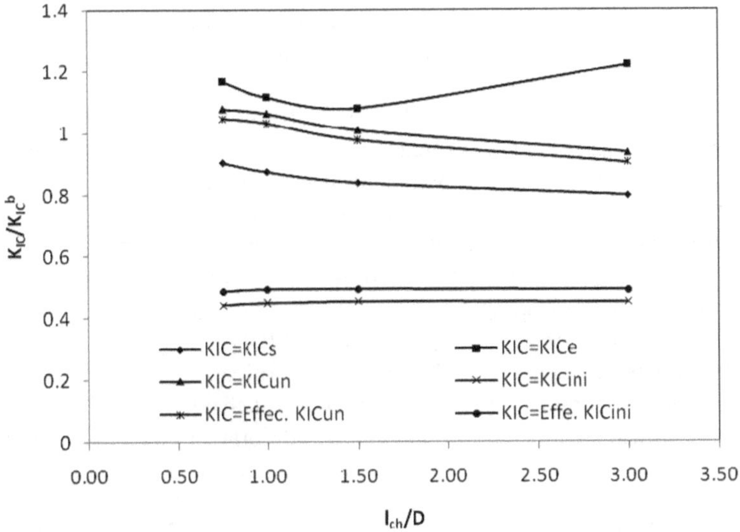

Fig. 6.4 Size-effect behavior of various fracture parameters at a_o/D ratio = 0.5

of 0.5. This deviation represented in the figure is probably due to the limitation of the applicability of the regression formula for determining the value of a_e in ECM.

It is also observed that the ratio of critical value of stress intensity factors predicted by ECM, DKFM, and DGFM to critical value of stress intensity factor predicted by SEM is close to 1. Furthermore, it is evident that the K_{IC}^{ini} and $\overline{K_{IC}^{ini}}$ are less dependent on the specimen size considered in the present study. This behavior was also observed in the previous studies of Chap. 4 (see Sect. 4.9.3). In the numerical study (Kumar and Barai 2008), it was shown that the parameter K_{IC}^{ini} is relatively less dependent on the specimen size ranging between 100 and 400 mm; however, beyond the size range 400 mm, a decrease in the value is observed. Similarly, the authors (Kumar and Barai 2009a) reported that the parameter K_{IC}^{ini} is almost independent of the specimen size ranging between 100 and 300 mm and beyond the size range 300 mm, a sharp decrease in the value is observed (see Sect. 4.10.2). In addition, it was observed that the $\overline{K_{IC}^{ini}}$ decreases with the increase in the specimen size. The discrepancy found in the results particularly with the $K_{IC}{}^{ini}$ and $\overline{K_{IC}^{ini}}$ may be possibly due to different softening functions employed in the calculation because the results of K_{IC}^{ini} and $\overline{K_{IC}^{ini}}$ are somewhat dependent on the softening function of concrete.

From Table 6.2 and Fig. 6.1, the ratios of K_{IC}^{s}/K_{IC}^{b}, K_{IC}^{e}/K_{IC}^{b}, K_{IC}^{un}/K_{IC}^{b}, $\overline{K_{IC}{}^{un}}/K_{IC}^{b}$, K_{IC}^{ini}/K_{IC}^{b}, and $\overline{K_{IC}^{ini}}/K_{IC}^{b}$ at a_o/D ratio 0.2 are found to be 0.611, 0.943, 0.941, 0.900, 0.425, and 0.492, respectively, for $D = 100$ mm and those are 0.830, 1.162, 1.091, 1.039, 0.400, and 0.481, respectively, for $D = 400$ mm. On the other extreme, from Fig. 6.4, the ratios of K_{IC}^{s}/K_{IC}^{b}, K_{IC}^{e}/K_{IC}^{b}, K_{IC}^{un}/K_{IC}^{b}, $\overline{K_{IC}^{un}}/K_{IC}^{b}$,

K_{IC}^{ini}/K_{IC}^{b}, and $\overline{K_{IC}^{ini}}/K_{IC}^{b}$ at a_o/D ratio 0.5 are found to be 0.799, 1.220, 0.936, 0.904, 0.453, and 0.492, respectively, for $D = 100$ mm and those are 0.903, 1.166, 1.079, 1.047, 0.443, and 0.487, respectively, for $D = 400$ mm.

The results indicate that the TPFM predicts the most conservative value of the critical stress intensity factor, whereas close results are predicted by the ECM, DKFM, and DGFM. The observation is consistent with the assumptions made for the development of various fracture models. In TPFM, the LEFM equations are applied for computation of different fracture parameters in which only elastic part of the total CMOD is considered for determining the critical effective crack length. The loading and the unloading are performed for the measurement of elastic part of the total CMOD. The inelastic part of that CMOD is neglected in calculation which possibly results in relatively lower value of critical effective crack length and K_{IC}^{s}. In ECM, the nonlinear P–δ (load–deflection) behavior before attainment of peak load is considered. Similar to compliance calibration method, the peak load and the corresponding midspan deflection (secant modulus) are used to evaluate the value of a_e, whereas the initial slope of the P–δ curve is used to determine the elastic modulus of concrete mix. In DKFM or DGFM, the linear superposition assumption (Xu and Reinhardt 1999) considering P-CMOD plot is used to obtain the critical effective crack length a_c. This assumption can be applied to determine the fictitious effective crack extension for complete analysis of fracture process in concrete. For critical condition, the effective crack length is determined using secant CMOD compliance at peak load, whereas the elastic modulus of concrete mix may be determined using initial compliance of P-CMOD plot. Hence the linear superposition assumption takes into account the nonlinearity effect in the P-CMOD curve before attainment of the unstable condition. This procedure seems to be similar to the method for calculating critical effective crack extension in ECM. From the above explanation it is clear that the K_{IC}^{e} may be the lowest value, whereas the fracture parameters K_{IC}^{e}, K_{IC}^{un}, $\overline{K_{IC}^{un}}$ should be in close agreement.

6.3.2 *Effect of Specimen Size on $CTOD_{cs}$ and $CTOD_c$*

The $CTOD_{cs}$ obtained using TPFM and the $CTOD_c$ evaluated using DKFM or DGFM are plotted with respect to the nondimensional parameter l_{ch}/D in Figs. 6.5 and 6.6, respectively.

It is observed from the figures that the $CTOD_{cs}$ and $CTOD_c$ maintain a definite relationship with the specimen size for a given value of a_o/D ratio and they increase as the specimen size increases. It is also observed from the figures that the $CTOD_{cs}$ and $CTOD_c$ depend on the a_o/D ratio for a given specimen size. The values of $CTOD_{cs}$ are more scattered particularly for smaller size of specimens when compared among the different a_o/D ratios, whereas the values of $CTOD_c$ are more closer and less scattered and appear to be in a narrow band for size range 100–400 mm considered in the study.

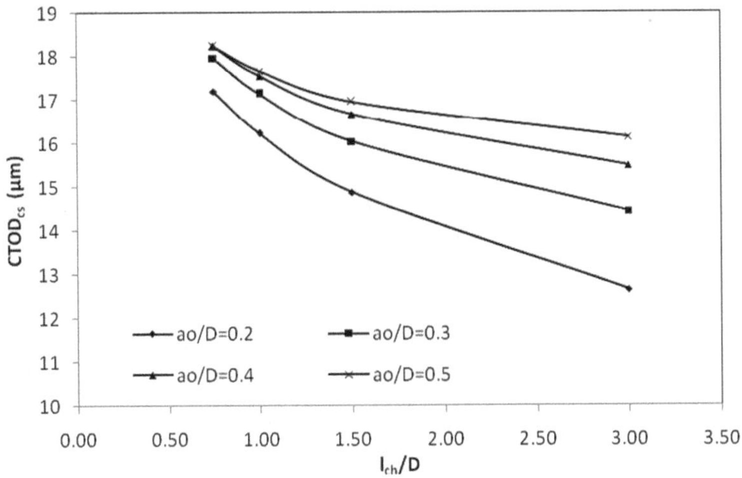

Fig. 6.5 Size-effect behavior of $CTOD_{CS}$ obtained using TPFM

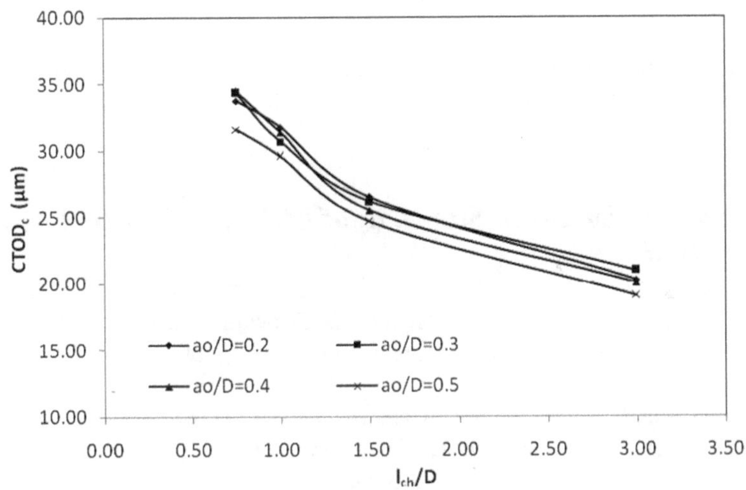

Fig. 6.6 Size-effect behavior of $CTOD_c$ obtained using DKFM

A relationship between $CTOD_{cs}$ and $CTOD_c$ is presented in Fig. 6.7 in which the ratio $CTOD_{cs}/CTOD_c$ is plotted with respect to the parameter l_{ch}/D.

It is seen from the figure that the ratio $CTOD_{cs}/CTOD_c$ maintains a definite relationship with the specimen size and the ratio decreases as the specimen size increases. For a_o/D ratio of 0.2, the value of $CTOD_{cs}/CTOD_c$ is 0.625, 0.561, 0.511 and 0.509 for specimen sizes of 100, 200, 300, and 400 mm, respectively, and the same for a_o/D ratio 0.5 is 0.847, 0.686, 0.597, and 0.577, respectively. Neglecting the effect of a_o/D ratio, the mean values of $CTOD_{cs}/CTOD_c$ for specimen size range 100 and 400 mm are determined and found to be 0.734 and 0.535, respectively.

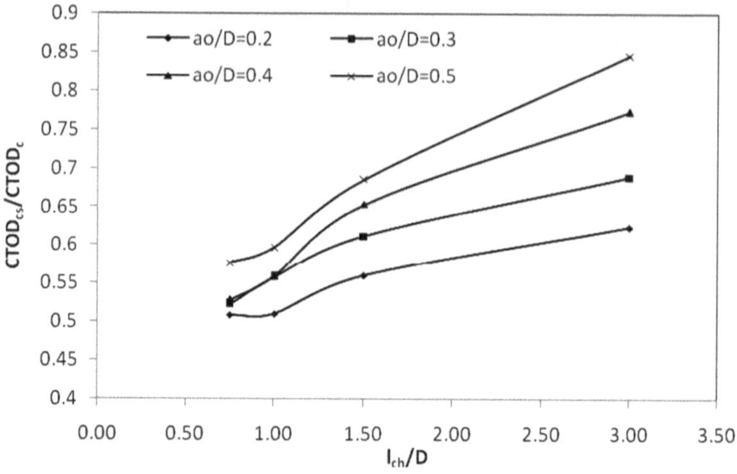

Fig. 6.7 Relationship between $CTOD_{CS}$ and $CTOD_C$ obtained using TPFM and DKFM

It means that the predicted CTOD at critical load using TPFM is relatively more conservative than that predicted by DKFM or DGFM.

6.3.3 Effect of Specimen Size on a_e of ECM and a_c of DKFM or DGFM

The critical effective crack extension ratio a_e/D obtained using ECM and a_c/D computed using DKFM or DGFM are plotted with l_{ch}/D in Figs. 6.8 and 6.9, respectively.

A similar trend on both the parameters a_e/D and a_c/D is observed from the figures. The values of a_e/D and a_c/D ratios are dependent on a_o/D ratio and specimen size. The assumption for determining both the parameters a_e/D and a_c/D is different. The secant compliance at critical load on P–δ curve is used for evaluation of a_e/D ratio, whereas the linear superposition assumption is applied on P-CMOD curve to determine the a_c/D value. In the present calculation, the regression equation (Karihaloo and Nallathamabi 1990) is used for evaluation of a_e/D ratio, while P-CMOD curve with linear superposition assumption is used for determining the a_c/D ratio.

Finally, an interrelation between a_e/D and a_c/D is plotted in Fig. 6.10.

It is interesting to observe from the figure that relationship between a_e/D and a_c/D ratios depends on the specimen size and geometrical factor. However, except for $D = 100$ at $a_o/D = 0.5$, the ratio a_e/a_c is very close to 1, that is, effective crack extension at critical load obtained using ECM and DKFM or DGFM is almost equivalent for the size range considered in the study.

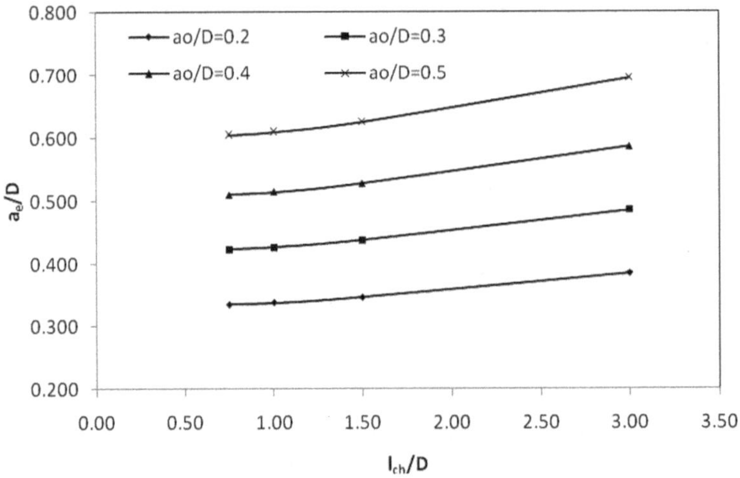

Fig. 6.8 Size-effect behavior of a_e/D obtained using ECM

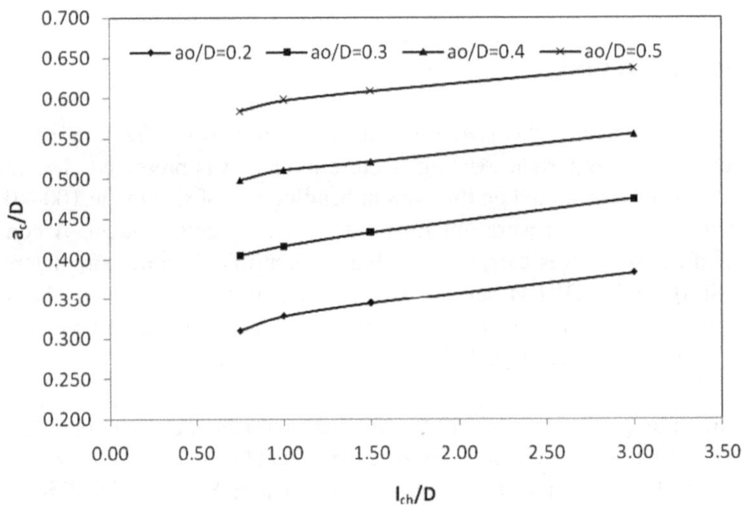

Fig. 6.9 Size-effect behavior of a_c/D obtained using DKFM

6.3.4 Relation Between c_f of SEM and $a_{cs\infty}$ of TPFM

From Table 6.2 it is seen that c_f slightly varies with the a_o/D ratio. For comparison purpose, the mean value of c_f is obtained as 36.51 mm and the mean value of $a_{cs\infty}$ is found as 22.40 mm. The ratio of a_{cs_∞}/c_f is 1.630, which shows that the effective crack extension for infinitely large structures predicted by TPFM is more conservative than the same predicted using SEM by about 38.64%.

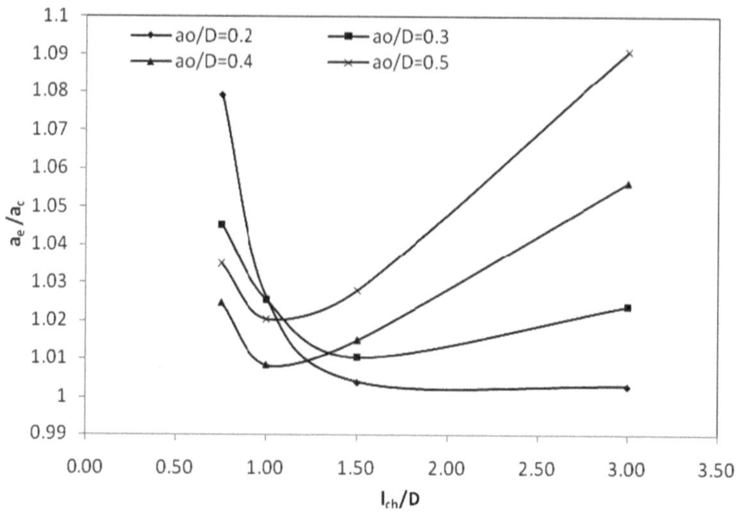

Fig. 6.10 Relationship of the equivalent critical crack extension obtained using ECM and DKFM

6.4 Closing Remarks

In the present chapter, the size-effect analysis of various fracture parameters obtained from the important existing fracture models was presented. The fracture parameters were determined on three-point bending test of size range 100–400 mm for which the input data were obtained from cohesive crack model. A comparative size-effect study was carried out using the possible fracture parameters from TPFM, SEM, ECM, DKFM, and DGFM. In general, it was observed that all the fracture parameters were dependent on geometrical factor and specimen size. From the present numerical study the following remarks can be highlighted:

- The fracture parameters of all the fracture models including double-*K* and double-*G* fracture parameters exhibited size-effect behavior.
- The critical stress intensity factors obtained using SEM, ECM, DKFM, and DGFM appear to be close to each other with an error range of $\pm 20\%$.
- TPFM predicted the most conservative critical stress intensity factor.
- The fracture parameters of double-*K* and double-*G* fracture models predicted the results very close to each other at initial cracking and unstable cracking loads.
- The crack-tip opening displacement at unstable fracture load predicted using TPFM was more conservative than that predicted using DKFM or DGFM by about in the range of 27–47%. This value was obtained on the basis of the mean values of crack-tip opening displacement at unstable fracture load from TPFM and DKFM or DGFM for specimen size 100 and 400 mm, respectively.
- The critical effective crack length obtained using ECM and DKFM or DGFM was very close to each other.

- The effective crack extension for infinitely large structures predicted by TPFM was more conservative than the same predicted using SEM by about 39%.

References

Karihaloo BL, Nallathambi P (1990) Size-effect prediction from effective crack model for plain concrete. Mater Struct 23 (3): 178–185.

Karihaloo BL, Nallathambi P (1991) Notched beam test: Mode I fracture toughness. Fracture Mechanics Test Methods for Concrete. Report of RILEM Technical Committee 89-FMT (Edited by SP Shah, A Carpinteri), London, Chapman & Hall: 1–86.

Kumar S (2010) Behaviour of fracture parameters for crack propagation in concrete. Ph.D. Thesis submitted to Department of Civil Engineering, Indian Institute of Kharagpur, India.

Kumar S, Barai SV (2008) Influence of specimen geometry on determination of double-K fracture parameters of concrete: A comparative study. Int J Fract 149: 47–66.

Kumar S, Barai SV (2009a) Equivalence between stress intensity factor and energy approach based fracture parameters of concrete. Eng Fract Mech 76: 1357–1372.

Kumar S, Barai SV (2009b) Size-effect of fracture parameters in concrete: A comparative study. Comput Concr Int J (under review).

Kumar S, Barai SV (2010) Size-effect prediction from the double-K fracture model for notched concrete beam. Int J Damage Mech 9: 473–497.

MATLAB Version 7, The MathWorks, Inc., Copyright 1984–2004.

Planas J, Elices M (1990) Fracture criteria for concrete: Mathematical validations and experimental validation. Eng Fract Mech 35: 87–94.

RILEM Draft Recommendations (TC89-FMT) (1990a) Determination of fracture parameters (K_{Ic}^{s} and $CTOD_c$) of plain concrete using three-point bend tests, proposed RILEM draft recommendations. Mater Struct 23 (138): 457–460.

RILEM Draft Recommendations (TC89-FMT) (1990b) Size-effect method for determining fracture energy and process zone size of concrete. Mater Struct 23 (138): 461–465.

Tada H, Paris PC, Irwin G (1985) The Stress Analysis of Cracks Handbook. Hellertown, PA, Del Research Corporation.

Xu S, Reinhardt HW (1999) Determination of double-K criterion for crack propagation in quasi-brittle materials, Part II: Analytical evaluating and practical measuring methods for three-point bending notched beams. Int J Fract 98: 151–177.

Xu S, Zhang X (2008) Determination of fracture parameters for crack propagation in concrete using an energy approach. Eng Fract Mech 75: 4292–4308.

Appendix

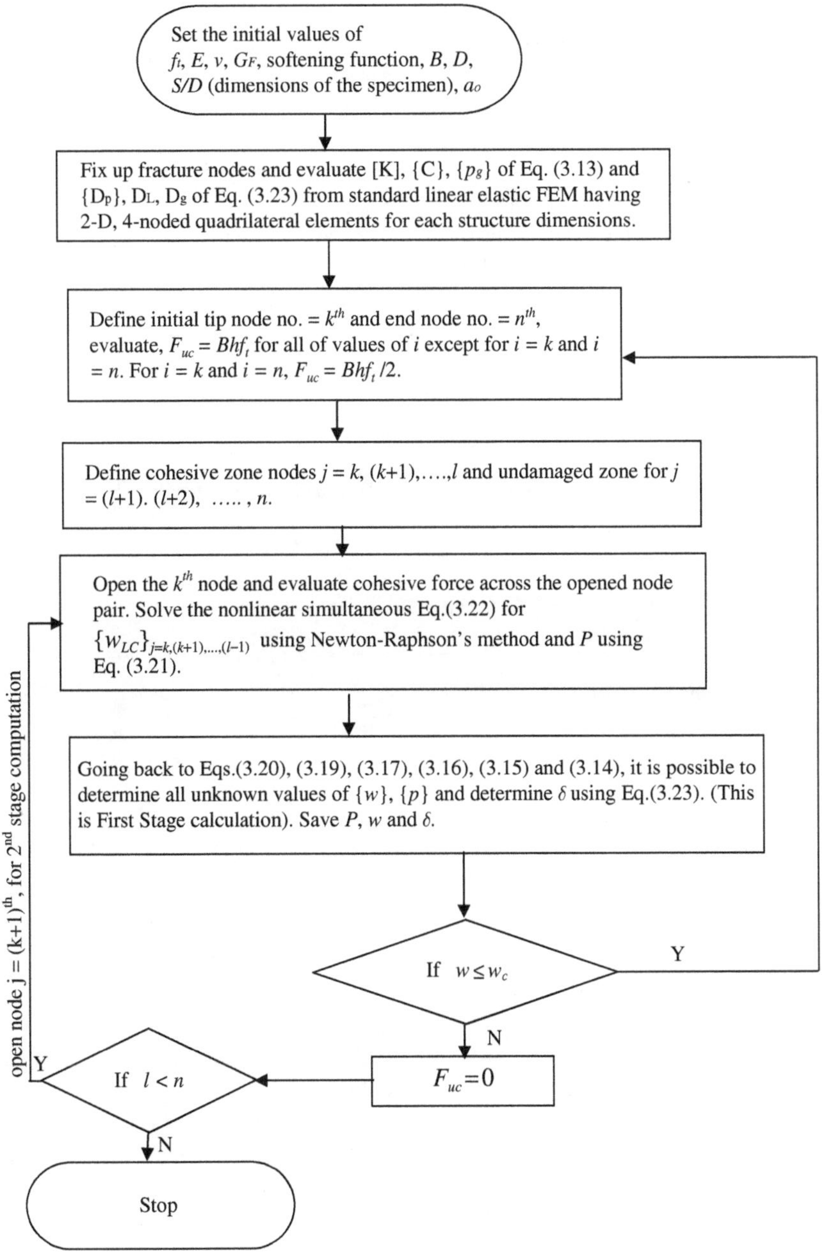

Fig. A.1 Flowchart of fictitious crack model

S. Kumar, S.V. Barai, *Concrete Fracture Models and Applications*,
DOI 10.1007/978-3-642-16764-5, © Springer-Verlag Berlin Heidelberg 2011

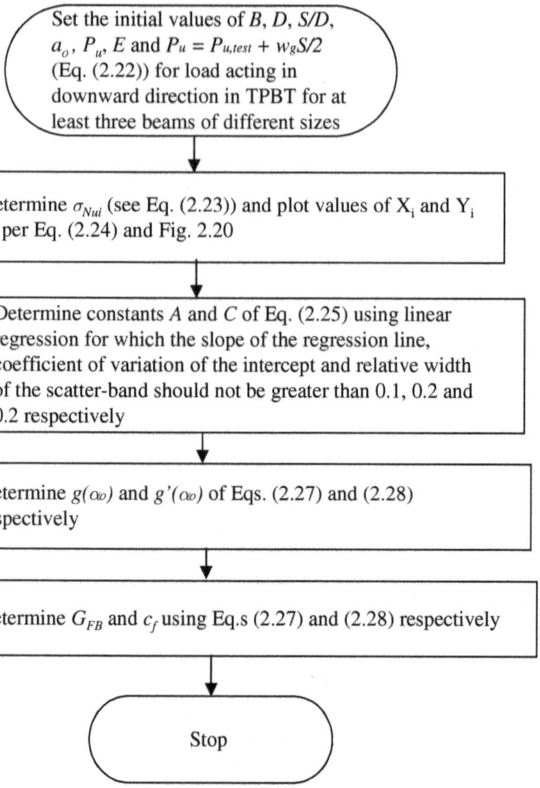

Fig. A.2 Flowchart of SEM

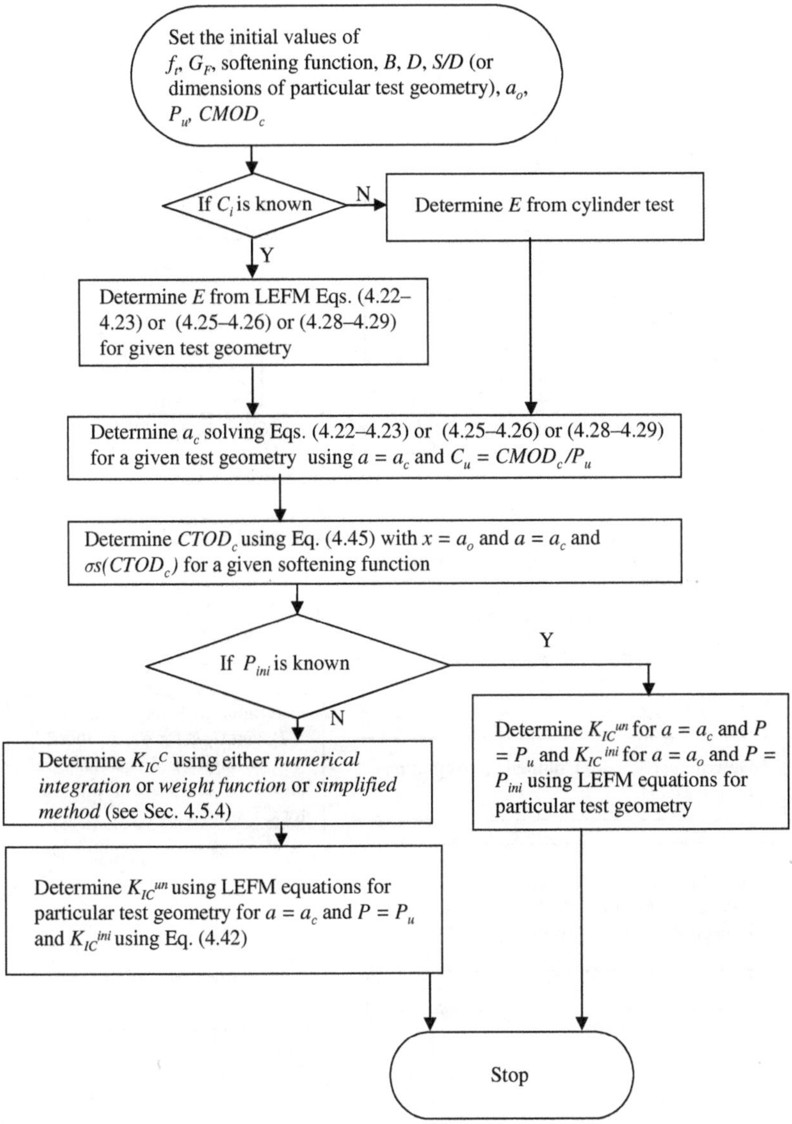

Fig. A.3 Flowchart of DKFM

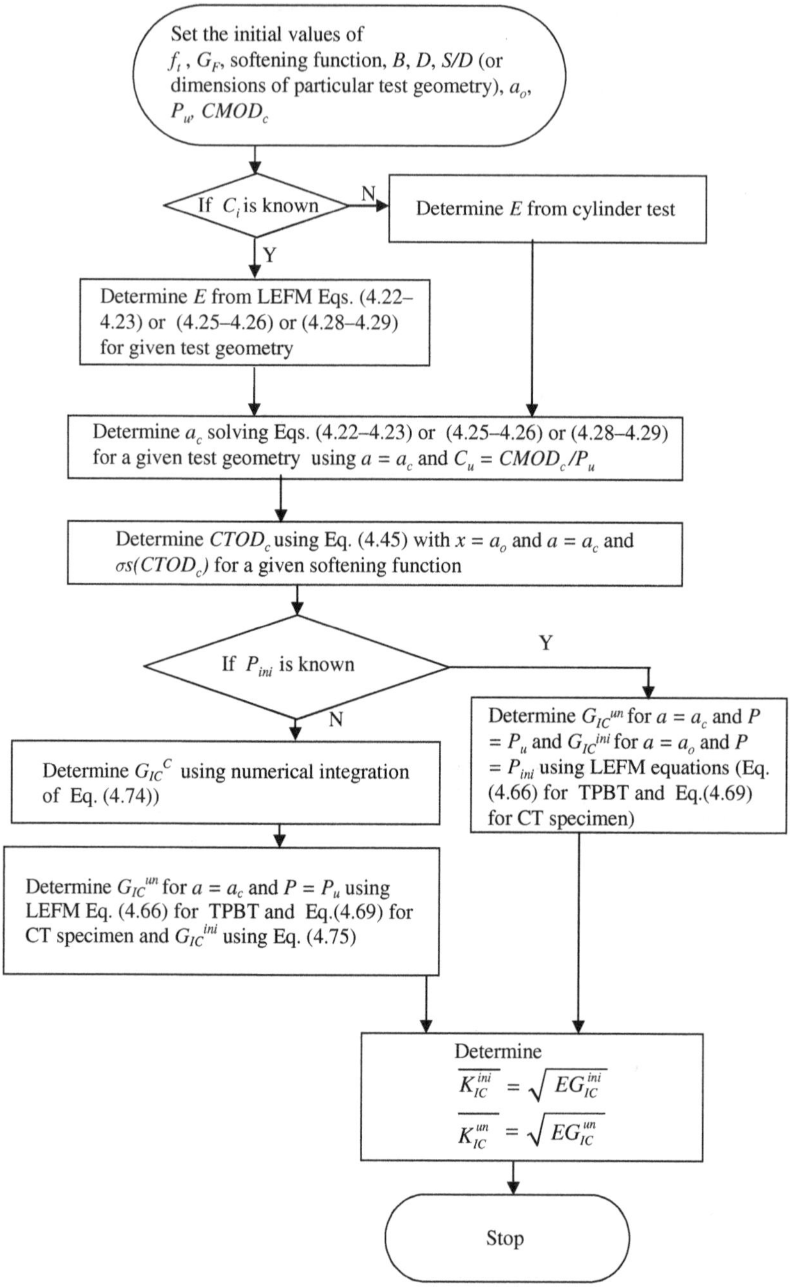

Fig. A.4 Flowchart of DGFM

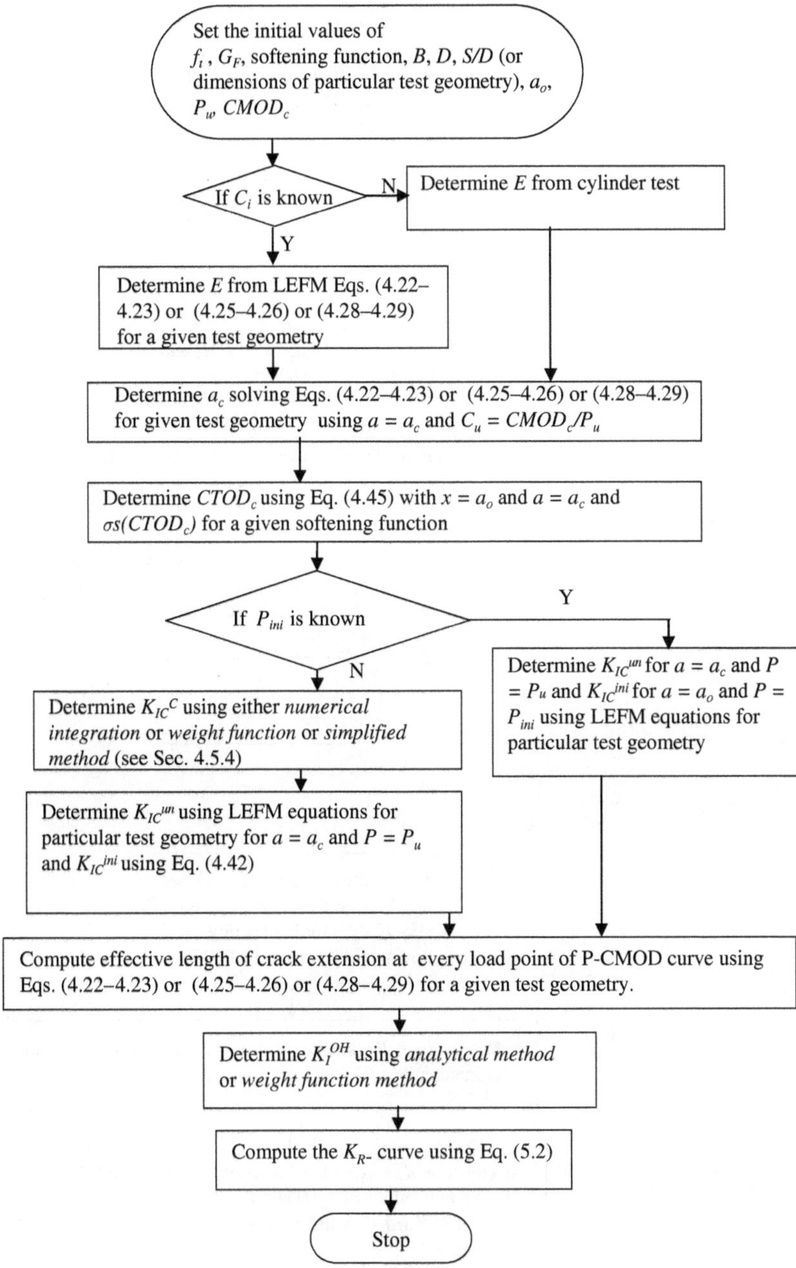

Fig. A.5 Flowchart of the K_R curve associated with cohesive stress distribution in FPZ

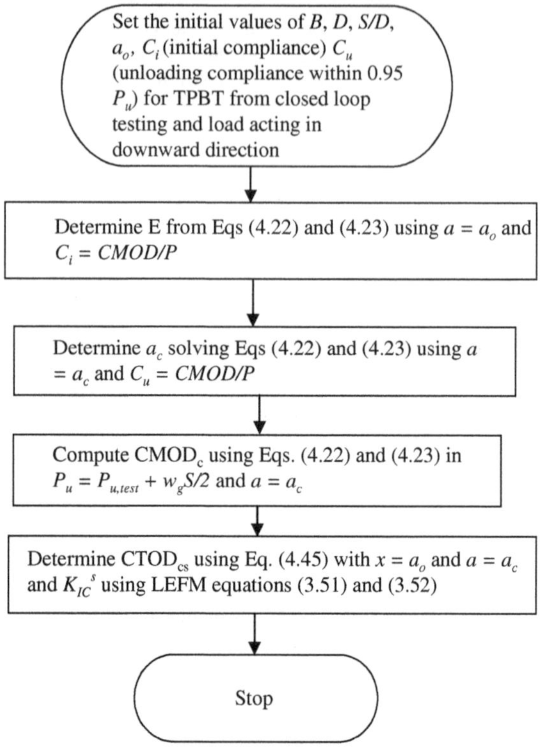

Fig. A.6 Flowchart of TPFM

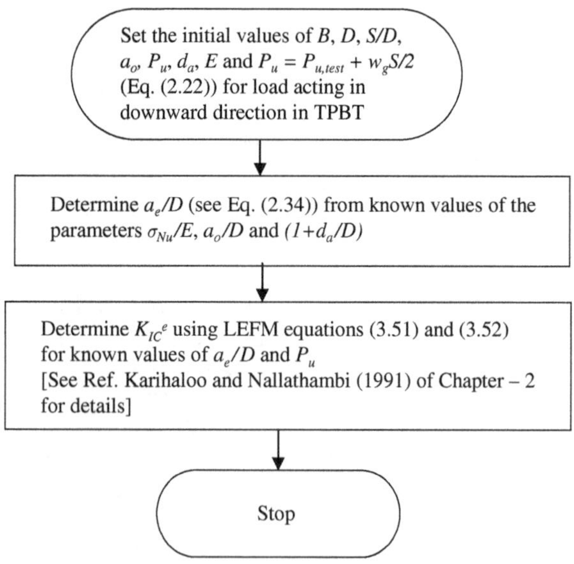

Fig. A.7 Flowchart of ECM using regression method

Index

D

Kolibri 360 GmbH
Ferdinand-Jühlke-Straße 7
99095 Erfurt, Deutschland
produktsicherheit@kolibri360.de